Lecture Notes in Economics and Mathematical Systems   551

Jacek Leskow
Martín Puchet Anyul
Lionello F. Punzo (Eds.)

New Tools of
Economic Dynamics

 Springer

Editors

Prof. Dr. Jacek Leskow
Econometrics Department
Nowy Sacz Graduate School of Business
National Louis University
33–300 Nowy Sacz, ul. Zielona 27
Poland

Prof. Dr. Lionello F. Punzo
Department of Political Economy
UNISI
Piazza S. Francesco 7
53100 Siena
Italy

Prof. Martín Puchet Anyul
Economics Faculty, UNAM
Av. Universidad 3000
Edif. de las Ofs. Admvas. 2, 1°
Ciudad Universitaria
México D. F. 04510
México

Library of Congress Control Number: 2005927235

ISSN 0075-8442
ISBN-10 3-540-24282-1 Springer Berlin Heidelberg New York
ISBN-13 978-3-540-24282-6 Springer Berlin Heidelberg New York

Springer is a part of Springer Science+Business Media

springeronline.com

© Springer-Verlag Berlin Heidelberg 2005
Printed in Germany

Typesetting: Camera ready by author
Cover design: *Erich Kirchner*, Heidelberg

Printed on acid-free paper   42/3130Di   5 4 3 2 1 0

Dedicated to *Richard M. Goodwin* and *Michio Morishima*

Preface

Innovation and advances in the techniques of analysis of economic dynamics have been dramatic in recent years. Taken together, they have formed a sort of second wave following the wave that has revolutionised macro dynamics in the 80s. Impact has been relevant to both theoretical and applied work and it has involved also econometrics, of course.

On one side, we have witnessed the birth of families of what could be broadly defined new growth and development models. They are quite new in comparison to those of traditional approaches the endogenous and exogeneous types- and can be collectively characterised by the fact that their much richer internal structure is capable of producing a richer, and more interesting, variety of dynamics. On the other hand, econometrics and especially time series analysis began looking more closely at the finer structure of our economies, a greater number of variables being attributed to different agents and represented in the models.

In either case, the result was first that the models economists got used to work with, had (often, many) more dimensions than the traditional ones. Moreover, the main force driving the economy's dynamics began to be identified with the various rules and forms of interaction among many heterogeneous agents (industries, firms, individuals). The engine of dynamics was seen to be fundamentally endogenous, rather than the the mere response to the exogenous shocks of New Classical dynamics. Thus, the whole analytical framework based upon the impulse-response mechanism had to be entirely overhauled, changing their relative weights: more was put into the internal structure of the economy, less in the complications of the shock profiles.

The emerging new modelling framework obviously demanded new analytical tools, too. These had to be (and have been) imported from elsewhere, this ranging over a very broad field, from statistics to mathematics to physics. Among such tools, more and more important became e.g. numerical simulations as an exploratory device with a theoretical dignity of its own. Sometimes, using simulations was a choice; more often, however, it was a necessity given the size of the model at hand. Taken together, at any rate, all those new tools were employed, at times, to search out the capabilities and to explore the structure of a given model. On other occasions, they were to depict alternative scenarios for growth and/or for policy actions. Even when quantitative results were expected - as in all econometrics and time series studies   still, some part of the added value of the research was in the qualitative nature of the information provided by its results. This calls for a comparison with the way the qualitative approach to dynamics entered into economics, and how it fared in the field since its discovery by the economists.

Qualitative analysis has been a key approach to dynamics since Poincar invented it at the end of 19th century and since its introduction into economics with the classical works of Frisch, Kaldor, Hicks and Goodwin, between the 30s and the 50s. It was born out of the incapability of handling certain non-linear dynamic models in a classical way, i.e. by explicitly finding their solutions. Going qualitative was a necessity, instead of a choice. It basically meant topological (hence, non numerical) analysis of individual models and the fundamentally topological theory of classes of models.

The qualitative approach that has been emerging recently is quite different, though it complements the classical one. Differences can be appreciated in many ways, but they all refer either to the intensive use of new, sometimes simulation and numerical techniques and the construction of models with greater dimensions than before, and/or to the deeper integration between theory and empirical evidence.

The New Tools project (and network) was born out of this challenge and it reflected the variety and heterogeneity of its aspects. Emphasis was however placed on the common ground, the exploration of tools rather than the construction of models around specific economic issues. The New Tools network now links researchers in various countries and universities of Europe and Latin America.

Most of the chapters collected in this volume are revised versions of research papers read in four workshops held consecutively at UNISI in Siena (December 2000), UDLA in Cholula (State of Puebla, Mexico, September 2001), CIMAT in Guanajuato (State of Guanajuato, October 2002) and in Nowy Sacz Graduate School of Business in Poland (September 2003). All papers were subjected to intense discussions during the network's meetings, with a varied public of researchers and students at different levels of their education. In fact, the purpose of the NT network is not only to promote research but also to enrich education, focusing on master and doctoral levels.

The broad areas in which the network's research activity fell so far, are reproduced in the volume's structure with the three sections: large interactive models of the economy; econometrics and time series; growth, development and structural change, Each section contains both theoretical and applied chapters as, in general, papers have been written with the need to look for such intersection in the authors'minds. In fact, it is a key hypothesis in the NT project that time is ripe for a reconciliation between the more theoretical and the more applied research lines in economic dynamics, ending thus a divorce and recomposing a unity that was at the birth of macro dynamics as envisioned by Ragnar Frisch and the founders of thee Econometric Society. We believe it is important to try in this direction by picking up the bits and pieces left from that divorce, in particular reconsidering the different tools that were developed then from the vantage point of the new ones. We now briefly go over the three sections, trying to highlight the novelty that is in the various applications of tools. Such novelty can often be appreciated more by the economists than by the praticitioner of those disciplines from which those tools have been imported. The main common ground can be identified with the study of various aspects of so called complex dynamics. As anticipated earlier, these aspects are hereafter investigated under the hypothesis that they spring from the endogenous mechanism more than from the characteristics of some exogenous forces. In other words, without denying the importance of the latter, often stochastic forces, it is the structure of the model economy, which is seen as the site of the basic explanation of its dynamics. Structure can be looked at in a variety of ways as shown in the various papers, and can also be seen in its evolution, dramatic or catastrophic as sometimes its discontinuous change is called (after the mathematical theory).

Thus, 6 out of 9 chapters in Section I are devoted to the analysis of the various effects and tools to analyse settings with heterogeneous agents, and to derive characteristics of the resulting (aggregate) dynamics. Thus, Aoki's Chapter 1 introduces the notion of classes or types of agents and deals with the issue of how to consider the uncertain appearance of new types along the economic trajectory. By looking at various schemes of local interaction between nearby firms, in Chpt. 2 Andergassen et als discuss the emergence of fluctuating growth and technological patterns shared by firms (trajectories in the evolutionary sense). On the other hand, through an explicit neural networks approach, Chapter 8 considers the emergence of firms and firms' networks as the result of processes of learning in an environments too complex to be handled efficiently by any individual, thus as the institutions adequate to solve the associated economic problems. Aoki's paper has implications for simulation techniques, which are heavily implemented in many of the other chapters. Through simulations, Chapter 2 tackles the problems of the emergence of different groupings of agents that are heterogeneous in their initial endowments and via bilateral exchange have to reach through an evolutionary process, equilibria implying different schemes of benefits sharing. Chapter 3 looks at a similar problem of social aggregation, there with a genetic algorithm approach, within a setting where the assumption of bounded rationality is central and a learning process is modelled. Chapter 4, on the other hand, innovates the conventional description of macroeconomic performance by studying (among other things) the effects of parameter perturbation over a system of equations tracing the time evolution of the first and second moments of the firms distributions (in terms of a chosen index of financial robustness). (Many of the implications of these analyses on growth and in particular on growth irregularity and fluctuations will come up again in Chapters of Section 3, while the statistical implications are practically dealt with in Section 2.) Finally, Chapter 9 reviews various easily available platforms for multi-agent simulations, thus providing a guide to the intriguing question of what to learn to do.

As said above, classical qualitative analysis basically meant topological methods applied to (classes of) models or of model predicted trajectories. This is very much the spirit in Chapter 6, where however the study of general equilibrium economies is carried on by means of much newer notions from Catastrophe theory. The point is the suggestion to focus upon the singular economies (that are structurally unstable) rather than on the structurally stables ones as is always the practice. (The theme of the importance of understanding instability comes up again in Section 3). But qualitative analysis can also be of a different type: an analysis where structural rather than functional dependence, and thus hierarchical and dominance relations are at the centre, as is in Chapter 7, where a pretopological approach is used to unveil the bare skeleton of an economy.

Virtually, all of the chapters mentioned above bear implications for observable dynamics, most of them do look also at empirical evidence. While this is a feature common with Section 3, empirical evidence and how to handle it is the very focus of Section II; as its title suggests. Here too, the common framework is one where effects of multidimensional economies and complex time evolution (including, uncertainty) are studied. This is the realm of econometrics, time series analysis and of course broadly defined simulation based-econometrics, a field fast growing specially in a version married with micro simulation.

The latter is basically the object of the two coordinated Chapters 13 and 14, and it appears in the topically related Chapter 15. In all three chapters, the study case of retirement choices is tackled for its own right, but also to demonstrate a variety of techniques to econometrically construct, handle and validate models with many agents, thus capable of exhibiting alternative outcomes through micro simulation experiments (in the former two chapters), or to endogenize choices (in this case, of retirement) as in the latter chapter. A critical review of outcomes of a bunch of econometric tools to evaluate monetary policy is presented in Chapter 16, with an application to a known difficult case, Mexico's highly volatility behaviour. The chapter makes a case for an informed policy decision-making process, whereby different scenarios produced by alternative techniques are systematically taken into account. This is again a link to themes in chapters of Section III, with their multiple illustrations of applications of complex dynamics tools to Latin America (and possibly to more recent events elsewhere). But before turning to that, we recall that the chapters opening this section, are all devoted to issues associated with detection of the driving force s behind seemingly irregular economic time series.

Thus, Chapter 12 reviews the recent advances of spectral analysis, a well-established technique in economics being associated with the still most favoured linear econometric framework, while in fact it has received major extensions through for instance the windowed filtering methods. The application to the well known Phillips' curve is a good link to Chapter 10, where time series are looked at as possibly embodying, next the more popular ingredients, also structural change. As a way to tackle such cases, the smooth transition formulation of an econometric model is exposed and some result shown. This is a rapidly expanding research in the filed of non linear econometrics, as much as is the modelling of financial markets, a sample to be found in Chapter 11 introducing an imported method of Value-at-Risk prediction (or VAR, not to be taken for the better known vector auto regression approach!), which promises to handle time series for which there are no multiple realizations, or it is safer not to assume it.

Section III deals with issues in what traditionally have been classified growth and development fields, until recently realm of well-defined theories with clearly understood predictions. The history of the last decades, and the theoretical reflections on it, has shown that the apparent consensus reached some time ago about their interpretation has definitely broken down. We are searching for an explanation, or more probably for various explanations for the series of events that have been happening in the various countries, explanations accounting of the variety of experiences and the evolution often dramatic shown by most of them. The so called convergence literature, enormously boosted by the growth debate and the availability of new statistical base in the mid of the 80s, has probably misled us by proposing the search for cross country uniform behaviours and long run stability towards some predicted equilibrium path. Neither prediction has proved to be reasonably tenable.

The critical implications of this failure are the common thread of the section, which opens with the revision of the notion of convergence in the light of its environmental implications. This leads to the unveiling of a double convergence hypothesis which is allegedly implicit in the growth re-interpretation of the so called environmental Kuznets curve, and to the rejection of the latter on the ground of the prevailing of different growth regimes across countries (the notion of regime recalls chapter 5 and 6 above). A re-examination of the growth findings in Chapter 18 focuses upon the much discussed issue of volatility in performance, an issue discussed by the authors resorting to distribution analysis with a Markov chain hypothesis. Chapter 19 re-examines in a detailed way a well known model of externally constrained growth, in the light of the Mexican and generally recent Latin American experiences, and it shows how it had to be to a large extent updated. The next two Chapters 20 and 21 dwell upon the uses of the notion of fractional brownian motion to explain, respectively, the time behaviour of indices of the Mexican stock market, where structural change is embedded as a result of the recent major organizational changes (including NAFTA), and the Argentinean high inflation before the parity with the US dollar as the evolution of self organized structures.

Optimisation is bread and butter for economists, and the intertemporal optimisation framework has become more so after the 80s. The book could not overlook this: Chapter 22 reviews the theory and various applications of a new field called semi-infinite programming, not as easy as standard programming, not so difficult as day to day, real life one.

Acknowledgement. The Editors of this volume would like to express their thanks to various government agencies, institutions and individuals. Our thanks go first to the University of Siena for its financial support also through a PAR project, and to the Italian Ministry of University and research (MIUR) with the financial support to a Project of National Interest.

We would also like to thank the Latin American institutions for providing a generous support. We are especially indebted to the Universidad Nacional Autonoma de Mexico (UNAM, Ciudad de Mxico), the Universidad de las Americas in Cholula (UdlA), to the Centro de Investigacion en Matematicas (CIMAT) for providing us excellent conference facilities and local support. Our fourth meeting was held in the conference center of the Nowy Sacz Graduate School of Business in Poland and we here gratefully acknowledge the support.

Last but not least, we want to express our gratitude to Alessandro Perrone who contributed to this volume as author and as technical problem solver. His work for the camera ready version of the manuscript made this project a finished product.

This acknowledgement would not be complete were we not to mention dedication and help of many individuals in Europe and Latin America that were generously offered in the course of our 4 year project. We would like to mention here Ana Elena Uribe from CIMAT in Guanajuato and Elzbieta Kubinska from Nowy Sacz. Without their help our efforts could not be as fruitful as they were.

March 2005

Nowy Sacz,   *Jacek Leskow*
Ciudad de México (DF)   *Martín Puchet*
Siena   *Lionello F. Punzo*

Contents

List of Contributors

Accinelli Elvio
U. de la República, Montevideo,
Uruguay and UAM-X,
México D.F.
elvio@fing.edu.uy

Ansergassen Rainer
Department of Economics,
University of Bologna
Italy
anderga@economia.unibo.it

Böhm Bernhard
Institute of Econometrics, Operations Research and Systems
Theory
University of Technology,
Wien, Austria
bboehm@e119ws1.tuwien.ac.at

Barroso-Castorena Mauricio
Universidad de las Américas, Puebla,
72820, Puebla, México,
grome@mail.udlap.mx

Bianchi Carlo
Dipartimento di Scienze Economiche
Università di Pisa
italy
cbianchi@ec.unipi.it

Bimonte Salvatore
Dipartimento di Economia Politica
Università degli Studi di Siena
Italy
bimonte@unisi.it

Blancas Andrés
Instituto de Investigaciones
Económicas.
UNAM, México.
neria@servidor.unam.mx

Catalán Horacio
Faculty of Economics
UNAM
México
catalanh@correo.unam.mx

Dabús Carlos
Departamento de Economía
Universidad Nacional del Sur
8000 Bahía Blanca, Argentina
cdabus@criba.edu.ar

Delli Gatti Domenico
ITEMQ,
Università Cattolica
Milano
domenico.delligatti@unicatt.it

Fiaschi Davide
Department of Economics,
University of Pisa,
Italy
dfiaschi@ec.unipi.it

Galindo Luis Miguel
Faculty of Economics,
UNAM
México
galindo@correo.unam.mx

Gallegati Mario
DEA,
Università di Ancona
Italia
gallegati@dea.unian.it

García Valdéz Carlos Alberto
Universidad de las Américas
Puebla, México,
grome@mail.udlap.mx

Huerta - Golzales Jorge
Universidad de las Américas
Puebla, México,
grome@mail.udlap.mx

Iacobucci Alessandra
OFCE, CNRS – IDEFI
06560 VALBONNE (France)
alessandra.iacobucci@sciences-po.fr

Lavezzi Andrea Mario
Dipartimento di Scienze Economiche
University of Pisa,
Italy
lavezzi@ec.unipi.it

Leskow Jaceck
Department of Econometrics,
The Graduate School of Business
NOWY SACZ, POLAND.
leskow@wsb-nlu.edu.pl

London Silvia
Departamento de Economía,
Universidad Nacional del Sur
8000 Bahía Blanca, Argentina
slondon@criba.edu.ar

Luna Francesco
International Monetary Fund
Washington DC,
fluna@imf.org

Masanao Aoki
Department of Economics
University of California,
Los Angeles
aoki@econ.ucla.edu

Moreno-brid Juan Carlos
Economic Commission for Latin
America and the Caribbean
(ECLAC)
México.
jcmoreno@un.org.mx

Napolitano Antonio
Dipartimento di Ingegneria Elettron-
ica e delle Telecomunicazioni,
Università di Napoli Federico II
Italy
antonio.napolitano@unina.it

Nardini Franco
Department of Mathematics for
Social Sciences,
University of Bologna
Italy
nardini@dm.unibo.it

Nicolas Garrido
Department of Economics
University of Trento
Italy
nico@black.economia.unitn.it

Pacini Pier Mario
Department of Economics
University of Pisa
Italy
pmpacini@ec.unipi.it

Perrone Alessandro
Department of Economics
University of Venice
Italy
alex@unive.it

Puchet Anyul Martín
Fac. de Economia
UNAM
México
anyul@servidor.unam.edu.mx

Punzo F. Lionello
Dipartimento di Economia Politica,
Università degli Studi di Siena
Italy
punzo@unisi.it

Ricottilli Massimo
Department of Economics,
University of Bologna
Italy
ricottilli@economia.unibo.it

Ricoy Carlos
Departamento de Fundamentos del
Análisis Económico,
Universidad de Santiago de Com-
postela,
Spain.
aericoy@usc.es

Romanelli Marzia
Sant'Anna School of Advanced
Studies
Laboratory of Economics and
Management (L.E.M.)
Pisa , Italy
romanelli@sssup.it

Romero-Meléndez Guillermo
Universidad de las Américas
Puebla, México,
grome@mail.udlap.mx

Rückmann Jan-J
Universidad de las Américas, Departamento de Física y Matemáticas,
Puebla, México
rueckman@mail.udlap.mx

Santigo-Bringas Manuel
Universidad de las Américas
Puebla, México,
grome@mail.udlap.mx

Solís Valentín
Facultad de Economía,
UNAM.
México
valentinsolis@mexico.com

Spataro Luca
Dipartimento di Scienze Economiche,
Università di Pisa,
Italy
l.spataro@ec.unipi.it

Tohmé Fernando
Departamento de Economía,
Universidad Nacional del Sur
8000 Bahía Blanca, Argentina
ftohme@criba.edu.ar

Vagliasindi Pietro
Dipartimento di Diritto, Economia e
Finanza Internazionale
Università di Parma
Italy
pvagli@unipr.it

Vázquez Francisco Guerra
Universidad de las Américas, Escuela
de Ciencias,
Puebla, México
fguerra@mail.udlap.mx

Part I

Large Interactive Economies

1

Modeling a Large Number of Agents by Types: Models as Large Random Decomposable Structures*

Masanao Aoki

Department of Economics
University of California,
Los Angeles
aoki@econ.ucla.edu

Summary. This paper introduces methods, based on decomposable random combinatorial analysis, to model a large number of interacting agents. This paper also discusses a largely ignored possibility in the mainstream economic literature that hitherto unknown types of agents may enter the models at some future time. We apply the notion of holding times, and introduce the results of the one- and two-parameter inductive methods of Ewens, Pitman and Zabell to economic literature. More specifically, we use the notion of exchangeable random partitions of a finite set to produce a simple rule of sucession, that is, the expressions for the probabilties for entries by new or known types, conditional on the observed data. Then Ewens equilibrium distriution for the sizes of clusters is introduced, and its use to examine market behavior is sketched, especially when a few types of agents are dominant. We suggest that the approaches of this paper and the notion of holding times are relevant to agent-based simulations because holding times can be used to randomly select agents that ''act'' first.

1.1 Introduction

Economists often face problems of modeling collective behavior of a large number of interacting agents, possibly of several different types. This paper discusses methods that are useful in this context. We indicate how the methods may explaine diverse phenomena such as equilibrium size distributions of clusters, that is subgroups formed by agents, market shares by different types of goods, changes in the adjustment speeds towards equilibria with model sizes, and emergence of macroeconomic regularities as the number of agents increases towards infinity, and so on.

We explicitly assume that there are several types of agents in our models, the number of which may not be known in advance, and that agents of new

---

* Prepared for the NT Book

types may enter the models at any time. We cannot assume in advance that we know all of them because new rules or new goods may be invented in the future. This is the so-called problem of unanticipated knowledge in the sense of Zabell, see Zabell (1992)[2]. In biology this problem is known as the sampling of species problem. In probability and statistics it is called laws of sucession, that is, how to specify the conditional probability that the next sample is a new type, given available sets of observation up to now. See Zabell (1982). In addition, agents may change their minds at any time about the decisions or behavioral rules they use. In other words, agents may change their types any time[3]. This paper presents some methods from the field of decomposable random combinatorial analysis. They are useful in modeling economic structures composed of a large number of possibly heterogeneous agents, components or basic units, and indicates some of their potential applications in macroeconomics and finance.

Large economic structures are regarded as decomposable random combinatorial structures. Of the many possible structures due to many possible configurations which a large number of components may assume, we wish to deal with "typical" structures of large sizes. By typical we mean structures which have high probabilities of being chosen, or observed from the set of all structures, when chosen at random in some sense from the set. This is the main reason for borrowing or adapting tools and concepts from combinatorial stochastic processes, and population genetics literature.

We interpret the word "types" broadly. This word may refer to some characteristics or rules that are used to partition the set of agents. Or, it may refer to some rules or behavioral patterns adopted by economic agents, or it may refer to some other characteristics to distinguish one subgroup of agents from other groups. We assume that the number of types are at most countable.

The methods described in this paper are not in the tool kit of traditionally trained economists or econometricians, but we have found them to be useful for understanding macroeconomic or financial phenomena from our new perspectives. Stochastic combinatorial tools are used to show how agents form clusters, and jump Markov processes are used to model how the clusters evolve over time through interaction among agents of several types. To describe dynamic phenomena, master equations (backward Chapman- Kolmogorov equations) are used to describe probability distributions over states. We use a new notion of states, called partition vectors, which is more appropriate in dealing with exchangeable or delabelled agents and category, i.e., type indices.

Some of the questions we examine are: How do we describe the process by which agents form clusters, that is, subgroups in modeling a collection of

---

[2] Zabell describes the problem faced by statisticians in classifying samples of insects collected in unexplored regions, since they may contain new species of insects, say. The naive Bayesian approach is not applicable. See, however, Antoniak (1969) on non-parametric Bayesian approach. He obtained the same distribution as the Ewens sampling formula, Ewens (1972).

[3] There is no lock-step behavior by agents.

interacting agents? What are the stationary distributions of sizes of fractions of agents of different types? What are the market shares of a typical largest cluster, two largest clusters, and so on? Distributions of cluster sizes matter, because a few of the larger clusters, if and when formed, approximately determine the market excess demands for whatever goods in the markets and the nature and magnitudes of fluctuations.

The methods mentioned in this paper have diverse origins. To discuss clusters and entries by agents of new types, such as new goods, new business models, new (sub)optimization procedures, and so on, we borrow from the literature of population genetics such as Ewens (1972, 1990), Watterson (1976), and Watterson and Guess (1977), and from statistics and stochastic processes such as Kingman (1978a, b), Arratia and Tavaré (1992), and Pitman (2002). See also Aoki (2002).

The Ewens sampling formula is an example of one-parameter inductive model. It is specified by a single parameter θ, which controls the rate of entries of new types, and correlations among agents of different types. We also describe its two-parameter extension by Pitman (1992), which is specified by two parameters, α and θ discussed later.

More concretely, we introduce the notion of partition vector as state vectors, which is different from the empirical distributions, and use the assssumption of exchangeable partitions induced by agents of different types in the models, in the technical sense of exchangeable random variables in the probability literature. We utilize the notion of holding times from the literature on continous-time Markov chain (also called jump Markov process) to decide randomly which agent acts first. We apply the equilibrium distribution to discuss the question of market shares, behavior of rates of returns, and volatilities of returns.

To conclude, we list some financial and economic applications: In finance we mention the work on power laws and volatility switching in Aoki (2002b, c). In economics, we briefly compare the approach of the traditional economists in allocating capital stock between two sectors as formulated by Dixit (1989), and our modeling procedure in terms of continuous-time Markov chains with a large but finite number of interacting agents in Aoki (2002a, Chapt.8). A new approach to the Diamond search model from our perspective is in Aoki and Shirai (2000). A new approach to growth model is in Aoki and Yoshikawa (2002).

1.2 New Concepts

Partition vectors

When it is known that there are n agents, and K distinct types of agents, a common choice of state vector is $\mathbf{n} = (n_1, n_2, \ldots, n_K)$ where n_i is the number of agents of type i, $i = 1, 2, \ldots, n$. This choice of state vector is natural, and appears to be satisfactory. There is, however, another choice of state which suits our needs better when identities of agents of various types are not the issue. In some cases, only the numbers of agents of different types may matter. Labels we assign to agents may be merely for convenience of reference. Permuting these labels often leaves nothing of substance changed. For example, agents may be labelled in the order we sample or examine them, or in the order they enter the market, but there is no essential meaning or substance to the labels. They are for mere convenience of referring to them. Permuting labels assigned to agents should not cause any essential changes in our conclusions about the models in such cases. When this holds true, agents are called exchangeable in the technical sense defined in probability literature[4]. To indicate this, exchangeable agents are sometimes called delabelled agents. For a collection of exchangeable agents their joint probability is invariant to permutations of indices assigned to agents in order to refer to them.

We regard collections of agents as exchangeable agents, and assume also that types are exchangeable. That is, agents are partitioned into distinct clusters. Labels of the clusters may again be for convenience. Category indices may again be for mere convenience of reference with no substance. If categories are delabelled, then the probability is also invariant with respect to permutations of category indicies. This is the notion of exchangeable partitions.

State of a population is described by the (unordered) set of type-frequencies i.e., fractions or proportions of different types without stating which frequency belongs to which type. In the context of economic modeling, this way of description does not require model builders to know in advance how many or what types of agents are in the population. It is merely necessary to recognize that there are K_n distinct types in his sample of size n, and that there are a_j types with j agents or goods in the sample. The vector **a** with these components is called partition vector by Zabell (1992), and we adopt this name. Note that $\sum_j a_j = K_n$, and $\sum j a_j = n$. The first equation counts the number of occupied boxes, and the second total number of agents. Partition vector is just the right notion to discuss models with delabelled agents and delabelled categories. The problem is the same as the occupancy problem of allocating unlabelled or indistinguishable balls and unlabelled or indistinguishable boxes. The same concept is known under different names in Kingman (1980), and Sachkov (1996).

Exchangeable random partitions

A partition of a finite set F into K blocks is an unordered collection of non-empty, disjoint sets $\{A_1, \ldots, A_K\}$ whose union is F. To be definite, we may use a convention that the blocks of partitions are listed in the order of appearance, that is, by the least elements of the blocks. Let $[n]$ denote a set of n elements, $\{1, 2, \ldots, n\}$. Let X_i, $i = 1, 2, \ldots n$ be random variables with values on $[K]$.

---

[4] See Feller (1968).

These X_s are grouped into subsets of K or less and induce partition of $[n]$. Any partition of $[n]$ defines a composition of n, which is a sequence of positive numbers with sum n. By using the size of the sets A_i, $n_i =| A_i |$, the set defines the composition $n = n_1 + n_2 + \ldots + n_K$. A partition Π_n of $[n]$ is exchangeable if its distribution is invariant with respect to permutations, i.e.,

$$Pr(\Pi_n = \{A_1, A_2, \ldots A_K) = p(| A_1 |, \ldots, | A_K |);$$

where $p(\cdots)$ is some symmetric function of the components.
Exchangeable random partition is such that

$$Pr(N^{ex}_{n,1}, \cdots, N^{ex}_{n,k}) = \frac{n!}{n_1! n_2! \cdots n_k!} \frac{1}{k!} \times$$
$$p(n_1, n_2, \cdots, n_k)$$

Two partitions with the same vector **a** are equiprobable when the partitions are exchangeable. Sequences associated with exchangeable random partitons are exchangeable sequences.

Exchangeable agents

This newer representation has roots in the exchangeable random partitions of a set of agents into clusters, which arises in examining clusters or subsets of agents of the same types. Stirling numbers of the first and second kind also appear in counting the configurations of clusters of agents of various types. These have roots in random combinatorial analysis. Probability distributions such as the Poisson-Dirichlet distributions and the multivariate Ewens distribution are not in the tool kit of conventionally trained economists or econometricians, but are important in examing the distributions of sizes of clusters of agents. We therefore present these as well as some others, as needs arise, to advance and support our views expressed in this book.

Let X_i, $i = 1, 2, \ldots, n$ be a sequence of random variables whose values are in the set of type indices. When an original sequence of random variables and that of permuted random variables, that is, $X_1, X_2, \ldots, X_n$ and $X_{\sigma(1)}, \ldots X_{\sigma(n)}$, where σ denotes permutation of the subscript, have the same probability depending on the empirical distribution $n_j =| \{X_i = typej\} |, j = 1, 2, \cdots, K$, that is $p(n_1, n_2, \cdots, n_K)$, where $p(\cdots)$ is symmetric of its arguments, then we call the sequence exchangeable. Two sequences with the same empirical distributions are equiprobable for exchangeable sequences of random variables.

Limit behavior of large fractions: Poisson-Dirichlet distributions

Suppose that a large number of agents interact in a market where each agent uses one of K available trading rules, where K is large. Then the set of agents is partitioned into at most K clusters. The number of clusters depends crucially on the correlations among agents which affects the probability that two randomly chosen agents in the market are using the same strategy. The sizes of clusters are arranged in non-increasing order as $n_{(1)} \geq n_{(2)} \geq \cdots$. With high correlations, a small number of large clusters tend to form. When correlations among agents are small, many smaller clusters are likely to emerge, as we later mention. We will also mention later that the sum of the sizes of the first two largest clusters, $n_{(1)} + n_{(2)}$, alone in some cases account for the majority, 70 per cent say, of the total number of participants. This observation is useful in characterizing aggregate behavior when it happens. See Aoki (2002b). Order statistics of the fractions have a well-defined limit distribution, called the Poisson-Dirichlet distribution, as the number of agents go to infinity. The probability density of the first few of the orders sizes of fractions can be used in our discussion of approximations of market excess demands or return dynamics of some traded assets.

1.3 Dynamics of Clustering Processes

Agents and goods are classified into clusters or subsets by associating types with strategies or choices of agents. Here we interpret the word types broadly.

As agents interact, new clusters form or some existing clusters break up into smaller ones. The transitions of these processes are captured by specifying how the partition vector $\mathbf{a}$ is transformed over a small time intervals. For example, transition rate specification

$$w(\mathbf{a}, , \mathbf{a} + e_1) = \lambda(n)$$

where e_i is a vector with the only non-zero component 1 at the ith component, refers to an event that a single agent enters the market without joining any existing cluster, while

$$w(\mathbf{a}, \mathbf{a} + \mathbf{e}_{j+1} - \mathbf{e}_j) = ja_j\lambda(n)$$

specifies one agent joints a cluster of size j, thereby increasing the number of clusters of size $j + 1$ by one, and reducing that of size j by one. The right-hand side specifies that the rates are proportional to some constant $\lambda(n)$, and ja_j which counts the total number of agents in the clusters of size j. There are many other possibilities, of course. At this stage of our exposition, let us pause to take stock of what we have done. Our model building procedures may be summarized as follows: We start with a collection of a large, but a finite number of microeconomic agents in some economic or financial context. We first select state space and specify a set of transition rates on it to model agent interactions stochastically. Agents may be households, firms, or countries depending on the context of models. Unlike examples in textbooks in simple probability, our transition rates are usually state-dependent, that is functions of states, to model effects of endogenously generated aggregate effects, called field effects in Aoki (1996), such as effects of aggregate behavior such as total outputs, crowding, fashion, group pressures on individual agents, and so on, as well as evaluations of consequences of specific choices subject to uncertainty or imperfect information that go into evaluations of value function maximization associated with alternative choices.

Then we describe by master equation the dynamics of the joint probability of the components of a state vector for the model which incorporates specified transition rates. Stationary or nonstationary solutions of the master equations are then examined to deduce model aggregate dynamic behavior.

In models which focus on the decomposable random combinatorial aspects, distributions of a few of the largest order statistics of the cluster size distributions are examined to draw economic consequences.

Ewens Distribution

Here we follow Aoki (1996, 1998, 2000a,b, 2002a, b,c) and sketch the basic ingredients for our modeling procedure without too much detail. The reader is asked to consult the cited references for detail.

Continuous-time Markov chains, also known as jump Markov processes, are completely specified by transition rates, when state spaces are at most countable.

For ease of explanation we use vector $\mathbf{n}$ some of the times and $\mathbf{a}$ at other times. Define a state vector X_t which takes on the value $\mathbf{n} := (n_1, n_2, \cdots, n_K)$, called frequency or occupancy vector, where n_i is the number of agents of type i, $i = 1, 2, \cdots, K$, $N = n_1 + n_2 + \cdots + n_K$.

In our model we need to specify entry rates, exit rates and rates of type changes. Over a small time interval Δ, rates are multiplied by the length of interval to approximate the conditional probabilities up to $O(\Delta)$. For example, entry rates by an agent of type j may be specified by

$$w(\mathbf{n}, \mathbf{n} + \mathbf{e}_j) = \phi_j(n_j, athbfn);$$

where $\mathbf{e}_j$ is a vector with the only nonzero element of one at component j.[5] Exit rates of an agent of type k is specified by

$$w(\mathbf{n}, \mathbf{n} - \mathbf{e}_k) = \psi_k(n_k, \mathbf{n});$$

and transition rates of type i agent changing into type j agent by

$$w(\mathbf{n}, \mathbf{n} - \mathbf{e}_i + \mathbf{e}_j) = \lambda_{i,j}\nu(n_i, n_j, \mathbf{n});$$

With transition rates between states specified, the dynamics for the probability is given by the following equation, where s, s', and s'' refer to some states

---

[5] For example, $w(\mathbf{n}; \mathbf{n} + \mathbf{e}_j)\Delta \approx Pr(X_{t+\Delta} = \mathbf{n} + \mathbf{e}_j \mid X_t = n)$.

$$dP(\mathbf{s},t)/dt = \sum_{s'} w(s',s)P(s',t) - \sum_{s''} w(s,s'')P(s,t)$$

This is called the master equation in phyics, ecology and chemistry, and we follow their usage of the name. A specific example of interest has the transition rates:

$$w(\mathbf{n},\mathbf{n}+\mathbf{e}_k) = c_k(n_k + h_k)$$

for $n_k \geq 0$,

$$w(\mathbf{n},\mathbf{n}-\mathbf{e}_j) = d_j n_j$$

$n_j \geq 1$, and

$$w(\mathbf{n},\mathbf{n}-\mathbf{e}_j+\mathbf{e}_k) = \lambda_{jk} d_j n_j c_k(n_k + h_k)$$

with $\lambda_{jk} = \lambda_{kj}$, and where $j,k = 1,2,\cdots K$. We assume that $d_j \geq c_j > 0$, and $h_j > 0$, and $\lambda_{jk} = \lambda_{kj}$ for all j,k pairs.

The first transition rate specifies entry rate of type k agents, and the second that of the exit or departure rate by type j agents and the last specifies the transition intensity of changing types by agents from type j to type k. In the entry transition rate specification $c_k n_k$ stands for attractiveness or disadvantage of larger group, such as network externality which makes it easier for others to join the cluster or group, or congestion which may induce avoidance of larger groups, as the case may be. The term $c_k h_k$ stands for the innovation effects which is independent of the group size. These transition rates for type changes are in Kelly (1979), for example. We need interactions or correlations among agents. It turns out that parameter θ, to be introduced in connection with (2) below, plays this role. See Aoki (2000a, 2002b). The jump Markov process thus specified has the steady state or stationary distribution

$$\pi(\mathbf{n}) = \prod_{j=1}^{K} \pi_j(n_j),$$

where

$$\pi_j(n_j) = (1-g_j)^{h_j} \binom{-h_j}{n_j} (-g_j)^{n_j}$$

where $g_j = c_j/d_j$

These expressions are derived straightforwardly by applying the detailed balance conditions to the transition rates. See Kelly (1979, Chapt.1), or Aoki (2002, p. 148) for example.

To provide simpler explanation, suppose that $g_j = g$ for all j. Then, noting that $\prod_j (1-g)^{h_j} = (1-g)^{\sum_j h_j}$, the joint probability distribution is expressible as

$$\pi(\mathbf{n}) = \left(\begin{matrix} -\sum_{n} h_k \end{matrix}\right)^{-1} \prod_{j=1}^{K} \left(\begin{matrix} -h_j \\ n_j \end{matrix}\right) \tag{1.1}$$

By a suitable limiting process this distribution goes to the Ewens distribution. To see this suppose that K becomes very large and h very small, while the product Kh approaches a positive constant θ. We note that the negative binomial expression

$$\left(\begin{matrix} -h \\ j \end{matrix}\right)^{\alpha_j}$$

approaches $(h/j)^{\alpha_j}(-1)^{j a_j}$ as h becomes smaller. Suppose $K_n = k \leq K$. Then, there are

$$\frac{K!}{a_1! a_2! \cdots a_n!(K-k)!}$$

many ways of realizing a vector. Hence

$$\pi(\mathbf{n}) = \left(\begin{matrix} -\theta \\ n \end{matrix}\right)(-1)^n \frac{K!}{a_1! a_2! \cdots a_n!(K-k)!} \prod_j (\frac{h}{j})^{\alpha_j} \tag{1.2}$$

Noting that $K! = (K-k)! \times h^k$ approaches θ^k in the limit of K becoming infinite and h approaching 0 while keeping Kh at θ, we arrive, in the limit, at the probability distribution, known as the Ewens distribution, or Ewens sampling formula very well known in the genetics literature, Ewens (1972).

$$\pi(\mathbf{n}) = \frac{n!}{\theta^{[n]}} \prod_{j=1}^{n} (\frac{\theta}{j})^{a_j} \frac{1}{a_j!},$$

where $\theta^{[n]} := \theta(\theta+1) \cdots (\theta+n-1)$. This distribution has been investigated in several ways. See Arratia and Tavaré (1992), or Hoppe (1987). Kingman (1980) states that this distribution arise in many applications. There are other ways of deriving this distribution. We next examine some of its properties.

The number of clusters and value of θ

Ewens sampling formula has a single parameter θ. Its value influences the number of clusters formed by the agents. Smaller values of θ tends to produce a few large clusters, while larger values produce a large number of smaller clusters. To obtain some quick feels for the influences of the value of θ, take $n = 2$ and $a_2 = 1$. All other as are zero. Then

$$\pi_2(a_1 = 0, a_2 = 1) = \frac{1}{1+\theta}$$

This shows that two randomly chosen agents are of the same type with high probability when θ is small, and with small probability when θ is large. In fact, θ controls correlation between agents' types or classification. Furthermore, the next two expreme situations may convey the relation between the value of θ and the number of clusters. We note that the probability of n agents forming a single cluster is given by

$$\pi_n(a_j = 0, 1 \leq j \leq (n-1), a_n = 1) = \frac{(n-1)!\theta}{\theta^{[n]}}$$

while the probability that n agents form n singleton is given by

$$\pi_n(a_1 = na_j = 0, j \neq 1) = \frac{\theta^{n-1}}{(\theta+1)(\theta+2)\cdots(\theta+n-1)}$$

With θ much smaller than one, the former probability is approximately equal to 1, while the latter is approximately equal to zero. When θ is much larger than n the opposite is approximately true.

We can show that

$$P_n(K_n = k) = \frac{1}{\theta^{[n]}}c(n,k)\theta^k$$

where $c(n,k)$ is known as the signless Stirling numbers of the first kind, and is defined by

$$\theta^{[n]} = \sum_1^n c(n,k)\theta^k.$$

See Hoppe (1987) for the derivation. Stirling numbers are discussed in van Lint and Wilson (1992, p.104) for example. Another class of interesting transition rates arise by applying what is called the Johnson's sufficientness postulate[6] in the statistical literature. In modeling industrial sector with ni being the number of agents of type i, the word type may refer to the kinds of goods being produced by firm i or n_i may refer to the size of the "production line", that is, a measure of capacity utilization by firm producing typ i good. Zabell (1982) proved that under the assumption of exchangeable partitions the functional form of f is specified by

$$f(n_i, n) = \frac{n_i}{n+\theta}$$

with some positive scalar parameter θ . Therefore, the entry rate of a new type is given by $\theta/(n+\theta)$. More generally, with K types, it is of the form

$$w(\mathbf{n}, \mathbf{n} + e_k) = \frac{\alpha + n_k}{K\alpha + n}$$

---

[6] Johnson's sufficientness postulate stipulates that the conditional probability that the next agent which enters is of type i, given the current state vector, is $f(n_i, n)$, that is, a function of the existing number of agents of type i and that of the total number of agents in the model. See Zabell (1982).

which reduces to (2) in the limit of α going to zero, and K to infinity while their product approaches θ, and

$$w(\mathbf{n}, \mathbf{n} - e_j) = \frac{n_j}{n}$$

See Costantini (1979, 2000), and Zabell (1982) for circumstances under which these transition rates arise. See Aoki and Yoshikawa (2001) and Aoki (2002a, Sec.8.6) for an application of this type of transition rates in models of economy or sectors of economy.

Densities of the large fractions

Aoki (2002a, Sec.10.6) gives expressions for the densities of the r largest fractions of clusters. In the case where the largest fraction x is greater than $1/2$, its density is given by

$$p(x) = \frac{\theta}{x}(1 - x)^{\theta - 1}$$

For the largest two fractions x and y such that $y \geq (1 - x)/2$, the joint density is

$$f(x, y) = \frac{\theta^2}{xy}(1 - x - y)^{\theta - 1}$$

These are used in Aoki (2002b) to discuss asset returns in an asset market . in which there are two dominant groups of agents, that is two largest clusters such that $x + y$ is about 0.7 or larger.

1.4 Gibbs Partitions

We can construct more complex combinatorial structures by introducing the notion of compound or internal states of a particular combinatorial structures. They could be called colors of structures, and may correspond to internal energies in the case of physical components.

The multiset $\{| A_1 |, \cdots | A_k |\}$, related to the sizes of k blocks of a partiton of $[n]$, of unordered sizes of blocks defines a partition of n. Earlier we have introduced two ways of representing or encoding these: non-increasing order of sizes, and partiton vectors.

Let V be some kind of combinatorial structures (called species in Pitman (2002)). As before $[n]$ is the set of n agents, objects or elements. Denote by $V([n])$ a set of V-structures such that the number of V-structures on $[n]$ is $| V([n]) |= v_n$.

Let W be another species of combinatorial structures. Let w_j be the number of W-structures on a set of j agents.

We construct the composite structure on $[n]$, $(VoW)([n])$, which is the set of all ways of partitioning $[n]$ into blocks $\{A_1, \cdots A_k\}$ for some $k = 1, 2, \cdots, n$, and assigning to each block A_i a W-structure. The number of composite structure is then

$$| (VoW)([n]) | := B_n(\mathbf{v}, \mathbf{w}) := \sum_{k=1}^{n} v_k B_{n,k}(\mathbf{w})$$

where

$$B_{n,k}(\mathbf{w}) := \sum \prod_{i=1}^{k} w_{|A_i|}$$

and where the sum is over all possible partitions of n agents into k clusters. Using the composition $n = n_1 + n_2 + \cdots + n_k$, we may write this as

$$B_{n,k}(\mathbf{w}) = \frac{n!}{k!} \sum \prod_i \frac{w_{n_i}}{n_i!}$$

where the sum is over $(n_1, n_2, \cdots, n_k)$. Since there are a_j of blocks of size j, and j agents can be arranged in $j!$ ways, and a_j blocks in $a_j!$ ways, we have

$$\left[\prod_j w_j^{a_j} \right] B_{n,k}(\mathbf{w} = \frac{n!}{\prod_j (j!)^{a_j} a_j!}$$

with $\sum_j = k \sum_j j a_j = n$. Here the notation $[x^j] f(x) = c_j$ means that the polynomial $f(x)$ has c_j as the coefficient of power x^j. The symbol $B_{n,k}()$ is known as the Bell polynomial in the combinatorics literature. We use the generating function

$$w(x) := \sum_{j=1}^{\infty} w_j \frac{x^j}{j!}$$

to write

$$B_{n,k}(\mathbf{w}) = \left[\frac{x^n}{n!} \right] \frac{w(x)^k}{k!}$$

For each partition of $[n]$

$$Pr(\prod_n = \{A_1, A_2, \cdots, A_k\}) = p(| A_1 |, \cdots (| A_k |]; \mathbf{v}, \mathbf{w}),$$

where

$$p(n_1, \cdots, n_k; \mathbf{v}, \mathbf{w}) = \frac{v_k \prod_i w_{n_i}}{B_n(\mathbf{v}, \mathbf{w})}$$

Here

$$B_n(\mathbf{v}, \mathbf{w}) = \sum_{k=1}^{n} v_k B_{n,k}((w)).$$

We call $\prod_n$ a Gibbs $(\mathbf{v},\mathbf{w})$ partition if the distribution of $\prod_n$ on the set of all partitions of $[n]$ is as given above. A random partition of n induced by a random partiton n of $[n]$ is represented by partition vector $\mathbf{a}$ with

$$Pr(|\,\pi_n\,|_j = a_j, j = 1, \cdots, n) = \frac{n!v_k}{B(\mathbf{v},\mathbf{w})} \prod (\frac{w_j}{j!})^{a_j} \frac{1}{a_j!}$$

where $|\,\prod_n\,|_j$ is the number of blocks of size j.

Economic interpretation

Our economic interpretation is as follows: Suppose that n agents are partitioned into clusters such that each agent belongs to a unique cluster. The collection of clusters is represented by a partition of $[n]$.

Assume that each cluster of size j can be in any one of w_j different internal "states", or "color" $\mathbf{w} := (w_1, w_2, \cdots)$, where w_i is a non-negative integer.

Configuration of the system of n agents is a partition of the set $[n]$ into clusters, plus the assignment of an internal state to each cluster. For each partition of $[n]$ into k blocks of sizes $n_1, n_2, \cdots, n_k$, there are $\prod_i w_{n_i}$ different configurations.

When $v_k = 1$ for some k and zero elsewhere, the Gibbs partition corresponds to those in which all configurations with k clusters are equiprobable. This is the microcanonical configurations in physics. A general weight sequence $\mathbf{v}$ randomizes k to allow all probabilistic mixture over k of these microcanonical states.

Ewens distributions are derived as a special case of the above in Pitman (2002). Because of our interest in underlying dynamics generating the distributions, we have earlierr provided a jump Markov process derivation for them. Whittle used reversible equilibrium distribution of a Markov process to construct particualr cases of Gibbs partitions, Whittle (1986).

Entry of new types

Our view on economic growth is that growth is sustained by continual introduction of goods of new types which stimulate demands for these new goods, not by R & D activities which re ne existing goods, Aoki and Yoshikawa (2002).

New entries could be newly invented or improved goods, new business models, new behavioral patterns and so on. Law of succession in the statistical literature address these questions as conditional probabilities of agents entering models from outside being new or one of existing types in the model. Here we rely on recent works by Kingman and Pitman. Their models can be approximated as birth-immigration models in the context of continous time branching processes and we introduce their resuls into our models. See Feng and Hoppe (1998)for the mathematical set-up.

Let $X_1, \ldots X_n \ldots$ be an infinite sequence of random variables taking on any of a finite numbe of values, say 1, 2, $\cdots$ k. The subscripts on X are thought of time index, or the order in which samples are taken or agents enter the system.

The sequence is said to be *exchangeable* if for every n, the cylinder set probabilities

$$Pr(X_1 = j_1, \ldots X_n = j_n) = Pr(j_1, j_2, \cdots j_n)$$

are invariant under all possible permutations of the time index. Two sequences have the same probability if one is a rearrangement of the other, or the probability is the function of the frequency vector, $\mathbf{n} = (n_1, n_2, \ldots n_k)$. The observed frequency counts, $n_j = n_j(X_1, X_2, \cdots X_n)$ are sufficient statistics for the sequence in the sense that probabilities conditional on the frequence counts depend only on the frequency vector

$$Pr(X_1, X_2, \ldots, X_n \mid \mathbf{n}) = \frac{n_1! n_2! \ldots n_k}{n!}$$

de Finetti theorem says that

$$Pr(X_1 = j_1, X_2 = j_2, \ldots, X_n = j_n) = \int p_1^{n_1} p_2^{n_2} \cdots p_1^{n_k} d\mu(p_1, p_2, \cdots, p_k),$$

over the simplex Δ_k of p_js which sum to one. Once the prior $d\mu$ is implicitly or explicitly specified, it is immediate that

$$Pr(X_{n+1} = j_1 \mid X_1, X_2, \ldots, X_n) = Pr(X_{n+1} = j_i \mid \mathbf{n}).$$

Such a conditional probability is sometimes called a rule of succession. Johnson's sufficientness postulate[7] is

$$Pr(X_{n+1} = i \mid \mathbf{n}) = f(n_i, N):$$

If $X_1, X_2, \ldots$ is an exchangeable sequence satisfying the sufficientness postulate, and $k \geq 3$, then assuming that the relevant conditional probabilities exist

$$Pr(X_{n+1} = i \mid \mathbf{n}) = \frac{n_i + \alpha}{n + k\alpha}:$$

See Zabell (1982).

How different are the estimates of probability of new types with the Ewens distribution and multinomial distribution? Ewens (1996) has some numerical examples which show that they are quite different. With the multinomial approach, a_1 (the number of singleton) is critical. With the Ewens formula, the total number of types, $\sum a_j$ is only relevant.

---

[7] so called by I. J. Good to avoid confusion with the notion of sufficient statistics

The Pitman two-parameter model

Pitman (1992) generalized the Ewens' distribution by using the transition rates

$$w(n, n + e_j) = \frac{n_j - \alpha}{n + \theta}$$

where $\theta + \alpha > 0$. In terms of the rule of sucession it becomes

$$Pr(X_{n+1} \mid \mathbf{n}) = \frac{n_i - \alpha}{n + \theta - K_n \alpha} :$$

for α between 0 and 1, and θ positive, the conditional probability for a new type is

$$Pr(X_{n+1} = \text{new}) = \frac{\theta}{n+\theta-K_n\alpha} :$$

Pitman (1995). With this, the conditional probability that a new type enters in the next Δ time interval is approximately given by $\frac{K_n\alpha+\theta}{n+\theta}\Delta$. Pitman also derived the equilibrium distribution for this two-parameter version.

The two-parameter Poisson-Dirichlet distribution, $PD(\alpha, \theta)$, for some α between 0 and 1 and $\theta > -\alpha$ is a probability distribution on the sequence of fractions V_n, with $V_1 > V_2 > \cdots$, and $\sum_n V_n = 1$. Let X_n, $n = 1, 2, \ldots$ be independent random variables with $Beta(1 - \alpha, \theta + n\alpha)$ distribution. Let U_n be the random variables with residual allocation, that is, $U_1 = X_1$; $U_2 = (1 - X_1)X_2$, ... Let V_n be the decreasing order statistics. This is Pitman's $PD(\alpha, \theta)$ distribution. When α is zero, it reduces to the Kingman $PD(\theta)$. This corresponds to the conditional probability

$$Pr \text{ (new type} \mid X_1, X_2, \ldots X_n) = \frac{\alpha+\theta}{n+\theta},$$

and

$$Pr \text{ (existing type} \mid X_1, X_2, \ldots X_n) = \frac{n-\alpha}{n+\theta},$$

In other words, when there are k clusters of types in the data, the probability of a new type appearing as the next observation is increased from the $\theta/(\theta+n)$ to $(\theta+k\alpha)/(\theta+n)$, and correspondingly the probability of observing type i next is reduced to $(n_i - \alpha)/(\theta + n)$. With the partition vector $\mathbf{a}$, the probability is

$$Pr(\mathbf{a} : \alpha, \theta) = \frac{1}{\theta^{[n]}} \prod_1^{K_n} (\theta + (i - 1)\alpha) \prod_1^n \left[(1 - \alpha)^{[j-1]} \right]^{a_j}$$

where $K_n = \sum a_j$, and $\sum j a_j = n$. There are interpretations in terms of what is called size-biased sampling of these, which is briefly mentioned below. See Kingman (1992), and Pitman (1992, 1995)

Urn models

It is most instructive to use urn models to describe the ways conditional probabilities of new types are constructed. Hoppe (1984) generalizes the Polya urn by introducing a special ball, the black ball say, of weight . All other balls of various colors have weight one. Initially, the urn contains just the black ball. So, at the first drawing, the black ball is drawn, and is returned to the urn, together with a ball of color 1, say. When a non-black ball is drawn in subsequent drawings it is returned together with another ball of the same color. When the black ball is drawn, it is returned togehter with a ball of color so far not seen. That is, a ball of new color signifies a new type of balls, that is agents. After n drawings, the urn contains balls of total weight $\theta + n$. Thus, the probability of a new type next is $\theta/(\theta + n)$, while that of drawing a ball of existing color is $n/(\theta + n)$. In (3), as α goes to zero, and $K_n \alpha$ approaches θ, Ewens (1996) discusses how the conditional probability assigned by the Ewnes distribution differs from the multinomial distributions.

Let a process $\{X_n\}$ generated by sampling from the urn. Let K_n be the number of colors in the urn.

Fix **a** and consider a possible sample path $\{X_1 = j_1, X_2 = j_2, \ldots, X_n = j_n\}$

$$Pr(X_1 = j_1, X_2 = j_2, \ldots, X_n = j_n) = \frac{\theta^{K_n} \prod_{j=1}^{n}(n_j - 1)!}{\theta^{[n]}}$$

Here $\theta^{[n]} := \theta(\theta + 1) \cdots (\theta + n - 1)$. Recall that the black ball has been selected K_n times, and that each ball of new color is followed $n_j - 1$ times by balls of the same color.

To count the number of samples, there are two constraints:

1. we observe that the first ball of color 1 precedes the first ball of color 2, which precedes the first ball of color 3, and so on.
2. We merely know that there are n_1 balls of some color, n_2 of another and so on.

To count the sample path subject to the above two constraints, arrange of occupancy numbers in decreasing order, $n_{(1)} \geq n_{(2)} \cdots n_{(K_n)}$. Suppose that there are p distinct integers in the set of the ordered frequencies. Define α_1 to be such that $n_{(i)} = n_{(1)}$, α_2 be such that $n_{(i)} = n_{(\alpha_1+1)}$, and so on. Finally, α_p is such that $n_{(i)} = n_{(K_n)}$.

There are $K_n!/\prod_{i=1}^{p} \alpha_i!$ ways of distributing $\{n_1, n_2, \ldots n_{K_n}\}$ among K_n types. For each such distribution of the occupancy numbers there are $n!/n_1! \cdots n_{K_n}$ permutations of labels agreeing with the occupancy numbers. In total, there are

$$\frac{K_n! n!}{\prod \alpha_i! \prod n_j!}$$

permutations which meets with constraint (2). Not all among these meet the constraint (1). Separate the permuta- tions into disjoint classes by the order of the first appearance of the digits $\{1, 2, \cdots K_n\}$. There are $K_n!$ such disjoint classes, all of the same cardinality by symmetry. Only one of them satis es (1). Thus dividing the above by $K_n!$, and multiplying by the probability of one sample, we have the probability of the random partition being given by **a**. This is the Ewens sampling formula.

When we let k go to infinity, and α to zero in such a way that $k\alpha$ goes to a positive limit, we have

$$Pr(X_{n+1} = i \mid \mathbf{n}) = \frac{n_i}{n + \theta}$$

Then

$$Pr(X_{n+1} = new \mid \mathbf{n}) = \frac{\theta}{n + \theta}.$$

This leads us to the Ewens sampling formula, well known in the population genetics literature.

The notion of partition exchangeability generalizes the notion of exchange-ble sequence by imposing exchangeability with respect to categories, as well as time indices. That is, a probability function P is partition exchangeable if the cylinder set probabilities $Pr(X_1 = j_1, \ldots X_n = j_n)$ are invariant under permutation of the time index and the type index. The role of frequency vec-tor is now played by the partition vector **a** where a_i denotes the number of types with i entries, that is, the number of n_j that are equal to i.

The predictive probabilities for partition exchangeable probabiities will have the form

$$Pr(X_{n+1} = i \mid X_1, X_2, \ldots, X_n) = f(n_i \mid \mathbf{a}).$$

Sampling, residual allocation models, and distributions of order statistics

Given K categories or types, and the associated random probability vector

$$\mathbf{p} = (p_1, p_2, \ldots, p_K);$$

let ν be a random variable having values $1, 2, \ldots K$ such that

$$Pr(\nu = r) = p_r;$$

$r = 1, 2, \ldots, K$. This probability p_μ is said to be obtained from **p** by sizebiased sampling. This process is repeated by renormalizing the remaining probability by dividing it by $1 - p$ and proceeding as before.

The components of p is rearranged as a vector q, the components of which are expressible by

$$q_1 = v_1, q_2 = (1 - v_1)v_2, q_3 = (1 - v_1)(1 - v_2)v_3, \ldots$$

The random variables $v_1, v_2, \ldots$ are independent. For example, vs may have beta distributions. This last construction is an example of a residual allocation model, also known as a broken stick model. With the one-parameter size-biased samples, when the random vari- ables q_s are arranged in descending order $q_1 \leq q_2 \leq \ldots$, that is, reordered into the descending order statistics is distributed according to the Poisson-Dirichlet distribution invented by Kingman, where the random variables qs are $Beta(1, \theta)$, other random variables are used to generate a two-parameter generalization of the Ewens distribution by Pitman.

Economic Applications

A traditional approach to model two-sector economy

In 1989 Dixit has analyzed several economic problems, such as that of how to optimally allocate capital stocks among two sectors, and of assessing the effects of exchange rate changes to induce entries or exits of firms in some export industry. In a setting of a two-sector economy what Dixit derives is the price schedule, that is, the price as a function of the number of firms existing in one sector. When the relative price of the two goods crosses the price schedule from above or below, a move by one more firm into or from one sector to the other is triggered. This, however, is analyzed as a problem for a central planner of the economy, not as problems for individual firm managers. He does not say how a firm manager knows that it is his turn to enter or switch sectors. In spite of random prices his approach is basically deterministic. What are some of the objections to this analysis? First, as we already mentioned, there is no explanation about which of the firms decide to move. Also, there is apparently no uncertainty as to which firm switches. Problems of imperfect or incomplete information and externalities among firms (agents) are cleverly hidden or abstracted away in his analysis.

An Alternative Approach: Basic Setup

Aoki (2002b) gives one potential applications of the Ewens sampling formula in finance.

Aoki (1996, 2002a) presents several examples of some alterntive approaches to that sketched above. Basically, our approach focuses on the random partitions of the set of firms into clusers induced by subsets formed by firms of the same types, and utilizes the conditional probability specifications for new entries and exits to derive equilibrium distributions, when they exist, for cluster sizes. We use the master equation (backward Chapman-Kolomorov equation) as the dynamic equation for the probabilities of state vectors.[8] Given the total number of agents, N, and the number of possible types, K, both of which are assumed in this paper to be known and finite for ease of explanation, we examine how the N-set, that is, the set $\{1, 2, \ldots N\}$ is partitioned into K clusters, or subsets. This partition is treated as a random exchangeable partition in the sense of Zabell (1992).

Holding times

Jump Markov processes stay at each state it visits for a while, called sojourn or holding time, before it jumps to another state. Holding times are exponentially distributed. Given a number of agents wishing to jump, one with the minimum holding time actually can jump. This notion is applied in Aoki (2002a, Chapt. 8) to a model of economy with several sectors. Each sector faces a fraction of the aggregate outputs of the economy as its demand. In his model some agents are in excess supplies and others are in excess demands. Those in the excess supply conditions wish to reduce their production, and those in excess demands wish to expand the production. The aggregate outputs a ect demand conditions each agent faces. Therefore, as soon as one agent adjusts its output first, that changes the aggregate output, and possibly the demand conditions each agent faces. Consequently, sets of agents with positive or negative aggregate demand conditions generally change as one agent actually jump to a new state. The notion of holding time is therefore useful as a conceptual device in choosing which of the agents actually can carry out their intended decisions in condiitons with externalities.

Power Laws and volatility switching

Aoki (2000, 2002) has examples of an asset market in which two dominant clusters of agents trade. This market exhibits power laws for returns. When two types of agents switch between two strategies, returns exhibit switchings of volatility as well.

Sluggish responses of dynamics as power laws, that is, decay of the form $t^{-\alpha}$ for some positive α rather than exponential decay $e^{-\lambda t}$ are found in models in which states are arranged as trees and transition rates between states are the functions of ultrametric distance as in Aoki (1996, Sec. 7.1). As the number of layers of tree nodes increase, reduce transition rates appropriately. In the limit we obtain power laws decays, not exponential decays. See Ogielski and Stein (1985), or Aoki (1996, p. 157).

---

[8] In a closed two-sector model the scalar variable of the number of firms in sector one, say, serves as the state variable. In an open model with K sectors, a K-dimensional vector is used.

1.5 Concluding Remarks

This paper proposes a finitary approach to economic modeling, that is to start with a finite number of agents with discrete choice sets, and with explicit transition rates. It discusses several entry and exit transition rates in economic models. In particular, it presented Ewens and related distributions as candidates for distributions of cluster sizes formed by a large number of economic agents who interact in a market. This distribution seems to be very useful in economic modelings, although we have only a few examples so far. However, see Arratia and Tavaré (1992), and Kingman (1980). These and other investigations strongly suggest that the Ewens' and related distributions are robust and ubiquitous.

Carlton (1999) discusses some estimation issues of two-parameter Poisson-Dirichlet distribution.

Although no application is described in this paper, Aoki (2002a, 2002b) has one simple application in which stocks of a holding company is traded by a large number of agents. With $\theta =: 3$, two largest groups are shown to capture nearly 80 per cent of the market shares and hence dominate the market excess demands for the shares, which in turn determine the stationary distributions of price. In this way it is also possible to relate the tail distribution of the market clearing prices with entry and exit assumptions.

References

1. Antoniak, C., (1969). "Mixtures of Dirichlet processes with applications to Bayesian nonparametric problems", Ph.D. dissertation, Univ. California, Los Angeles. Arratia, R., and A.D. Barbour and S. Tavar e, " Poisson process approximations for the Ewens sampling formula", Ann. Appl. Probab. 2 519-35, 1992.
2. Arratia, R., and S. Tavaré, "Independent process approximations for random combinatorial structures ", Adv. Math. 104, 90-154, 1994.
3. Aoki,M.,(1996) New Approaches to Macroeconomic Modeling: Evolutionary Stochastic Dynamics, Multiple Equilibriua, and Externalities as Field effects, Cambridge University Press, New York.
4. ——. (1998). "A simple model of asymmetrical business cycles: Interactive dynamics of a large number of agents with discrete choices", Macroeconomic Dynamics,2, 427-442
5. ——, "Open Models of Share Markets with Several Types of Participants", presented at 1999 Wheia Conference, Univ. Genova, Genova, Italy, June, 1999.
6. ——. (2000a). "Cluster size distributions of economic agents of many types in a market," J. Math. Analy. Appl. 249, 32-52.
7. ——. (2000b). "Herd behavior and return dynamics in a share market with many types of agents," in M. Tokuyama, and H. Stanley (eds). Statistical Physics, Amer. Inst. Phys., Melville, New York.
8. ——. (2002a) Aggregate Behavior and Fluctuations in Economics. Cambridge Univ. Press, New York.

9. —. (2002b). "Open models of share markets with two dominant types of participants," J. Econ. Behav. Org., 49 199-216.

10. —, (2002c), "A simple model of volatility fluctuations in asset markets", pp.180-85, in H.Takayasu (ed.) Empirical science of financial fluctuations: The advent of econophysics, Springer, Tokyo, and New York.

11. —, and M. Shirai. (2000). "Stochastic business cycles and equilibrium selection in search equilibrium." Macroeconomic Dynamics 4, 487-505.

12. —, and H. Yoshikawa. (2001). " A simple quantity adjustment model of economic fluctuatuions and growth, " presented at the 2001 Wehia annual meeting, Maastrict, June, 2001.

13. —, and — (2002). "Demand saturation-creation and economic growth", J. Econ. Behav. Org. 48

14. Carlton, M. A., (1999). "Applications of the two-parameter Poisson - Dirichlet distribution", Ph.D. dissertation, Univ. California. Los Angeles.

15. Costantini, D, and U. Garibaldi (1979) " A probabilistic foundation of elementary particle statistics" Stud. Hist. 28, 483-506

16. Costantini, D. (2000) "A probability theory for macroeconomic modelling", Mimeo, Dep. Stat. University of Bologna.

17. Dixit, A., (1989). "Entry and exit decisions of firms under fluctuating real exchange rates," J. Pol. Econ. 97, 620-637.

18. Ewens, W. J., (1972). "The sampling theory of selectively neutral alleles", Theor. Pop. Biol. 3 87-112.

19. —, (1990) "Population genetic theory-The past and the future'" in Mathematical and Statistical Developments of Evolutionary Theory, edited by S. Lessard, Kluwer Academic Publishers, London

20. —, (1996) "Remarks on the law of sucession", in Athens Conf. on Applied Probab. on Time Series Analysis

21. Feller, W. 1968 Introduction to Probability Theory, Vol.I, Wiley, New York.

22. Feng, S., and F. M. Hoppe (1998). "Large deviation principles for some random combinatorial structures in population genetics", Ann. Appl. Probab. 8, 975-94.

23. Hoppe, F., (1984) "Polya-like urns and the Ewens sampling formula" J. Math. Biol. 20 91-94

24. — (1987) "The sampling theory of neutral alleles and an urn model in population genetics," J. Math. Biol. 25, 123-59

25. Kelly, F., (1979)Reversibility and Stochastic Networks, J.Wiley, New York.

26. Kingman, J.F.C., (1992) Poisson Processes, Oxford Clarendon Press.

27. — (1980), Mathematics of Genetic Diversity, SIAM, Philadelphia

28. Pitman, J., (1992) "The two-parameter generalization of Ewens' random partition structure", Tech. Report 345, Department of Statistics, UC Berkeley, reissued with addendum in (2002).

29. —, (2002). "Combinatorial Stochastic Processes," Lecture notes, St.Flour lecture course.

30. — (1995) "Partially exchangeable random partitions" Probab. Theory and Related Fields 102 145-58. van Lint, J. H., and R.M.Wilson, (1992) A Course in Combinatorics, Cambridge Univ. Press, New York

31. Sachkov, V. N., Probabilistic Methods in Combinatorial Anaysis, Cambridge Univ. Press, Cambridge, 1996.

32. Watterson, G.A., (1976) "The stationary distribution of the infinitely-many neutral alleles diffsusion models", Journal of Applied Prob. 13, 639-651

33. —, and Guess,(1977) H. A.,"Is the most frequent allele the oldest?' Theor.Popul. Biol. II, 141-60
34. Van Lint, J., and R. M. Wilson (1992) A Course in Combinatorics Cambridge Univ. Press, New York.
35. Whittle, P (1986) "Systems in Stochastic Equilibrium" Wiley, New York
36. Zabell, S., " Predicting the Unpredictable", Synthese 90 205-32, 1992.
37. —- (1982). "W. E. Johnson's sufficientness postulate," Ann. Stat. 10, 1090-1099.
38. —-(1997) "The continuum of inductive methods revisited" in Earman, J., and J.D. Norton (eds) Cosmos of Science, Univ. Pittsburg Press.

An ABM-Evolutionary Approach: Bilateral Exchanges, Bargaining and Walrasian Equilibria*

Nicolás Garrido[1] and Pier Mario Pacini[2]

[1] Department of Economics - University of Trento (Italy)
nico@economia.unitn.it
[2] Department of Economics - University of Pisa (Italy) pmpacini@ec.unipi.it

Summary. This paper analyzes, via intensive use of simulation techniques, the effects of the introduction of direct exchange relationships through bilateral trades in a simple general equilibrium pure exchange economy. Agents are heterogeneous in their endowments and repeatedly match in random pairs bargaining on how to split the advantages of a trade; possibly they can agree to exchange at the known market clearing prices. Simulations of this evolutionary process show that while walrasian outcomes emerge in the interaction among people with similar outside opportunities, people of different groups converge to accept an equilibrium in which agents with the best outside opportunity extract the greater part of the surplus out of an exchange. On other hand the acceptance of market mediation (i.e. walrasian outcomes) is more probable when either the parties try to exploit too much from the opponent or when there is anonymity in the trading process. The results show evidence that the acceptance of decentralized, personalized contracting (apart from efficiency considerations) increases the probability of amplifying the asymmetries in the initial distribution beyond what is produced by the pure market mechanism.

Key words: Bargaining, Bilateral trades, Social conventions, Walrasian allocations, Learning, Numerical simulations

2.1 Introduction

There has been recently an increasing interest in incorporating elements of direct interaction among agents in economic modeling in order to take into account the effects of the externalities produced by individual behaviour on

* We thank D. Fiaschi and two anonymous referees for helpful comments and all the participants in the workshops held at the University of Siena within the research project on New Tools for Qualitative Economic Dynamics and the CIMAT (Guanajuato - Mx). We acknowledge financial support from MURST PRIN 2000 New Tools for Qualitative Analysis of Economic Dynamics.

others behaviour. The ways and modalities with which interaction can enter the scene are multiform and extensively reviewed and analyzed in the economic literature ([4], [11] and [5] for an extensive survey and motivation); it is commonly stressed that incorporating wider forms of interaction (also beyond the well known strategic interaction of game theory) in which either actions, constraints, preferences or expectations on the side of one agent affect others actions, constraints, preferences or expectations may produce dramatic and some time unexpected changes in the aggregate and systemic behaviour (an example of this is [7], see also [10]).

On the other hand general equilibrium theory, as the prototype representation of an economy in which the contrasting interests of a population of heterogeneous agents are composed by a price system, stands immune to this form of interaction and traditionally it is based on an abstract way of market interaction among people, i.e. global and anonymous, as it is implied by the standard equilibrium condition by which the sum of individual excess demand has to vanish. The abstraction of this form of interaction makes it more compelling the old and often claimed desire to decompose market equilibrium [...] into specific bilateral transactions. This tendency arises from our concrete experience, especially as customers in retail markets, of market exchanges as bilateral transactions, and from the legal requirement that all transfers of property be between identifiable individual owners (see [6], p. 13). Beside that it is also common experience that in many and important markets (e.g. labour markets, financial and insurance markets) transactions are less and less coordinated by a general price signal, but exchanges are rather the consequence of an unmediated and often conflicting relation between two parties.

This paper is an attempt at incorporating direct exchange relations into a standard general equilibrium pure exchange economy and analyzes their effects on the distribution of commodities rising out of bilateral trades, in which agreements to exchange at walrasian prices are just one of the possible options. To this purpose we set up a standard (very simplified and in many respects ad hoc) economy with a number of types of agents differing as to their own endowments, but identical in preferences. An equilibrium price system for the economy as a whole (a walrasian price system) exists and is assumed to be known to everyone. Agents match in random pairs and, when paired, bargain on how to divide efficiently the surplus of the exchange; one of the possible agreements is to trade at the known walrasian prices, but they can also agree on two other different efficient divisions. In choosing a course of action in a bargaining session, everyone responds at best to the expectations that he has about the behaviour of the partner he is occasionally paired with. In this setting, the acceptance of trading at walrasian prices is a social convention supported by individual beliefs that everyone-else will behave accepting that trades will be regulated by walrasian equilibrium prices.

By means of ABM simulations we show that the dynamics of this process a la Axtell-Epstein-Young (see [1]) converges to a persistent pattern of behaviour in which the walrasian outcome is accepted in the exchange between similar parties, i.e. agents with complementary endowments and similar outside opportunities, while bilateral bargaining between agents with different outside opportunities will determine allocations in which the stronger agent gets the greater part of the surplus in the exchange.

We investigate also whether making the trading process more anonymous will restore (at least partially) the possibility of observing the walrasian outcomes to emerge as a long run convention for this society. Indeed we observe that the more the trading process becomes anonymous (despite the fact that it remains bilateral) the more probable is the fact that the socially accepted convention will prescribe walrasian allocations.

The paper is organized as follows: Section 1 describes the characteristics of the pure exchange economy that will be dealt with. Section 2 introduces the structure of interaction based on bilateral trades and defines the various settings in which it can take place according to the amount of information that agents are assumed to have when engaged in a bargaining session. Sections 2.2 and 2.3 present the details of the interaction in the various case together with basic results and the dynamics in the corresponding case. Section 3 compares the cases previously examined analyzing the characteristics of the emergent socially accepted convention and identifying its relationships with the underlaying distribution of endowments. Conclusions close the paper.

2.2 The Basic Model

We consider a standard pure exchange economy $\mathcal{E}\left(\Im, X_i, \succsim_i, \omega_i\right)$ with a population $\Im$ of agents indexed by i, each of them characterized by a consumption set $X_i \in \mathbb{R}_+^\ell$, individual preferences $\succsim_i$ and initial endowments $\omega_i \in int\left(X_i\right)$. The preference preordering is assumed to satisfy standard assumptions, i.e. the relation $\succsim_i$ is complete, reflexive, transitive, convex, monotone and continuous. For the sake of simplicity we make the following assumption:

Axiom 1 *With respect to the consumption space and preference ordering we assume that*

1. *$\ell = 2$;*
2. *all agents in $\Im$ have identical preferences that can be represented by a Cobb-Douglass utility function of the form $U_i\left(x_1, x_2\right) = \sqrt{x_1 \cdot x_2}$.*

This assumption permits on the one hand to work with well known theoretical results[3] that can be easily achieved in this context and, on the other hand, entails that any heterogeneity in the population $\Im$ can be traced back

---

[3] Notice that the aim of the paper is to explore using an evolutionary learning algorithm what eventually is the selected convention adopted by the agents to exchange their endowments, using as reference point a well known theoretical result based on real analysis. We are not concerned with how the Walrasian price system arises in our economy.

just to the distribution of endowments. Indeed heterogeneity is introduced differentiating agents by *types*, each type being identified by a particular vector of initial endowments. In order to simplify things, agents types are defined in the following way:

Axiom 2 *Take three values* $(\underline{\omega}, \overline{\omega}, \gamma)$ *such that* $\gamma > 0$ *and* $0 \leq \underline{\omega} < \overline{\omega}$; *then* $\Im$ *can be partitioned in 4 groups* $T_1, \ldots T_4$ *such that*

$T_1:$ $\forall i \in T_1,$ $\omega_i = (\overline{\omega}, \underline{\omega})$;
$T_2:$ $\forall i \in T_2,$ $\omega_i = (\underline{\omega}, \overline{\omega})$;
$T_3:$ $\forall i \in T_3,$ $\omega_i = (\overline{\omega} + \gamma, \underline{\omega} + \gamma)$;
$T_4:$ $\forall i \in T_4,$ $\omega_i = (\underline{\omega} + \gamma, \overline{\omega} + \gamma)$.

This assumption together with Assumption 1 entails that $U_i(\omega_i) = U_j(\omega_j)$ for any pair of agents i and j such that either (1) $i \in \{T_1 \cup T_2\}$ and $j \in \{T_1 \cup T_2\}$ or (2) $i \in \{T_3 \cup T_4\}$ and $j \in \{T_3 \cup T_4\}$. In other words agents of the first two types get the same level of welfare in the case they would not agree to make a transaction and the same is true for agents of types 3 and 4; however it is also true that $U_i(\omega_i) < U_j(\omega_j)$ for any pair of agents i and j such that $i \in \{T_1 \cup T_2\}$ and $j \in \{T_3 \cup T_4\}$, i.e. agents of the latter two types have better outside opportunities than agents of type 1 and 2 actually have. The situation is portrayed in the following picture

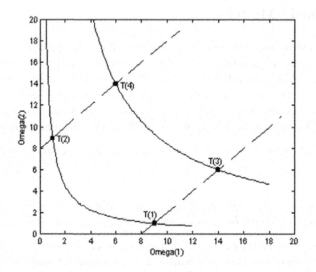

Fig. 2.1. Distribution of endowments and agents types

For the sake of convention we will term agents of types 1 and 2 on the one side and of type 3 and 4 on the other side as agents of the same *class*, while agents of type 1 and 3 on the one side and 2 and 4 on the other will be agents of the same *species*. The class and the species are sorts by *tags* that agents carries and that can eventually be observed by the counterpart in an exchange according to the different information settings that we will examine in Section 2. Finally, and again for the sake of simplicity, we assume that the number of agents of any type is the same so that $|T_n| = \frac{3}{4}$ for $n \in \{1, 2, 3, 4\}$. This assumption will remove any asymmetry in the realizations of the economy due to the predominance in the population of a certain type of agents. Walrasian equilibrium prices and allocations for the economy described can be easily computed. Indeed, given Assumption 1 and taking commodity 2 as the numeraire, the equilibrium price of commodity 1 is given by

$$p_1^* = \frac{\omega_2}{\omega_1}$$

where $\omega_j = \sum_{i \in \Im} \omega_{ij}$ is the aggregate endowment of the j-th commodity. Given the assumed distribution of endowments (as in Assumption 2), we have

$$p_1^* = \frac{\frac{3}{2} \cdot (\underline{\omega} + \overline{\omega} + \gamma)}{\frac{3}{2} \cdot (\underline{\omega} + \overline{\omega} + \gamma)} = 1$$

Equilibrium allocations are identical for agents of the same class, i.e.

$$x_i^* = \left(\frac{\underline{\omega} + \overline{\omega}}{2}, \frac{\underline{\omega} + \overline{\omega}}{2} \right), \ i \in \{T_1 \cup T_2\}$$

and

$$x_i^* = \left(\frac{\underline{\omega} + \overline{\omega}}{2} + \gamma, \frac{\underline{\omega} + \overline{\omega}}{2} + \gamma \right), \ i \in \{T_3 \cup T_4\}$$

From these equilibrium configurations we can easily compute the patterns of equilibrium net trades of agents of different types, i.e.

$$\Delta x_i^* = \left(\frac{\underline{\omega} - \overline{\omega}}{2}, \frac{\overline{\omega} - \underline{\omega}}{2} \right), \ i \in T_1$$

$$\Delta x_i^* = \left(\frac{\overline{\omega} - \underline{\omega}}{2}, \frac{\underline{\omega} - \overline{\omega}}{2} \right), \ i \in T_2$$

$$\Delta x_i^* = \left(\frac{\underline{\omega} - \overline{\omega}}{2}, \frac{\overline{\omega} - \underline{\omega}}{2} \right), \ i \in T_3$$

$$\Delta x_i^* = \left(\frac{\overline{\omega} - \underline{\omega}}{2}, \frac{\underline{\omega} - \overline{\omega}}{2} \right), \ i \in T_4$$

We can see that, in a walrasian equilibrium, all agents of the same species have exactly the same excess demand vector; more precisely agents of types 1 and 3 supply commodity 1 to agents of the opposite species from which they obtain a net positive amount of the numeraire commodity. Therefore there is *no intra-species transaction* in equilibrium, but all transactions are of the inter-species type.

2.3 The Mechanisms of Exchange: Bilateral Trades

To frame the interaction mechanism suppose that trading takes place in time and in every period $s, s = 1, 2, \ldots$, there is a *bargaining session* where agents meet in random pairs. Any pair is formed by a couple of agents of different type; two partners in an exchange can at most recognize the type of the opponent, but cannot determine his specific identity. In other words an agent, when paired with a random partner, can receive information on the *tags*, i.e. class and species, of the opponent (the information may not be complete as it will be clarified in the following); after receiving this information, he can condition his own behaviour on what he knows about the past behaviour of the agents with those tags (or that tag), though he does not know the precise identity of the counterpart. Therefore any agent is playing against groups (indeed types) of agents and not against single persons. Once a pair of agents is formed, they both simultaneously make a proposal of net trade to the counterpart; if the two proposals are compatible, i.e. the sum of the two net trades vanishes, exchange takes place, the so obtained allocations are consumed and payoff are distributed[4]; if the two proposals are not compatible the two agents leave the bargaining session without trading and enjoy their own outside opportunity (initial endowments). Then we pass to the subsequent period in which a new session of bilateral bargaining and exchange opens with the same rules. From period to period individual characteristics (i.e. preferences and endowments) remain unchanged.

2.3.1 The Nature of Matching and Bilateral Trades

In this framework we can distinguish basically two different bargaining situations, according to the information that an agent has about the characteristics of the agent which he is paired with. In particular this information concerns the recognition of the *tags* of an opponent.

In the first case we assume that agents, when they form a trading pair, can recognize all the characteristics (the class and the species, though not the identity) of the opponent and hence they can condition their own behaviour on what they know about the precise type of the opponent. This situation tries to represent our common experience in retail markets, capturing the usual circumstance in which exchange takes the form of a contract between two perfectly identifiable parties. This is the situation that we will refer to as *bilateral trade with perfect identifiability*.

However it may be objected that this situation is not so conformable to the one commonly observed in real markets, indeed the formation of multilateral coalition of traders is common in markets (see [6]). In the present setting this means that an agent, when paired with another one, knows that he will make

---

[4] This entails that the outcomes of an exchange cannot be accumulated and transferred to future periods.

an exchange with a trader out of a coalition of traders, but he is not able to identify the opponent as an agent of a particular type and hence he is not able to condition his behaviour to the opponents type; in other words he bargains against a coalition of types.

Thus, the second case is the one in which a less refined amount of information is available prior to the exchange; an agent can recognize at least the species of the opponent and he can condition his own behaviour to the behaviour of the other species. This setting tries to capture the situation in which an agent bargains and trades with a coalition of other complementary agents (precisely agents of the opposite species), but the trading is multilateral in the sense that it takes the form of one against many. We will refer to this situation as bilateral trade *with imperfect identifiability*[5].

In the following we will examine these bargaining situations separately describing what are the information, expectations, actions and payoffs available to an agent when he bargains with an opponent.

2.3.2 Bilateral Trade with Perfect Identifiability

In the present case, agents in any pair form a sub-economy with complete information on the respective endowments and also the bargaining space is determined with certainty. Given the assumed distribution of endowments, the equilibrium prices and allocations of the formed sub-economies are identical to those that would be obtained in the walrasian equilibrium of the grand economy; this indeed serves the purpose to eliminate any asymmetry between the behaviour in the grand economy and in any subeconomy that can form.

Strategies and Payoffs

Bargaining takes the form of an ultimatum game in which the two partners simultaneously make proposals on how to allocate efficiently among themselves the available aggregate quantities of the two commodities. To this purpose we assume that a proposal is the prospect of a net trade z_i that agent i makes to his opponent, say j, and such that the pair of allocations $(w_i + z_i, w_j - z_i)$ is *feasible* and *efficient* for both parties; if the simultaneous proposals of the two agents (i.e. z_i and z_j) are compatible (i.e. $z_i = -z_j$) then the exchange takes place, otherwise they can only enjoy the outside opportunity u_i^w obtained by

---

[5] Actually there is a third case that we can refer to as bilateral trade with no iden- tifiability. This represents a further relaxation of the trading relation in which an agent has no information at all on the opponents type and agents of the same species but different classes can possibly be partners in a trading session. However agents of the same species will not trade together, so every time they face they will obtain just their outside opportunity (i.e. just a scaling down of the expected payoffs by a constant). By this fact this relaxation produces a quantitative change in expected individual payoffs, but will not induce any qualitative change compared to Case 2. For this reason this case will not be examined further.

consuming the initial endowments. For the sake of simplicity we assume that agents have only three strategies at their disposal:

z_i^2: agent i proposes to the counterpart to accept walrasian allocations, i.e. he proposes to behave as if their exchanges were regulated by the walrasian equilibrium price system of the economy as a whole.

z_i^1: agent i asks a δ-*sacrifice* to the opponent j, in the sense that he proposes to the opponent a net trade which is a convex combination with parameter δ of the walrasian equilibrium trade and of the trade that would guarantee himself all the advantages from exchange: i.e. $z_i^1 = (1 - \delta) \cdot z_i^2 + \delta \cdot \bar{z}_i$ where $\bar{z}_i$ is such that $U_j\left(\omega_j - \bar{z}_i\right) = U_j\left(\omega_j\right) = u_j^\omega$.

z_i^3: agent i accepts a δ-*sacrifice* proposed to him by the opponent j.

According to the described mechanism of exchange (agents play an ultimatum game) and denoting with u_i^ω and u_j^ω the payoffs of consuming the initial allocations, we have that the payoff matrix corresponding to the bargaining situation with perfect identifiability of the opponent can be written as

	z_j^1	z_j^2	z_j^3
z_i^1	u_i^ω, u_j^ω	u_i^ω, u_j^ω	$\overline{u}_i, \underline{u}_j$
z_i^2	u_i^ω, u_j^ω	u_i^*, u_j^*	u_i^ω, u_j^ω
z_i^3	$\underline{u}_i, \overline{u}_j$	u_i^ω, u_j^ω	u_i^ω, u_j^ω

Table 2.1. Payoff matrix in the perfect bilateral trade case

Clearly, keeping δ fixed, the values of $\overline{u}_i$ and $\underline{u}_i$ change according to the type of the opponent, but, for every $h = i, j$, the following relation among individual payoffs holds

$$\overline{u_h} = U_h(\omega_h + z_h^1) \geq u_h^* = U_h(\omega_h + z_h^2) \geq \underline{u_h} = U_h(\omega_h + z_h^3) \geq u_h^{omega} = U_h(\omega_h) \tag{2.1}$$

In the bargaining situation described by the payoff matrix in Table 1 and provided $0 < \delta < 1$, there are only three possible agreements (i.e. Nash Equilibrium) between two traders (i, j) of different species and we will indicate them respectively as

e_1: i proposes a δ-sacrifice to j and j accepts; in this case the obtained allocation is represented as $(\overline{x}_i, \underline{x}_j)$ with payoffs $(\overline{u}_i, \underline{u}_j)$.

e_2: i and j accept to trade at the -local and global- walrasian allocations; in this case the obtained allocation is (x_i^*, x_j^*) with payoffs (u_i^*, u_j^*).

e_3: j proposes a δ-sacrifice to i and i accepts; in this case the obtained allocation is represented as $(\underline{x}_i, \overline{x}_j,)$ with payoffs $(\underline{u}_i, \overline{u}_j)$.

Individual Choices

Agents condition their behaviour on the acquired information about the behaviour of the types of agents met in the past. Indeed the actions played by the opponents are kept in memory $M_i^L(s,t)$, which is a private record kept at time s by agent i of the strategies played by opponents of type T in the previous L direct meetings. The basic assumption about individual behaviour is that every agent uses his memory to compute the relative frequencies with which opponents of a given type played their actions; for instance agent $i \in T_1$ uses $M_i^L(s,t)$ to compute the probabilities $p_i^{T_2}(z^a)$ and $p_i^{T_4}(z^a)$ at time s, where $a = 1..3$.

Agents, when called upon to make a choice play the best reply to the expected behaviour of the opponents type, where the latter one is represented by the probabilities $p_i^T(z^a)$ described in the previous paragraph. More precisely we assume a stochastic fictitious play (see [12] and [8]) where agents will use the best reply to their own memory with a probability $(1 - \frac{2}{3}\epsilon)$ where ϵ is the probability of committing a mistake and is taken small; with probability $\frac{2}{3}\epsilon$ they will choose either one of the actions that are not the best reply to his own memory.

Actions, once played, are observed by the opponent; hence everyone can update the part of the memory corresponding to the type of the opponent, payoffs are distributed according to the rules specified in paragraph 2.2 and then we pass to the next period, when the story repeats with a new sequence of random matching.

Realizations and Dynamics

To observe a single realization of the process take a situation characterized by the following parameters: the number of agents is $\mid \Im \mid = 80,$, initial endowments are $(\underline{\omega}, \overline{\omega}) = (1, 9)$, the difference in the endowments of agents of the same species is $\gamma = 11$, the sacrifice asked in the bargaining is determined by a value of δ, equal to 0.6, the length of the record of previous actions (length of memory) is $L = 12$, the probability of a mistake in choosing the best reply is $\varepsilon = 0.1$ and the dynamic process takes place over $S = 500$ periods.

The results are represented on a simplex with three differently shaded regions, each of them representing the basin of attraction of any of the three actions. The beliefs of an agent i ($\forall i \in \Im$) about an opponent of type T_n in any given period s can be represented as a point in the unitary simplex; the coordinates of this point are nothing but the probabilities $\left[p_i^{T_n}(z^1), p_i^{T_n}(z^2), p_i^{T_n}(z^3) \right]$, as resulting from i's memory $M_i^L(.,...)$, with which i expects an agent of type T_n to play any given action $z^k, k \in \{1, 2, 3\}$. For instance in Figure 2; if an agents beliefs, represented by a point, fall in the white area, then his best response will be z^1; if the point lies in the light shaded area then his best response is z^3, and finally, if the point lies in the darker area, then his best reply will be the walrasian action z^2. In Figure 2 we show the outcomes of the interactive bilateral exchange process in two particular subeconomies; simplexes in the first row represent the state of the system in the interclass subeconomy $\{T_1, T_4\}$ formed by the agents of type 1 and 4; the state of the intraclass relation between agents of type 1 and 2 is represented in the second row[6].

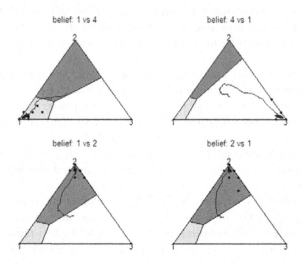

Fig. 2.2. Final states and evolution of beliefs in the $\{T_1, T_4\}$ and $\{T_1, T_2\}$ subeconomies in the perfect identifiability case

As we can see, in the subeconomy $\{T_1, T_2\}$ formed by agents with the same outside opportunity, the system converges to a state in which any agent adopts the walrasian strategy z_2 because he observes agents of the opposite type to adopt that strategy almost surely, so that his best reply is in turn to play the walrasian action. Indeed the walrasian outcomes emerge as a social convention supported by the beliefs that anyone else will play according to that convention (this is indeed the meaning of the clusters of points around the vertex labeled as 2 in the second row of simplexes). On the contrary, when outside opportunities are different, i.e. the subeconomy $\{T_1, T_4\}$, agents with higher outside opportunities play more aggressively and the agents with lower outside opportunities accept to bear the sacrifice asked to them by the rich opponent. This is shown in the first row of simplexes; in the left simplex we observe all points scattered around the vertex labeled as 1, which means that agents of type 1 observed agents of type 4 playing action z_1 very often in the past so that they are ready to expect them to play it again in the future with a probability close to 1. Given these expectations, the best that agents of type 1 can do is to play action z_3, and this is confirmed in the right simplex

---

[6] Clearly the outcomes represented in Figure 2 are the results of a single run (one experiment) of the model with the given parameters, but they indeed represent a rather robust behaviour in this setting as it will appear clearer later.

of the first row, representing the state of the memory of agents of type 4 as to the behaviour of agents of type 1: as expected they observed agents of type 1 playing their own best reply, i.e. actions z_3, so that equilibrium e_3 is actually the emergent pattern of behaviour in the interclass relation.

In order to understand the dynamics driving the system[7] to these outcomes, in Figure 2 we plotted, together with the basins of attraction and the final beliefs of the agents, a line representing the time evolution of the average beliefs that agents of type i have about the behaviour of agents of type j. In the case of the sub-economy formed by agents of type 1 and 4 (first row of simplexes), notice that, notwithstanding the initial motion toward z_2, the beliefs of the agents of type 4 remain inside the basin of attraction of the action z_1 (white area) and indeed they keep playing this action. As agents of type 4 keep playing action z_1, the beliefs of the agents of type 1 are revised in the direction of a further increase in the expectation of observing action z_1 being played in the future and indeed the trajectory moves further in the direction of the vertex labeled 1 in the left simplex up to the point in which it abandons the basin of attraction of actions z_2 and enter the region where the best response is to play z_3. From this point onward the line representing the evolution of the beliefs of the agents of type 4 starts bending toward the vertex labeled 3 thus reinforcing agents of type 4 to play action z_1. Thus the system is trapped into the regions in which both types of agents play the best response to the opponents actions, though the state of the system is characterized by an asymmetric behaviour in which agents with the better outside opportunities ask a sacrifice to the agents with the lower outside opportunity and the latter ones accept.

In the case of interaction between agents of the same class the dynamics is simpler, because the basins of attraction of the different actions are the same for the two types of agents. Thus, when the process starts, both of them play z_1 so that the trajectories representing average beliefs start moving in the leftward direction up to the point in which they both enter in the basin of attraction of the walrasian action z_2 (the darker area); from that point on both types start playing z_2 reinforcing the emergence of a walrasian convention.

2.3.3 Bilateral Trade with Imperfect Identifiability

In the case of bilateral trade with imperfect identifiability an agent trades with the coalition of agents of the opposite species, since he is not able to identify the precise characteristics of the subject he is bargaining with in any particular session, but only the fact that he is complementary to him and hence a possible partner. In this case he knows only that there is a $\frac{1}{2}$ probability of meeting either one of the type of traders of the opposite species. Strategies are a course of action that an agent chooses prior to receiving information

---

[7] Speed of convergence has not been studied in this paper; an excellent source for the reader interested in studying this property of the process is [9].

about the specific opponent type and whose results are determined ex-post conditionally on the specific type of the partner one is bargaining with, as it will be described in the following paragraph.

Strategies and Payoffs

Again the bargaining takes the form of an ultimatum game in which the two parts simultaneously make proposals on how to allocate efficiently among themselves the available quantities of the two commodities; one of the possible strategies, denoted by z_2, is again to trade at the walrasian equilibrium prices.

The other two strategies consist again in asking or accepting a ?-sacrifice, however the precise allocations can be determined only ex-post, once the characteristics of the opponent become known. To determine the ex-ante payoff of an action, say z_2^i (the same would be true as to z_3), for an agent i we have to take into account the fact that this action can be crossed with probability $\frac{1}{2}$ with an action z_j^n ($n \in \{1,2,3\}$) of an opponent j that is of *is* own class and with probability $\frac{1}{2}$ with an action z_h^n ($n \in \{1,2,3\}$) of an opponent h who is of the other class. In other words the expected value of a given action z^a for the agent i is computed as the population weighted value of the expected value of playing separately against agent j and h. Formally,

$$E_i[z^a] = \omega_j E_i[z^a \mid j] + \omega_h E_i[z^a \mid h] \qquad (2.2)$$

where $E_i[z^a \mid t]$ represent the expected payoff of agent i at playing action z^a against an agent of type t and ω_t is the population share of agent of type t in its own specie; in this case $\omega_j = \omega_i = \frac{1}{2}$. With imperfect identifiability agent does not know ex-ante the characteristics of the opponents, but he can recognize them ex-post; therefore he can upgrade his own memory of the opponents behaviour conditional of his type, just as in the previous case, simply recording the last L actions in his memory $M_i^L(s,t)$ that has the same meaning as in paragraph 2.2.

Individual Choices

The assumption about individual behaviour remains unchanged in the present case and we assume that agents play the action that maximizes their expected payoff given expectations about the behaviour of the opposing coalition. Again we assume that choices are subject to random mistakes that occur with a small probability ε. After actions are played, the distribution of payoffs allows any agent to recognize the type of the opponent in that bargaining session: then the proper part of the memory is updated as in the previous case and we pass to the next bargaining session, when the story repeats with a new sequence of random matching.

Realizations and Dynamics. The outcomes of a single experiment are shown in Figure 3, using the same parameters as in the perfect identifiability case[8]. Now the representation in two dimensions of the basins of attraction of the different actions is more complex, so that we plot on the simplex only the

---

[8] Precisely we set the basic parameters to the following values: $\mid \Im \mid = 80,; (\underline{\omega}, \overline{\omega}) = (1,9); \gamma = 11, \delta = 0.6, L = 12, \varepsilon = 0.1$ and $S = 500$.

final state of the agents memories and the trajectories representing the time
evolution of the beliefs for every type.

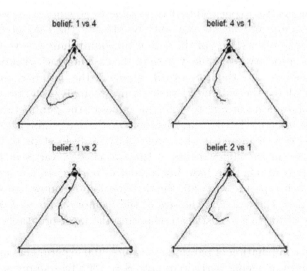

Fig. 2.3. Final states and the evolution of beliefs in the $\{T_1, T_4\}$ and $\{T_1, T_2\}$
subeconomies: imperfect identifiability case.

While in the perfect identifiability case it is possible to analyze the behaviour of the independent subeconomies to understand their dynamics, this
is not possible when there is imperfect information about the agents. Uncertainty makes the social network of interaction become global, so any convention has to be globally consistent. According to this, there are only three
globally consistent Nash equilibria in pure strategies; $E_1 = \{z_1, z_3, z_1, z_3\}$,
$E_3 = \{z_3, z_1, z_3, z_1\}$ and finally the walrasian convention $E_2 = \{z_2, z_2, z_2, z_2\}$.
In figure 3 the beliefs of agents of type T_1 against agents of type T_2 and T_4,
represented by the dots, show how in all cases agents agree to trade under
the walrasian convention. The time evolution of beliefs out of the individual
interaction are represented by the smooth line. In order to understand the
sort of mechanism at work here, take the case of agents of type T_1 and T_4:
the formers start playing z_2 increasing $p_4^1(z_2)$ i.e. the probability with which
agents of type T_4 expect agents of type T_1 to play the middle action; on other
hand agents of type T_4 play aggressively, i.e. z_1, making the trajectory for
agent 1 to bend toward the label 1. However in this movement the trajectory
enters into the set of points in which playing z_2 becomes the most profitable
action and hence they switch to this pattern of behaviour.

In the case of agents of the same class the dynamics of the trajectory is
not different from that observed in the case of perfect identifiability.

2.4 Comparison of the Perfect and Imperfect Identifiability Cases

In the previous two experiments it is possible to advance that, when bilateral exchange occurs and agents are able to identify the characteristics of the opponents, the observation of the walrasian equilibrium as a widely accepted convention in society is less likely than in the case in which agents are not able to distinguish ex ante the type of the opponent they are bargaining with. In order to reinforce the validity of such claim, we analyze the results of Monte Carlo simulations according to different values of the parameters δ and γ.

The parameter related to aggressivity in the bargaining session between two agents, δ, is explored in the range $[0.02, 0.98]$ in steps of 0.02 while γ, the difference in the endowments of the two classes, varies in the range $[1, 25]$ with steps of 0.5. The base line model in both cases is composed by 20 agents of each type, i.e. $\Im = 80$, the distribution of endowments is based on the values $\underline{\omega} = 1$ and $\overline{\omega} = 9$, the size of the memory is set to $L = 18$ and the probability of making a mistake in choosing the most profitable action is set to $\varepsilon = 0.1$.

For any combination of parameters values, 100 experiments are run and in each of them the agents participate into $S = 1000$ bargaining sessions. From the experiments we compute the probabilities with which the whole society will accept the social convention to trade at any of the three societywide consistent equilibria; $E_1 = \{z_1, z_3, z_1, z_3\}$, $E_3 = \{z_3, z_1, z_3, z_1\}$ and $E_2 = \{z_2, z_2, z_2, z_2\}$.

In the perfect identifiability case there are four independent subeconomies, and each of them happens to converge toward any of three possible local Nash equilibria. As consequence in this case society could converge to any of the 81 possible combinations of local equilibria, E_1, E_2 and E_3 being just three out of that multiplicity. Figure 4 portrays the probabilities with which any of the three social convention E_1, E_2 and E_3 will be observed in our economy. As we can observe the probabilities of the asymmetric conventions E_1 or E_3 are very low in this case (on average around 0.13%), whereas the probability of observing a society-wide convention in which all agents trade at walrasian prices is substantially higher (close to 22%)[9]. On the other hand, under imperfect identifiability, E_1, E_2 or E_3 are the only possible social conventions and the probability with which each of them will be observed as the one regulating trades is reported in Figure 5. The probability of observing the social convention in which all agents accept market mediation and trade at walrasian prices is almost doubled in passing from perfect identifiability to a situation in which there is uncertainty about the characteristics of a partner in a trading session.

---

[9] Almost 77% of the other cases converged toward equilibria in which the high endowment class takes advantages over the lower endowment class.

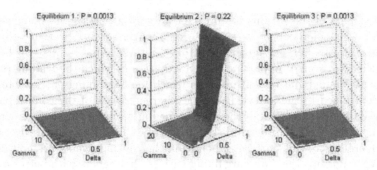

Fig. 2.4. Probability of observing equilibrium E_1, E_2 and E_3 under perfect identifiability

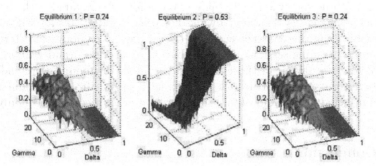

Fig. 2.5. Probability of observing equilibrium E_1, E_2 and E_3 under imperfect identifiability

A way of rationalizing this systemic behaviour is the following: for any agent i the increase in uncertainty makes him face any trading sessions as if his opponents endowments were from a new type computed as the average of the endowments of the agents of the opposite species. Thus all the trading sessions are carried out on the basis of the perceived average endowment gap between classes, which is smaller than the actual gap in the distribution of initial endowments. On the other hand, from the experiments in the case of perfect identifiability, we know that if the difference between the classes is low (low values of γ) and the value of δ is high (i.e. agents play aggressively), then the probability of observing walrasian equilibrium as the accepted convention increases. This observed increase in the likelihood of the walrasian convention as a consequence of the introduction of uncertainty about the trading partners can be interpreted in two ways; the first one is close to the standard core story, since we observe that the more the exchange becomes multilateral and the economy becomes fully connected, the more likely is the fact that agents agree to trade at walrasian prices. As outlined above the basic explanation of this change in the system-wide behaviour is that agents (namely those of the upper class), while still have incentives to play aggressively and extract the greatest part of the surplus of an exchange with a partner, do not insist too much in this behaviour because they are no longer sure to meet an agent that will accept this kind of agreement. However these same results can be read the other way round: in passing from a situation in which the trading partner is ex-ante anonymous to a situation in which the bargaining process can be at all conditioned to the knowledge of the characteristics of the counterpart we observe the abandonment of the market mediation in favour of the emergence and acceptance of asymmetric distributions of welfare. Basically the reason for the emergence of this pattern of behaviour lies in that agents are now sure of the type of agents they are bargaining with so that agents with better outside opportunities can exert all the bargaining power stemming from the ownership of better endowments in the face to face contracting with agents that are less endowed. While the system efficiency is not affected compared with the case in which exchange is regulated by market prices (indeed trades will always take place along the contract curve of the two agent economy that forms in the bilateral trading session), the distribution of resources (and welfare) now clearly bends in favour of those who right from the beginning were endowed with better outside opportunities and extract a surplus from the counterpart exceeding what they could obtain in a purely market mediated economy.

2.5 Conclusion

The aim of the paper is to investigate how the inclusion of direct exchange relations into a standard general equilibrium pure exchange economy affects the adoption of the walrasian price system as the convention to trade the heterogeneous endowments owned by the agents. To this purpose we set up a standard (very simplified and in many respects ad hoc) economy with a number of types of agents differing as to their own endowments, but identical in preferences. An equilibrium price system for the economy as a whole (a walrasian price system) exists and is assumed to be known to everyone. Agents match in random pairs and, when paired, they bargain on how to divide efficiently the surplus of the exchange; one of the possible agreements is to trade at the known walrasian prices. When called upon to make a choice, agents use fictitious play to decide their best response.

By means of ABM simulations we show that the dynamics of this process a la Axtell-Epstein-Young (see [1]) converges to a persistently asymmetric pattern of behaviour in which (i) walrasian outcomes are accepted in the bilateral exchange between similar parties, i.e. between agents with complementary endowments and similar outside opportunities, while (ii) bilateral bargaining between agents with different outside opportunities (i.e. of different class) will end up in allocations in which the stronger agent gets the greater part of the surplus out of the exchange. This result is stronger the greater is the difference between classes in terms of outside opportunities, thus showing that the coupling of direct interaction and the presence of heterogeneity in endowments makes the distribution of commodities to bend in favour of the strongest part in the contract. Also the strength with which agents conduct the bargaining process plays a role: indeed social conventions in which an agent asks the counterpart to renounce to a very large part of the benefits of an exchange will never be accepted; on the contrary asymmetric conventions are accepted for low and moderate values of the parameters determining the degree of conflictuality in the bargaining session.

We investigate also if these conclusions are robust to a weakening of the assumptions imposed on the structure of interaction. In particular we examined whether making the trading process more anonymous will restore (at least partially) the possibility of observing the walrasian outcomes as the long run convention in this society. Indeed we observe that the more the trading process becomes anonymous (despite the fact that it remains bilateral) the more probable is that the socially accepted convention will prescribe walrasian allocations: the impossibility of conditioning individual behaviour to the complete knowledge of the characteristics of the opponent makes more likely the emergence of social conventions in which commodities are traded at walrasian prices. These results on the one hand support the common tenet that anonymous large markets will be regulated by walrasian equilibrium prices; however they can be read also the other way round, i.e. the more the abstract market interaction is abandoned in favour of a more decentralized, personalized contracting, the more we can observe the emergence of asymmetric behaviours amplifying the asymmetries in the initial distribution beyond what is produced by the pure market mechanism.

As a future line of work we propose to explore whether there exist a structure of interaction or an institution for the bargaining process that restores the Walrasian outcome as a convention for the trading process among the agents in a pure exchange economy.

References

1. Axtell R., Epstein J., and Young, P.H. (2000). The Emergence of Classes in a Multi-Agent Bargaining Model. Center on Social and Economic Dynamics. February 2000.
2. Blouin M. and Serrano R. (2001), A Decentralized Market with Common Values Uncertainty: Non-Steady States. Review of Economic Studies. 68, 323- 346.
3. Brock W. and Durlauf S., (2001), Discrete Choice with Social Interaction. Review of Economic Studies. 68, 235-260.
4. Glaeser E. and Scheinkman J. (2001). Measuring Social Interactions in Social Dynamics. D.C.: Brookings Institution Press; Cambridge and London: MIT Press. Ed. Durlauf S. and Young P.H.
5. Durlauf S. and Young P.H. (2001). Social Dynamics. Washington, D.C.: Brookings Institution Press; Cambridge and London: MIT Press.
6. Foley D. (1999). Statistical Equilibrium and Financial Arbitrage. Mimeo. Paper presented at the XII Workshop on General Equilibrium: Problems, Prospects and Alternatives. Siena.

7. Follmer H. (1974). Random Economics with many interacting agents. Journal of Mathematical Economics. 1, 51-62 1974
8. Fudenberg D. and Levine D. (1998). Learning in Games. The MIT Press.
9. Kandori, Michichiro, George Mailath and Rafael Rob. (1993) Learning, Mutation and Long-Run Equilibira in Games. Econometrica. 61:2956.
10. Kirman A.P. (1999) Aggregate Activity and Economic Organization. Mimeo. Paper presented at the XII Workshop on General Equilibrium: Problems, Prospects and Alternatives. Siena.
11. Manski C. F. (2000). Economic Analysis and Social Interaction, Journal of Economic Perspectives, 14(3), pp. 115-136.
12. Young P.H. (1998). Individual Strategy and Social Structure. Pricenton University Press. Princeton, New Jersey.
13. Young P. (1993), An Evolutionary Model of Bargaining. Journal of Economic Theory. 59, 145-168.

A Genetic Algorithms Approach: Social Aggregation and Learning with Heterogeneous Agents*

Davide Fiaschi[1] and Pier Mario Pacini[2]

[1] Department of Economics, University of Pisa, Italy dfiaschi@ec.unipi.it
[2] Department of Economics, University of Pisa, Italy pmpacini@ec.unipi.it

Summary. We analyze an economy in which increasing returns to scale incentivate social aggregation in a population of heterogeneous boundedly rational agents; however these incentives are limited by the presence of imperfect information on others' actions. We show by simulations that the equilibrium coalitional structure strongly depends on agents' initial beliefs and on the characteristics of the individual learning process that is modeled by means of genetic algorithms. The most efficient coalition structure is reached starting from a very limited set of initial beliefs. Furthermore we find that (a) the overall efficiency is an increasing function of agents' computational abilities; (b) an increase in the speed of the learning process can have ambiguous effects; (c) imitation can play a role only when computational abilities are limited.

Key words: Coalition formation, Learning, Genetic Algorithms, Increasing returns to scale, Numerical simulations

3.1 Introduction

Aggregation among agents to cooperate and produce is a typical economic phenomenon. Teams, clubs, cooperatives (coalitions) are all instances of this behaviour. Within the framework of non-cooperative games, [5] was the first to analyze the emergence of coalitional structures (**CS**) as the outcome of a bargaining process in which agents cannot be committed to binding pre-play agreements.[3] Following this approach we analyze an economy where a

---

* We thank two anonymous referees and all the participants to the workshops "New Tools for Qualitative Economic Dynamics" held at the CIMAT, Guanajuato - Mex (October 2002) and "VII WEHIA Workshop" held in Trieste (May 2002) for helpful comments. We acknowledge financial support from MURST PRIN 2000 'New Tools for qualitative analysis of economic dynamics'.

[3] This type of games well adapts to represent situations concerning the decentralized supply of public goods (see [8]), the formation of cooperative firms (see [5])

population of individuals, persecuting the same goals, structures in coalitions under two competing forces: on the one hand an increasing returns to scale technology that incentivates aggregation, on the other the non-monitorability of actions in formed coalitions that incentivates free-riding behaviours, thus reducing the incentive to form and act in groups.[4] Generally a situation like this admits a multiplicity of Nash equilibria (see [6]). Traditionally this indeterminacy has been reduced appealing to a sound refinement proposed by [3], i.e. *coalition proofness*. However coalition-proofness requires the possibility of a pre-play communication stage, in which agents can agree at no cost to correlate their strategies (for example a simultaneous exit from a coalition to form a new one), a stage that is difficult to justify when the population is large, as in the case we are going to study. Therefore we take another point of view and substitute the implicit pre-play communication phase with an explicit dynamic process in which anonymous agents repeatedly interact without the possibility of making jointly agreed deviations and learn from experience. In this way we try to provide an answer to the following questions: when the population is so large that it is not reasonable to assume that pre-play agreements can be reached, does a system in which agents act rationally find an equilibrium position? And, provided that it does, what are the characteristics of the equilibrium positions? (see also [11]).

The analysis is performed by means of computational experiments reproducing an artificial society in which adaptive and heterogeneous agents interact. Their behaviour results from two simultaneous learning processes:[5] the first one deals with the revision of beliefs about the possible realizations of an action, so that agents learn what will be the plausible outcome of a strategy conditional on others' behaviour; this revision of expectations will take the form of a simple *adaptive learning* process. The second one deals with the choice of the course of action: given a certain set of beliefs, an agent finds his best strategy through a process of *learning* and *imitation* on the set of possible strategies; this decision process is modelled by means of *genetic algorithms (GAs)* which tends to capture the idea that agents, when called upon

---

4 For example, take the case of the formation of work cooperatives (research groups as well). The larger the group the greater the potential output because of the integration of different competencies and the cutting of administrative costs; but the larger the group the more a single has an incentive to shirk, benefiting of collective output and saving on the private effort directed to the coalitional goal.

5 The learning process is a crucial aspect in the modelling of any evolutionary game (for an excellent review see [12]). A distinguishing feature of the present approach is that the dynamics of the system is not driven by the payoffs that strategies receive when played, but by the configuration of beliefs that change over time as new realizations of the system occur; these affect the expected payoffs and, by this way, the strategies actually played.

or multilateral bargaining; more generally they can be applied to all those situations in which there is a problem of coordination and imperfect information (as to this point see also [7]).

to make a choice in a complex environment, do not make explicit optimization, but rather operate on a set of rules that they continuously modify reacting to the effects of their own behaviour.[6] The latter point is the main difference from [6], where actions are chosen by an explicit maximization process. Our main finding is that agents play the strategies sustaining a coalition proof Nash equilibrium (CPNE henceforth), which corresponds to the maximum efficient allocation compatible with individual incentives, only if the population start out with a priori beliefs that are close to those sustaining coalition-proof outcomes. There are two main sets of resting points (attractor sets in the following); the first one is made up of coalitional structures "close" to CPNE, i.e. there is a very limited number of agents refusing aggregation, while the large part of the population behave according to the CPNE prescriptions. The other attractor set is characterized by coalitional structures in which coalitions are, on the average, of smaller size, so that the overall result is certainly less efficient. The latter shows a greater basin of attraction than the first one.

The robustness of our results is then tested to various parametrizations of learning process; we find that the efficiency of the aggregation process is an increasing function of agents' computational abilities, while an increase in agents' learning speed can have an ambiguous effects. Finally imitation appears to be relevant only when computational abilities are limited.

The paper is organized as follows: Section 3.2 describes the basic characteristics of the model. Section 3.3 describes how coalitions form, how expectations are revised and decisions are taken. Section 3.4 reports the results of numerical simulations. Conclusions close the paper.

3.2 The Model

In the following we report only the basic characteristics of the model and refer to [6] for more details. There is a population $\mathfrak{S}$ of I agents indexed by i. Agents are identical in all physical characteristics and are endowed with 1 unit of time that they can use either working $(l_i - 1)$ or as leisure time $(l_i = 0)$. They receive utility from the consumption of a commodity y and leisure time, according to the following utility function

$$U_i\left(y_i, l_i\right) = y_i + \left(1 - l_i\right) \cdot \omega,$$

where ω measure the pleasure of not working.

Agents can form coalitions, but an agent can participate into one and only one coalition. A coalition is a group of agents that agree to share the

---

[6] The use of *GAs* to represent individual behaviour has been motivated by important contributions from the theory of cognitive processes (see [9]). Reference [4] provides some examples of applications to evolutionary games. Most contributions using *GAs* in economic analysis focus on macroeconomic models with rational expectations and multiple equilibria (see e.g. [2]), but some authors also tackle explicitly the decision problem (see e.g. [10]).

output they produce by means of the labour input that they provide. Labour is the only productive input and the technology for the production of the coalitional output Y is given by the production function $Y = L^{\alpha}$, where L is the number of labour units and α is assumed to be greater than 1. The production within a coalition has no external effect on the production of other coalitions; commodity Y deteriorates in a single period.

We assume that the agreement within a group entails an *equal sharing* distribution of the coalitional output, i.e. in a coalition of N agents anyone receives an amount $y_i = \frac{Y}{N}$ of the produced output Y. While participation into a group is publicly observable, the contribution of the individual working time to the coalitional production process cannot be monitored, so that defection ($l_i = 0$) cannot be punished.

Within this setting agent has to make two decisions: (i) which coalition to participate into and (ii) which action to perform in a coalition, once formed. The first action determines the formation of a **CS** σ, i.e. a partition of the population $\Im$ in coalitions S_k, while the second determines the individual payoffs within a coalition S_k.

Strategies

In this framework a strategy for an agent i must concern (i) the formation of a coalition and (ii) the action to perform in a coalition once formed. Therefore any strategy θ_i is made up of two components, i.e. $\theta_i = \{\theta_i', \theta_i''\}$: θ_i' is a *signal* indicating the maximum cardinality of the coalition that i is ready to form;[7] θ_i'' is a complete contingent plan indicating the action l_i that i is willing to take conditional on the cardinality of the coalition he may happen to belong to. As an example a strategy θ_i can take the form

$$\theta_i = \{\theta_i', \theta_i''\} = \{N, [1, 1, 1, 0, \dots]\}$$

meaning that agent i is ready to participate into any coalition of cardinality up to N and cooperate ($l_i = 1$) in all coalitions with cardinality less or equal to 3, while he will not cooperate ($l_i = 0$) in coalitions of greater cardinality. Finally, we assume that agents cannot play mixed strategies.

Coalition Formation

Once strategies are announced, agents randomly match. If two agents i and j match and $\min\{\theta_i', \theta_j'\} \geq 2$ then the coalition $S = \{i, j\}$ forms; this coalition is ready to accept another (randomly chosen) agent h provided $\min\{\theta_i', \theta_j', \theta_h'\} \geq 3$ otherwise h will be the first member of a new coalition S' and so on. A **CS** σ is obtained when $\Im$ is partitioned in groups S in such a way that $\min\{\theta_i'\}_{i \in S} \geq |S|$ ($\forall S \in \sigma$) and there is no couple of groups S

---

[7] The fact that θ_i' is the *maximal* acceptable cardinality implies that an agent, signalling θ_i', refuses to participate in all coalitions larger than θ_i' but is ready to belong to coalitions of smaller cardinality.

and S' in σ such that $\min\left\{\min\{\theta_i'\}_{i \in S}, \min\{\theta_i'\}_{i \in S'}\right\} \geq |S| + |S'|$, i.e. a **CS** σ is formed whenever no agent is compelled to participate into a coalition of greater cardinality than the one he is willing to accept and no two groups are compelled to remain separated when they could join without the objection of any participant.

Equilibria

A configuration of strategies $\theta^* = \{\theta_i^*\}_{i \in \Im}$ is a Nash equilibrium of the game if it gives rise to a **CS** σ^* and a corresponding profile of actions l^* such that no agent has an incentive to deviate from his strategy. Indeed this game may have many Nash equilibria: any partition of society in groups $\{S_1, \ldots, S_K\}$ such that $1 \leq |S_k| \leq \bar{N}(\alpha)$ where $\bar{N}(\alpha)$ is the cardinality of the maximal coalition in which cooperation is incentive compatible for all its members,[8] is a Nash equilibrium for a suitable configuration of strategies; for example the **CS** $\sigma' = \{S_1', \ldots, S_K'\}$, $1 \leq |S_k| \leq \bar{N}(\alpha)$, is certainly a Nash equilibrium if i's strategy is $\theta_i = \{\theta_i', \theta_i''\} = \{|S_k'|, [1, \ldots, 1, 0, \ldots]\}$, $S_k \ni i$, where the last 1 in the conditional action part of θ_i is in the $|S_k'|^{th}$ position.

Among all Nash equilibria the **CS** in which all, but possibly one, coalitions have cardinality $\bar{N}(\alpha)$, is particularly interesting because it implies the maximum aggregate output and furthermore it is a *coalition-proof Nash equilibrium* (see [7]). However the implementation of this refinement presumes a lot of communication and coordination capacities on the side of agents.

Here we analyze this equilibrium selection problem by adopting an evolutionary approach. We suppose that agents repeatedly play the game; at each stage they choose their strategies on the basis of their beliefs on the other agents' behaviour, having as time-horizon only one period (i.e. they play a series of one-shot games). Then, at the end of every period, agents revise their beliefs on the basis of experience and imitation. While the fact that agents will finally play Nash equilibria is an expected outcome, our focus will be on the question on which Nash equilibrium agents will play.

The analysis is performed by computational experiments; the next Section describes the details of the artificial setting used for the simulations.

---

[8] Provided $\frac{1}{2} < \omega < 1$, $\bar{N}(\alpha)$ is the integer part of the value of N solving the equation $N^\alpha - (N-1)^\alpha - \omega \cdot N = 0$. $\bar{N}(\alpha)$ is monotonically increasing in α, so that all agents merge and cooperate in the grand coalition provided $\alpha \geq \alpha^* > 1$, where α^* solves $I^{\alpha^*} - (I-1)^{\alpha^*} - \omega \cdot I = 0$. Furthermore, since $\bar{N}(\alpha) < 2$ for $\alpha \leq 1$, it follows that increasing returns are a necessary condition in order to observe cooperation in groups of at least 2 agents. Finally, for $\alpha \in (1, \alpha^*)$ cooperation can emerge just in coalitions that are proper subsets of $\Im$; this is the case we will deal with in the rest of the paper.

3.3 The Design of the Simulation

The basic idea behind the simulation is to represent the evolution of the game as a sequence of periods. At the beginning of each period every agent plays a strategy stating the maximum cardinality of the coalition he intends to belong to and the action he will take, conditional upon the cardinality of the realized coalition. The choice of a strategy is the outcome of a maximizing process *conditional on the current period beliefs* about the actions of the others; the decision process is implemented by *GAs*, that, as we will see, provide a very intuitive way to model learning and imitation, two crucial aspects of any decision process.

The strategies $\theta = (\theta_1, ..., \theta_I)$ played by the agents and the random matching process described in the previous Section determine the **CS** σ and the agents' actions $l = (l_1, ..., l_I)$ of the current period. Given σ and l, every player uses this new information to upgrade his beliefs and receives his payoff. This concludes the period. The simulation goes on until a persistent pattern in the **CS** emerges. In the following we describe the various components of this procedure in more details.[9]

3.3.1 Initial Beliefs and Agents' Type

In this game agents play their "best" strategy *given* their expectations on others' behaviour; according to the two parts of a strategy, agents have two forms of expectations:

1. the probability $p_i^t(N|S)$ that an agent i assigns to the event in which a coalition of size N forms in period t if he communicates his willingness to participate in coalitions of size at most S.
2. the probability $Q_i^t(n|N)$ with which an agent i, in period t, expects to find n other cooperators $(0 \leq n \leq N-1)$ in a coalition of N persons $(0 \leq N \leq I)$; to keep things as simple as possible, we assume that:
   a) in any period t, an agent i assesses the probability $q_i^t(N)$ with which another randomly chosen agent will be a cooperator in a coalition of cardinality N, and
   b) in any period t, the probability $Q_i^t(n|N)$ is the value at n of the binomial distribution with probability $q_i^t(N)$ and $N-1$ trials.[10]

It is straightforward to show that the more optimist an agent is (i.e. the higher the value of q_i for whatever N) the greater is the cardinality of the

---

[9] The code used for the simulation is available from the authors upon request.

[10] This assumption is an intuitive way of modelling the probability distribution of cooperators within a coalition, when agents are identical; in short we assume that the probability distribution of cooperators is always binomial. Clearly more complex distributions (e.g. gamma) could be considered, but for the sake of simplicity, we restrict ourselves to the simpler case of the binomial.

maximal coalition in which i is ready to cooperate (clearly always within the limit of $\bar{N}(\alpha)$). If we take the following values of the parameters: $\alpha = 1.428$ and $\omega = 0.635$, it is easy to verify that an agent will cooperate in any coalition up to cardinality 3 when $1 \geq q_i^t(N) \geq 0.77$ ($\forall N$), while he will sustain cooperation in coalitions of cardinality up to 2 when $0.77 > q_i^t(N) \geq 0.39$ ($\forall N$); finally for lower values of $q_i^t(N)$ will cooperate only in the singleton coalition. We will use this correspondence between beliefs about others' cooperative attitude and individual behaviour to classify agents in three different types, i.e.:

- *optimistic* agents: agents of this type assign high confidence to the fact that other people are cooperators in a formed coalition; in particular we assume that an agent has optimistic beliefs when $1 \geq q_i^t(N) \geq 0.77 \,\forall\, N$. As we have seen, agents of this type are ready to support cooperation in larger coalitions, so that we can say that optimistic beliefs sustain a *strongly associative* (SA) behaviour.
- *mildly optimistic* agents: agents of this type assign a lower confidence to the cooperative attitude of the other agents and, in particular, beliefs are mildly optimistic when $0.77 > q_i^t(N) \geq 0.39 \,\forall\, N$. As we have seen, agents of this type sustain cooperation only in lower coalitions beliefs, i.e. they will give rise to what will be termed a *weakly associative* (WA) behaviour.
- *pessimistic* agents: agents of this type assign a very low probability to the fact that anyone else will be a cooperator in a formed coalition; in particular we assume that beliefs are pessimistic when $0.39 > q_i^t(N) \geq 0 \,\forall\, N$. Again we know that for these values of the probabilities of cooperation anyone will cooperate only in the singleton coalition, i.e. he will sustain only a *non associative* (NA) behaviour.

In the sequel the terms (i) optimistic and strongly associative, (ii) mildly optimistic and weakly associative, (iii) pessimistic and non associative will be used as synonyms when referred to agents.

3.3.2 Update of Beliefs

Individual expectations are revised through time. Starting from prior beliefs, posterior distributions are formed taking into account observations; the latter are relative to the local experience of an agent, i.e. he can only observe the cardinality of the coalition he happens to be into and the profile of actions of the other members. We assume that the mechanism governing the process of revision of expectations is of a very simple type and takes the form of *adaptive learning*.

Consider the expectations about the cardinality of a coalition and let $H_i^t(N|S)$ be the relative frequency with which agent i observed the formation of a coalition of cardinality N up to period t as a consequence of playing a strategy θ in which $\theta_i' = S$; then for any $t > 0$

$$p_i^{t+1}(N|S) = \begin{cases} \left(1 - \delta^{CARD}\right) \cdot p_i^t(N|S) + \delta^{CARD} \cdot H_i^t(N|S) & \text{if } \theta_i'^t = S \\ p_i^t(N|S) & \text{otherwise,} \end{cases}$$

(3.1)

where δ^{CARD} is a parameter measuring the importance of experience in the formation of beliefs. Equation (3.1) states that agents revise the prior on the possible events ensuing from S on the basis of the relative frequencies with which they occurred in the past.

In the same manner consider the expectations about the cooperation within a coalition and let $C_i^t = \frac{n_i^t}{N_i^t}$ be the proportion of cooperators in the coalitions of size N which agent i belonged to in period t; then for any $t > 0$

$$q_i^{t+1}(N) = \begin{cases} \left(1 - \delta^{COOP}\right) \cdot q_i^t(N) + \delta^{COOP} \cdot C_i^t & \text{if } N = N_i^t \\ q_i^t(N) & \text{otherwise} \end{cases}$$

(3.2)

Finally, we assume that at period 0:

$$P_i^0(N|S) = \frac{1}{S}, \forall S \in [0, ..., I]$$

(3.3)

and

$$q_i^0(N) = \begin{cases} 0.9 \text{ for optimistic agents} \\ 0.6 \text{ for mildly optimistic agents} \\ 0.3 \text{ for pessimistic agents.} \end{cases}$$

(3.4)

We point out that, given this construction, there are two sources of heterogeneity in individual beliefs: the first one is a difference in initial beliefs, the other one is the possible difference in the history agents have experienced up to a certain period.

3.3.3 The Decision Process

GAs is the engine by which strategies are selected and reproduced to arrive to a strategy that is "optimal" given individual beliefs; in this process also new strategies are created and evaluated. This procedure, where trial-and-error learning and imitation are two crucial aspects, seems well suited to our purposes and it agrees with the evolutionary approach.[11]

Building on the idea of natural selection, GAs start working on a set of candidate strategies for a given period and select the strategies with the highest fitness, calculated as the expected payoff given the beliefs on others' strategies, and "recombine" their single components (building blocks) to produce new ones. An intuitive interpretation of these building blocks is to consider

---

[11] Moreover, by a computational point of view, GAs has been proved to be very efficient in searching for a solution in highly dimensional spaces, which is particularly useful in our case, where agents are looking for the best strategies in a solution space that, according to our encoding procedure (see later), is the $\left(I + \frac{\log(I)}{\log(2)}\right)$-dimensional hypercube.

a strategy as the result of different components; for example a component could be the action in a coalition for a certain cardinality; if this action is particularly efficient (and this is measured by the fitness of the string), then *GAs* will tend to use this component in the formation of new strategies. The performance of *GAs* is further aided by what has been termed *imitation*: in this context imitation means that an agent uses the observable part of the strategies played by others in his coalition as new building blocks (genetic material) to produce better strategies. In the following we briefly describe the working of the *GAs* procedure.

Encoding

The first step is to represent the strategy in a way that can be handled by *GAs*. This is done by coding a strategy in a binary alphabet as a string of bits; the first part indicates the maximum cardinality of the coalition agent intends to belong to (this encodes θ'_i) and the second part the contingent plan of actions (θ''_i) conditional on the cardinality of possible coalitions. For example take an economy with 4 agents; in such a case a possible strategy is a string like the following: $\boxed{0}\boxed{1}\boxed{1}\boxed{1}\boxed{1}\boxed{1}$. The first two bits indicates the maximum cardinality and the last four bits the conditional actions. Therefore with this strategy an agent intends to form a coalition with a maximum acceptable cardinality of 3 (0 1 in binary alphabet) and cooperate in every coalition he can belong to (the first three bits of the second substring are 1, which means cooperation).[12]

Therefore the set of rules (strategies) is a $J \times L$ binary matrix, where J is the number of strategies coexistent in the population (the "mind" of an agent) and L is the number of bits necessary to express a **CS** in binary form.

GAs work sequentially and their procedure is made up by three basic steps: (i) selection, (ii) recombination and (iii) mutation. Each of them will be dealt with separately.

Selection

In the biological evolution, the greater is the ability of a species to adapt and cope with the environment, the higher is the probability for it to survive and reproduce; similarly *GAs* privilege those strings (strategies) with the highest fitness, giving them the highest chance to survive and reproduce. In order to model this mechanism we assume that J strings are drawn from the available population, where the probability of a string to be drawn is positively correlated to its fitness indexes $F_i^t(j)$. In this way the highest is the fitness index the highest is the probability that the corresponding strategy is selected and passed to the next step of the procedure; conversely strategies with a low fitness index are candidate to be eliminated soon, since they proved inefficient.

---

[12] We use a string with a fixed length to simplify algorithm implementation, although, in this case, the 4th bit in the second substring is not necessary (indeed the maximum cardinality for this strategy is equal to 3).

Cross-over

The mere selection of the fittest strategies serves the purpose to refine the set of current schemes but does not allow for the discovery of better ones. This further step is accomplished by recombining the building blocks of the selected proposals, as in the natural process of procreation and consequent exchange of genes. By "mixing" the building blocks (genes) of the fittest strategies we get new strings with an hopefully enhanced capacity of adaptation to the environment; such a mixing is called *cross-over*. There are several ways to model the crossing over of strings. We adopt the most commonly used in the literature: first a couple of strings is picked up at random from the set of selected strings (they are candidate to be parents) and, with a certain probability, they mix their genes. This means that each of them is partitioned into two substrings of length v and $L - v$ respectively, where v is a random integer drawn from an uniform distribution over the interval $[2, L - 1]$; finally two substrings of equal length are interchanged and, by recombination, we get two new strings.

Mutation

A crucial element in any evolutionary process is chance. The presence of chance in our context is taken into account by adding a further step in the *GAs* procedure in which every single bit in the set of strings is subjected (with a low probability) to a random mutation of its state. By this trick we avoid the lock-in phenomenon and we help search to escape from local inefficient optima.

Imitation

An important aspect of any learning process is to emulate the most successful strategies played by the opponents. To make more realistic the analysis we consider that agent i can imitate only what he can observe, i.e. only the cardinality of his coalition, that provides new information on the maximum acceptable cardinality, and the actions *actually* played by the agents belonging to his coalition (not their full strategies, that are unobservable). Therefore he will replace some of the candidate strategies with new ones obtained modifying the formers both in the maximal cardinality part and conditional action part so as to incorporate the new accruing information.

Fitness

The fitness of a string is nothing but its expected payoff conditional on the individual beliefs $p_i^t(.,.)$ and $Q_i^t(.,.)$ in the period in which it is to be evaluated. In particular, if agent i is going to play a strategy $\theta = \{\theta', \theta''\} = \{\theta', [\theta_1'', \ldots, \theta_N'', \ldots \theta_I'']\}$, $Q_i^t(n, N) \cdot p_i^t(N, \theta')$ is the probability with which i expects to find himself in a coalition of N people with other n cooperators and $Q_i^t(n, N) \cdot p_i^t(N, \theta') \cdot \left(\frac{(n+\theta_N'')^\alpha}{N} + (1 - \theta_N'') \cdot \omega \right)$ is the expected payoff in that situation. Therefore the expected payoff of playing θ in period t, i.e. the fitness $F_i^t(\theta)$ to the strategy $\theta = \{\theta', \theta''\}$ is given by

$$F_i^t(\theta) = \sum_{N=1}^{\theta'} \left[p_i^t(N|\theta') \cdot \sum_{n=0}^{N} Q_i^t(n|N) \cdot \left(\frac{(n + \theta_N'')^\alpha}{N} + (1 - \theta_N') \cdot \omega \right) \right]$$

where θ_N'' is the Nth component of the conditional action part of the strategy θ.

3.4 Computational Experiments and Results

Our computational experiments can be divided in two main stages; in the first we analyze how, setting the parameters concerning learning speed and computational abilities to appropriate values, the aggregation of a population composed by agents with heterogeneous initial beliefs evolves through repeated interactions by learning and experimentation. In this respect we identify equilibria with positions of the system that do not show "significative" changes over appropriate number of periods. In the second we analyze how our findings are robust to changes in the degree of computational abilities, learning speed and imitation.

In any stage we run several simulations changing the seed of random numbers for any distribution of initial beliefs. This is necessary to eliminate possible random disturbances, deriving from the fact that agents' initial set of candidate strategies are randomly generated and GAs make use of random numbers.

3.4.1 Parametrization

For simplicity we limit our attention to an economy with just 16 agents; even if the economy is so simple, the set of possible strategies is very large being made up by 2^{20} elements. Given the parameters' values reported in Section 3.3, the cardinality of the greatest coalition capable of sustaining cooperation is $\bar{N}(\alpha) = 3$, so that every partition of 16 agents in coalitions with 3 or less agents with full cooperation are Nash equilibria. The CPNE is given by 5 coalitions of 3 agents and 1 of one agent; it is easy to check that this configuration implies maximum aggregate output, subject to the individual incentive constraints. As to GAs, we set to 30 the number of strings forming the individual population of rules (that is $J = 30$), the parameter for the selection to 0.8,[13] the crossover probability to 0.6 and the mutation probability to 0.01.

---

[13] Heuristically this parameter measures the probability of the string with the highest fitness to be selected for the crossover.

3.4.2 Equilibrium Selection

In this section we study the profile of equilibrium strategies with respect to the agents' initial beliefs. We set the speed of learning process equal to 0.25 both for cardinality and cooperation (i.e. $\delta^{CARC} = \delta^{COOP} = 0.25$), the number of iterations of GAs per period equal to 25 and consider 153 different compositions of initial population defined on the basis of agents' initial beliefs on cooperation (any possible permutation of 16 agents for the possible three types of initial beliefs on cooperation). For every possible composition of initial population we ran 50 simulations, modifying the seed of random numbers, so that we have $153 \times 50 = 7650$ observations.

Figure 3.1 reports in a three dimensional simplex the frequency of the rest points of the simulations. Every vertex corresponds to a population composed by only one type of agents; in particular the top vertex corresponds to a population with only NA agents, the right-bottom vertex to a population of only WA agents and finally the left-bottom vertex to a population with only SA agents. Every point in the simplex corresponds to a mixture of these three types of agents and the darker is the colour of that point the higher is the frequency with which the corresponding combination of types occurs in the simulation.

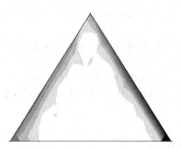

Fig. 3.1. Rest points of the simulation in terms of the type of composition of the population

Figure 3.1 shows that most rest points are characterized by a population of WA and NA agents and only few simulations converge to a population with a significative amount of SA agents (in particular the shadow zone on the left represents more or less the 15% of total rest points, while the shadow zone on the right the 80%). It is worth observing that populations composed only by SA and NA agents (all the rest points near to the edge on the left) and those with only WA and NA agents (all the rest points near to the edge on the right) are the most frequent, while the coexistence of SA and WA agents is hardly observable (all the rest points near to the edge on the bottom).

However notice that Fig. 3.1 does not show how the rest points are related to the initial composition of the population. To see this we calculate the number of times a simulation converged to every points in the grid (all the possible rest points) and consider only those for which this number is greater than 1/100 of the total number of simulations (that is 7560/100 = 75.6); we find that most simulations converge to two main sets, one on the right, denoted as A, and one on the left, denote as B. Moreover we characterize the basin of attraction of these two sets by calculating for every possible starting point the percentage of simulations converging to one of the two sets. Figure 3.2 reports the results.

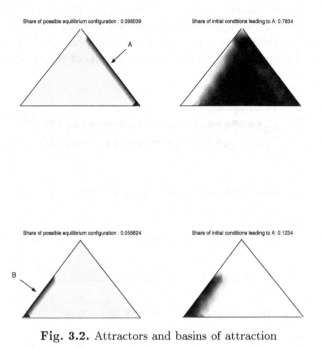

Fig. 3.2. Attractors and basins of attraction

On the left of Fig. 3.2 we represent the two attractor sets (A and B), while on the right the two basins of attraction. The depth of gray in any point of the right simplex is proportional to the probability with which the corresponding initial population converged to the corresponding attractor set on the left; we can notice that the basin of attraction of the set A is absolutely greater than those corresponding to the set B (respectively 78.3% and 12.3% of the total number of simulations).[14]

The basic finding is that the presence of few WA agents can be sufficient to lead to a rest point where WA agents are the majority. This result is due to the free-riding behaviour of WA agents in coalitions of size 3 (given their beliefs, this is their best action), which causes SA agents to revise their beliefs

---

[14] Notice that the two attractor sets count for more than the 90% of the total number of simulations.

downturn. On the other hand, NA agents quickly learn that cooperation is the best action in a coalition of size 2, so becoming WA agents.

3.4.3 Computational Abilities

There are important contributions in the literature showing the importance of agents' computational abilities in the selection of an equilibrium (see [1]). In our model agents' computational abilities can affect the rest points of the simulation, since they determine the accuracy of the individual maximization process. We ran 50 simulations for 6 different types of initial populations (see Fig. 3.3 for the exact composition of each one), setting $\delta^{CARD} = \delta^{CARD} = 0.25$ and varying the number of iterations of GAs performed in every period from 1 to 30. Figure 3.3 reports the average payoffs of the agents (in log) for every type of initial population.[15]

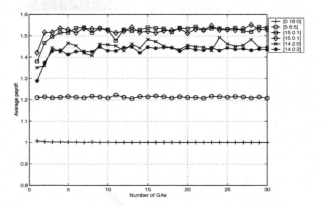

Fig. 3.3. Computational abilities and average payoffs

Figure 3.3 shows that a number of iterations of GAs equal to 25, the one considered in the previous Section, guarantees that the rest points are not substantially affected by the agents' computational abilities, independently of the type of population we consider (in fact this already holds for a number of GAs greater than 5). For a low number of iterations of GAs, that is in presence of "imprecisions" in agents' decision process, the average payoffs are remarkably lower.

3.4.4 Learning Speed

Another key factor in evolutionary games is agents' learning speed (see [12]). In order to test the importance of this factor we ran several experiments

---

[15] The type of the population is shown in the legend on the right where the first number represents the number of SA agents, the second the number of WA agents and the last the number of NA agents.

varying the parameter δ for 6 different types of initial populations and setting to 25 the number of iterations of GAs per period; in particular we consider 31 different values of δ in the range $[0, 1]$ and for any value of δ we ran 30 simulations. Figure 3.4 reports the average payoffs for each type of population:

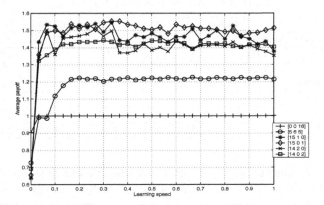

Fig. 3.4. Learning speed and average payoffs

We observe that for $\delta^{CARD} = \delta^{COOP} \in [0.1, 1]$ the properties of the rest points do not show substantial changes, so that the choice of $\delta^{CARD} = \delta^{COOP} = 0.25$ as in our previous experiments seems to be appropriate. However average payoffs show first an increase then a decrease as δ^{CARD} and δ^{COOP} increase for at least two types of initial distribution.

For low value of δ^{CARD} and δ^{COOP} agents do not learn from experience; they play strategies indicating an acceptable coalitional cardinality greater than that compatible with full cooperation and, at the same time, they do not cooperate, given the persistent (incorrect) beliefs that other agents will be cooperating. As learning speed increases, agents learn that this strategy is also adopted by other agents, so that they decrease the acceptable coalitional cardinality and play cooperation (this happens for values of δ^{CARD} and δ^{COOP} lower than 0.1). For higher values of δ^{CARD} and δ^{COOP} the average payoffs may have a slight decrease because a higher "reaction" to experience can easily lead to the destruction of optimistic beliefs in presence of possible "mistakes" in the formulation of strategies. For example if some agent does not cooperate in a coalition with cardinality equal or lower than 3 (we know that this is not the best action for an SA agent), this can lead other SA agents to revise drastically their beliefs, so that they will not cooperate in future in coalitions of cardinality equal to 3.

3.4.5 Imitation

To measure the full effect of imitation on our results we consider the case where agents' computational abilities are the minimum, i.e. the number of GAs is set equal to 1. As usual we perform 50 simulations for 6 types of initial population. Table 3.1 reports the average payoffs and the expected composition of final population for each type of initial population: Imitation generally does not

Table 3.1. Average payoffs and imitation

	Imitation				No imitation			
Init. pop.	Av. payoffs	SA	WA	NA	Av. payoffs	SA	WA	NA
[0 0 16]	1.00	0	0.32	15.68	1.00	0	0.42	15.58
[5 6 5]	1.21	0	10.52	5.48	1.21	0	10.4	5.6
[15 1 0]	1.38	4.08	11.92	0	1.36	3.82	12.12	0.06
[15 0 1]	1.42	8.68	6.32	1	1.36	4.40	10.6	1
[14 2 0]	1.35	1.84	14.04	0.12	1.35	2.16	13.62	0.22
[14 0 2]	1.29	2.86	10.16	2.98	1.30	3.04	10.00	2.96

seem to affect substantially the results, but when there is just one WA or one NA agent. In fact in this case imitation allows the SA strategy to persist and not be abandoned in favour of the WA strategy, because every agent tends to adopt the most performing observed strategies that, in a population of almost all SA agents, will be actually a SA strategy.

3.5 Conclusions

In this paper we have shown that, if agents adjust their behaviour on the basis of their experience, then they play the CPNE strategies only if the population starts out with a priori beliefs that are close to those sustaining coalition-proof outcomes; in this case the role of the learning process will be just to size down the dimension of the equilibrium coalitions to the one consistent with individual incentives to cooperate. The composition of initial population strongly affects the selection of equilibrium. The computational experiments highlight how there are two main attractor sets, the first one characterized by a population composed by a large part of SA agents and a minority of NA agents, while the second one is characterized by a population composed by a large part of WA agents and a minority of NA agents. This suggests that equilibria with a population of SA and WA agents are not "sustainable" because of the free-riding behaviour in large coalitions of the WA agents. Moreover, we observe that the possibility that WA agents become SA is limited to the case in which WA agents are a strict minority in a population with a large share of SA agents. On the contrary, if the number of WA agents is sufficiently high, we observe convergence toward an equilibrium in which there are no SA agents. These findings suggest that these two attractor sets can be preserved even in face of small random perturbations in agents' beliefs. Therefore a straightforward extension of this work would be to allow for random mutations in agents characteristics in order to test the robustness of the attractor sets from a purely evolutionary point of view. Finally we test the robustness of our results to various parametrizations; we find that the overall efficiency is an increasing function of agent's computational capabilities, while an increase in agents' learning speed can have an ambiguous effect. Imitation appears to be crucial only if the number of WA and /or NA agents is small.

References

1. Anderlini L, H Sabourian (1995) Cooperation and Effective Computability. Econometrica 63:1337-1369
2. Arifovic J (1994) Genetic Algorithm Learning and the Cobweb Model. Journal of Economic Dynamics and Control 18:3-28
3. Bernheim BD, Peleg B,Whinston MD (1987) Coalition-Proof Nash Equilibria: Concepts. Journal of Economic Theory 42:1-12
4. Birchenhall CR (1996) Evolutionary Games and Genetic Algorithms. In: Gill' M. (ed) Computational Economic Systems. Kluwer Academic Publishers
5. Farrell J, Scotchmer S (1988) Partnerships Quarterly. Journal of Economics 103:279-97
6. Fiaschi D, Pacini PM (2003) Coalition Formation with Boundedly Rational Agents. Forthcoming in: Kirman A,Marsili M, Gallegati M. (eds) The Complex Dynamics of Economic Interaction. Lectures Notes in Economics and Mathematical System. Springer, Berlin
7. Fiaschi D, Pacini PM (1998) Endogenous Coalition Formation with Identi- cal Agents. Working paper n. 308 (Progetto dAteneo: Leconomia italiana e la sua collocazione internazionale: una redifinizione delle politiche di wel- fare edelloccupazione per una pi'u efficiente crescita economica). University of Bologna
8. Guesnerie R, Oddou C. (1988) Increasing Returns to Size and Their Limits. Scandinavian Journal of Economics 90:259-73
9. Holland JH, Holyoak KJ, Nisbett RE (1986) Induction: Processes of Inference, Learning and Discovery. MIT Press, Cambridge, USA
10. Holland JH, Miller JH (1991) Artificial Adaptive Agents in Economic Theory. American Economic Review Papers and Proceeding 81:365-70
11. Mailath GJ (1998) Do People Play Nash Equilibrium? Lessons From Evolutionary Game Theory. Journal of Economic Literature 36:1347-1374
12. Marimon R (1996) Learning from Learning in Economics. Working Paper N. 96/12, European University Institute, Florence
13. Young, H. Peyton and Dean Foster (1991) Cooperation in the Short and in the Long Run. Games and Economic Behavior 3:145-56

4

Structure and Macroeconomic Performance: Heterogeneous Firms and Financial Fragility*

Domenico Delli Gatti[1] and Mauro Gallegati[2]

[1] ITEMQ, Università Cattolica, Milano domenico.delligatti@unicatt.it
[2] DEA, Università di Ancona gallegati@unian.it

Summary. In this paper we adopt a new macrodynamic tool, i.e. a system of non-linear difference equations describing the evolution over time of the first and second moments of the distribution of firms' degrees of financial robustness captured by the ratio of the equity base to the capital stock - the *equity ratio* for short - which affects supply and capital accumulation decisions. For particular configurations of parameters the dynamic patterns of the average equity ratio and the variance generate irregular and asymmetric time series in which growth and fluctuations are jointly determined (*fluctuating growth*).

4.1 Introduction

According to the mainstream view, fluctuations are the equilibrium adjustment process which follows a shock; therefore, they have to be analyzed using the impulse-propagation approach. This view has been recently criticized as '' ad-shockery (Cochrane, 1991; Zarnowitz, 1998). Moreover, it is very difficult for the theory to explain well known stylized facts concerning the business cycle such as persistence, asymmetry and the occurrence of large fluctuations due to small shocks. One candidate for explaining those puzzles is the role of financial factors in business fluctuations, which in turn is rooted in the notion of asymmetric information (Greenwald and Stiglitz, 1988, 1990, 1993; Bernanke and Gertler, 1989, 1990; Bernanke et al., 1994, 1998; Kiyotaki and Moore, 1997).

The notion of asymmetric information of the New Keynesian Economics (Mankiw and Romer, 1991), in turn, is inconsistent with the representative agent hypothesis (Stiglitz, 1992). In fact, asymmetric information is relevant

---

* We also thank participants to seminars held at the meeting of the European Economic Association in Istanbul, LATAPSES at the University of Nice, CENDEF at the University of Amsterdam, University of Ancona and University of Bergamo for useful comments and criticisms. Special thanks to the referees for their insightful comments and criticisms. The usual disclaimer applies.

inasmuch as agents are different from one another: it is only in this context that phenomena such as adverse selection or moral hazard may exist. Moreover, empirical evidence at the microeconomic level suggests that heterogeneity matters and aggregation of heterogeneous agents is of central relevance in macroeconomic modelling, since there is systematic evidence of individual differences in economic behavior (Stoker, 1993). Several contributions show that aggregate dynamics and individual heterogeneity are intertwined, because neglecting individual heterogeneity in aggregate equations generates spurious evidence of dynamic structure (Lippi and Forni, 1996).

According to us, therefore, a consistent theory of the business cycle has to take agents' heterogeneity seriously: the evolution over time of the distribution of heterogeneous agents must be the cornerstone of business cycle theory.

In this paper, therefore, we build a simple model of production and investment of financially constrained heterogeneous firms in an uncertain environment with capital market imperfections due to asymmetric information. Each firm is characterized by the degree of *financial fragility*, inversely related to the ratio of the equity base or net worth to the capital stock the *equity ratio* for short which affects supply and investment decisions. The links among financial conditions, supply and capital accumulation are explained in section 3. The dynamics of the macrovariables which describe the business cycle (such as the rate of change of the aggregate capital stock) are determined by the evolution over time of the entire distribution of firms according to their degree of financial fragility. In particular, in section 4 we derive the laws of motion of the mean and the variance of the distribution of firms' equity ratio. In section 5 we explore the properties of the two-dimensional nonlinear map which describes the dynamics of the first and second moments of the distribution. For particular (but not implausible) configurations of parameters, the dynamic patterns of the mean and the variance of the distribution of firms' equity ratio generate irregular and asymmetric time series in which growth and fluctuations are jointly endogenously determined (*fluctuating growth*).

4.2 Background Assumptions

We model a closed economy without a public sector. In order to simplify the argument, we assume that firms carry on production in a competitive environment by means of a linear technology (uniform across firms) which uses only capital as an input[3]. In particular we distinguish between "hard capital", i.e. machinery and equipment, which is a fixed factor of production in the short run and becomes variable only in the long run by means of investment activity, and "soft capital", i.e. circulating capital such as inventories, maintenance

---

[3] Greenwald and Stiglitz (1993), on the contrary, assume that the only factor of production is labour.

and "organization" which is a variable factor of production[4]. Output is a homogeneous good which can be used for consumption or investment purposes. In order to invest, firms must raise funds on the credit market[5]. We assume that the interest rate on bank loans (r) is uniform across firms. Firms are differentiated by the individual price they face. Following Greenwald and Stiglitz we assume that each firm is not sure at which price it will sell its goods. Uncertainty can be captured by defining the selling price of the i-th firm at time t as follows: $p_t^i = u^i P_t$ where u_i is a stochastic idiosyncratic shock and P_t is the general (average) price level at time t. We will denote by $f(u^i)$ the p.d.f. and by $F = \int f(u^i) du^i$ the c.d.f. of u^i. By assumption, the distribution of the shock is time-invariant. Moreover, we assume that $E(u^i) = 1$. Therefore, the expected value of the individual selling price is equal to the average price level i.e. $E(p_t^i) = P_t$. Given the relatively narrow focus of our analysis we are not concerned here with the way in which the average price level is determined. As a consequence, in the following we treat P_t as an exogenous variable. This assumption is acceptable for small scale macroeconomic cycles which do not involve major price oscillations. In order to bring to the fore the impact of firms' financial conditions on supply and investment we overlook the monetary side of the model. Profit in nominal terms is the difference between revenue and cost, which in turn is equal to the remuneration of capital[6]:

$$\Pi_t^i = p_t^i Y_t^i - r(P_t K_t^i + \sigma P_t Y_t^i) =$$
$$= P_t \left[u^i Y_t^i - r(K_t^i + \sigma Y_t^i) \right]$$

where Y_t^i is output K_t^i is hard capital and σ is the (given and constant) soft capital-output ratio, so that σY_t^i is soft capital in real terms. Dividing by P_t, we obtain profit in real terms:

$$\Pi_t^i / P_t = u_i Y_t^i - r(K_t^i + \sigma Y_t^i) =$$
$$(u^i - r\sigma)r(Y_t^i - rK_t^i)$$

Dividing by capital K_t^i we get the (real) *profit rate*:

$$\pi_t^i = \Pi_t^i / P_t K_t^i = (u^i - r\sigma)y_t^i - r$$

where $y_t^i = Y_t^i / K_t^i$ is the *output ratio*. Since $E(u_i) = 1$, the *expected profit rate* is:

----

[4] See Gertler and Hubbard (1988).

[5] We rule out the issue of new equities as a viable means of raising funds due to equity rationing. This assumption can be justified by asymmetric information on the Stock market (Myers and Majluf, 1984; Greenwald, Stiglitz and Weiss, 1984).

[6] We are implicitly assuming that both internally and externally financed capital (i.e. borrowed and non-borrowed funds) are remunerated at the current interest rate. This is a simplifying shortcut. In an asymmetric information context in which a financing hierarchy can be envisaged, internal finance has a cost advantage over bank loans and the issue of new equities (Fazzari, Hubbard and Petersen, 1988; Bernanke and Gertler, 1991).

$$E(\pi_t^i) = (1 - r\sigma)y_t^i - r$$

Bankruptcy occurs if the individual price happens to be "too low", so that actual revenue is not sufficient to remunerate capital, a loss occurs and the loss is so big as to deplete the equity base inherited from the past. The bankruptcy condition in intensive form can be expressed as follows:

$$r + (r\sigma - u^i)y_t^i > a_{t-1}^i \tag{4.1}$$

where $a_{t-1}^i = A_{t-1}^i/K_{t-1}^i$. The ratio $a_t^i = A_t^i/K_t^i$ will be referred hereafter as the *equity ratio*. The lower the equity ratio, the higher the financial fragility of the firm. The bankruptcy condition can be rearranged as follows:

$$u^i < r\sigma + \frac{r - a_{t-1}^i}{y_t^i} = \overline{u_t^i} \tag{4.2}$$

that is: bankruptcy occurs if the relative price is lower than a critical threshold $(\overline{u_t^i})$. The probability of bankruptcy, therefore, is: $Pr(u^i < \overline{u_t^i}) = F(\overline{u_t^i})$.

In order to simplify the argument, we assume that the distribution of the shock u^i is uniform with support $(0,2)$, so that $E(u^i) = 1$. Thanks to this assumption, the probability of bankruptcy is an increasing linear function of the critical threshold:

$$Pr(ui < \overline{u_t^i}) = F(\overline{u_t^i}) = \frac{\overline{u_t^i}}{2} = \frac{1}{2}(r\sigma + \frac{r\sigma - a_{t-1}^i}{y_t^i})$$

that is an increasing function of the interest rate and the soft capital output ratio and a decreasing function of the equity ratio and the output ratio.

We assume that bankruptcy is costly. First of all, there are the legal and administrative costs of default (Gordon and Malkiel, 1981; Altman,1984; White, 1989). Second, managers may suffer a loss of reputation as a consequence of bankruptcy (Gilson, 1990; Kaplan and Reishus, 1990).

Following Greenwald and Stiglitz, we assume that the cost of bankruptcy is increasing with the scale of production: managers of bigger firms face greater costs (a bigger loss of reputation, for instance) than owners/managers of smaller firms. In our framework which is built on variables in intensive form the scale of production is measured by the output ratio[7]. We make the additional technical assumption that the cost of bankruptcy is a quadratic increasing function of the output ratio:

$$c_t^i = c(y_t^i)^2 \tag{4.3}$$

---

[7] 6Moreover, since the stock of hard capital is fixed in the short run, the output ratio changes with the level of production. Changes of the scale of production measured in intensive form, therefore, are equivalent to changes in the scale of production measured in levels.

In conjunction with the specification of the distribution of the idiosyncratic shock, this assumption allows to derive a ''nice (linear) closed form solution to the problem of the firm i.e. the determination of the optimal output ratio as we will see in the next section. Admittedly, it is hard to believe that the rate of change of bankruptcy costs might be increasing with the scale of production as in (4.3). A more realistic assumption, however, would have made computations more cumbersome without adding much to the results.

4.3 Financial Conditions, Supply and Investment

The firm's objective function is the difference between the expected profit rate and the cost of bankruptcy in case bankruptcy occurs. Therefore, the i-th firm's problem is:

$$max_{y_t^i} V_t^i = E(\pi_t^i) - c(y_t^i)^2 Pr(u^i < \overline{u_t^i}) = \qquad (4.4)$$

$$(1 - r\sigma)y_t^i - r - \frac{c}{2}(y_t^i)^2(rg + \frac{r - a_{t-1}^i}{y_t^i})$$

The First Order Condition for a maximum is:

$$1 - r\sigma - \frac{c}{2}(r - a_{t-1}^i) - cr\sigma y_t^i = 0$$

The Second Order Condition:

$$-cr\sigma < 0$$

is obviously satisfied. Solving FOC for the *optimal output ratio* yields:

$$y_t^i = \alpha + \beta a_{t-1}^i \qquad (4.5)$$

where

$$\alpha = \frac{1}{cr\sigma} - \frac{1}{c} - \frac{1}{2\sigma}$$

$$\beta = \frac{1}{2r\sigma}$$

The output ratio, therefore, is linear in the equity ratio of the previous period. The stock of (hard) capital can be modified (in the ''long run) by means of investment activity. Investment is proportional to the difference between the ''target capital stock which in turn is proportional to output and the stock of capital inherited from the past:

$$I_t^i = \lambda(\gamma Y_t^i - K_{t-1}^i) \qquad (4.6)$$

where γ is the hard capital-output ratio and λ is the stock adjustment coefficient. Both parameters are uniform across firms. Moreover, $0 < \lambda < 1$. In order to simplify the calculations, we assume that the capital-output ratio is the same for soft and hard capital, i.e. $\lambda = \sigma$. Ruling out depreciation and taking into account (6), the law of motion of the capital stock can be written as follows:

$$K_t^i = K_{t-1}^i + I_t^i = K_{t-1}^i(1 - \lambda) + \lambda\sigma Y_t^i$$

Dividing by capital and rearranging we get:

$$\frac{K_{t-1}^i}{K_t^i} = \frac{1 - \lambda\sigma y_t^i}{1 - \lambda} \qquad (4.7)$$

In the following, in order to simplify the analysis, we will assume that $\alpha \cong 0$. Taking this technical assumption into account, substituting (5) into (7) and rearranging we get:

$$\frac{K_{t-1}^i}{K_t^i} = \frac{1}{1 - \lambda} - \frac{\lambda}{2(1 - \lambda)r}a_{t-1}^i \qquad (4.8)$$

From a simple algebraic manipulation of (8), we derive the equation of the rate of capital accumulation (the *growth rate* for short) of the i-th firm:

$$g_t^i := \frac{K_t^i - K_{t-1}^i}{K_{t-1}^i} = \frac{\lambda(a_{t-1}^i - r)}{2r - \lambda a_{t-1}^i} \equiv g(a_{t-1}^i, r, \lambda) \qquad (4.9)$$

From a quick inspection of (9) it is clear that the partial derivative of the growth rate with respect to the equity ratio of the previous period

$$g_1 = \frac{\partial g}{\partial a_{t-1}^i} = \frac{2r(1 - \lambda)}{(2r - \lambda a_{t-1}^i)^2}$$

is positive. The second derivative:

$$g_{11} = \frac{\partial^2 g}{\partial(a_{t-1}^i)^2} = \frac{4r\lambda(1 - \lambda)}{(2r - \lambda a_{t-1}^i)^3}$$

is positive if $a_{t-1}^i < 2r/\lambda$, negative otherwise. In words: the growth rate of the i-th firm is an increasing convex function of the equity ratio if $a_{t-1}^i < 2r/\lambda$, it is an increasing concave function of the equity ratio if $a_{t-1}^i > 2r/\lambda$.

In period t, each firm is characterized by a degree of financial robustness a_t^i. We will denote by $\phi_t(a_t^i)$ the p.d.f. of the distribution of the equity ratio across firms a_t^i. This distribution is time dependent as we will see. Let

$$a_t = E(a_t^i) \qquad (4.10)$$

$$V_t = E(a_t^i - a_t)^2 \qquad (4.11)$$

be the *average equity ratio* and the *variance of the equity ratio* respectively in period t. Applying Taylors formula to (9) we can write the second order approximation of the growth rate of the i-th firm around the average equity ratio as follows:

$$g_t^i \cong g(a_{t-1}) + g_1(a_{t-1}^i - a_{t-1}) + \frac{1}{2}g_{11}(a_{t-1}^i - a_{t-1})^2 \qquad (4.12)$$

where g_1 and g_{11} are the first and second partial derivatives of the growth rate with respect to the equity ratio *evaluated at the average equity ratio* of the previous period. The expected value of (12) is the *average growth rate*:

$$g_t = E(g_t^i) \cong g(a_{t-1}) + g_1 R(a_{t-1}^i - a_{t-1}) + \frac{1}{2}g_{11}E(a_{t-1}^i - a_{t-1})^2 \qquad (4.13)$$

Recalling that

$$E(a_{t-1}^i - a_{t-1}) = 0$$

and denoting by

$$V_{t-1} = E(a_{t-1}^i - a_{t-1})^2$$

the variance of the firms equity ratio at time t-1, from (13) we derive the following equation of the average growth rate:

$$g_t = g(a_{t-1}) + \frac{1}{2}g_{11}V_{t-1} \qquad (4.14)$$

From (14), it is clear that the average growth rate is an increasing function of the average equity ratio of the previous period. Moreover, it is increasing (resp. decreasing) with the variance if $a_{t-1}^i < 2r/\lambda$ (resp.$a_t^i > 2r/\lambda$). From the discussion above, it is clear that if the relationship between variables (the growth rate and the equity ratio of the previous period) at the individual (micro) level is non-linear, the relationship between variables at the aggregate (macro) level is a function of the variance. For instance, the relationship among the equity ratio and the growth rate at the macro level (14) is a function of the variance of the distribution of firms according to their financial conditions[8].

4.4 Laws of Motion

Assuming, for the sake of simplicity, that the firm does not distribute dividends, the change of the equity base, that is internal finance or "cash flow, is equal to output less the remuneration of capital, that is profit. Therefore, the law of motion of the equity base for the i-th firm in real terms is:

$$A_t^i = A_{t-1}^i + (\Pi_t^i/P_t =$$
$$A_{t-1}^i + (u^i - r\sigma)Y_t^i - rK_t^i$$

---

[8] Since the microeconomic relationship is non-linear, the knowledge of the distribution of firms helps explaining the response of the system to an adverse shock (see also Caballero et al., 1997).

Dividing by K_t^i and recalling (8) and (5), assuming for the sake of convenience, that $acong = 0$, we obtain the following *law of motion of the equity ratio for the i-th firm*:

$$a_t^i = \Gamma_1^i a_{t-1}^i - \Gamma_2^i (a_{t-1}^i)^2 - \Gamma_0 \qquad (4.15)$$

$$\Gamma_1^i = \frac{1}{1-\lambda} + \frac{u^i - r\sigma}{2r\sigma}$$

$$\Gamma_2 = \frac{\lambda}{2(1-\lambda)r}$$

$$\Gamma_0 = r$$

(15) is a non linear stochastic difference equation described by a quadratic map subject to a stochastic shock.

Lets consider now aggregate (average) variables. We assume that the idiosyncratic shocks ui are i.i.d. and each shock is not correlated to the degree of financial robustness of the firm. Summation and averaging over firms, assuming that the law of large numbers applies and recalling that $V_{t-1} = E(a_{t-1}^i - a_{t-1})^2 = E(a_{t-1}^i)^2 - a_{t-1}^2$, yelds:

$$a_t = \Gamma_1 a_{t-1} - \Gamma_2 a_{t-1}^2 - \Gamma V_{t-1} - \Gamma_0 \qquad (4.16)$$

$$\Gamma_1 = \frac{1}{1-\lambda} + \frac{u^i - r\sigma}{2r\sigma}$$

$$\Gamma_2 = \frac{\lambda}{2(1-\lambda)r}$$

$$\Gamma_0 = r$$

Equation (16) is the *law of motion of the average equity ratio*, which can be interpreted as the equity ratio of the "average firm. It differs from the law of motion of the equity ratio of the individual firm (see equation (15)) because of the absence of the stochastic shock, which is replaced by the average $E(u^i) = 1$ thanks to the law of large numbers assumption, and the presence of the variance.

Equation (16) is another instance of the claim already exemplified above with reference to the growth rate according to which if a relationship between variables at the individual level is non-linear, the relationship between variables at the aggregate level is a function of the variance. In this case, the relationship among the average equity ratio in period t and the average equity ratio in period $t-1$ is affected by the variance in period $t-1$. In particular, since the individual relationship is concave, the average equity ratio in period t is a decreasing function of the variance in period $t-1$. Lets consider now the variance of the equity ratio in period t:

$$V_t = E(\Gamma_1^i a_{t-1}^i - \Gamma_2 (a_{t-1}^i)^2 - \Gamma_0 - a_t)^2$$

After some tedious calculations, we get:

$$V_t = \Gamma_2^2(\theta - 1)V_{t-1}^2 - \Gamma_1\Gamma_2\mu^3 + (2\Gamma_2 a_{t-1} - \Gamma_1)^2 V_{t-1} + 4\Gamma_2^2\mu^3 a_{t-1} \quad (4.17)$$

where μ^3 is the third moment from the mean of the distribution of firms equity ratio which measures the degree of asymmetry and θ is a coefficient which measures the degree of kurtosis[9]. We treat these parameters as exogenous. We end up, therefore, with a two-dimensional dynamical system (16)(17) which describes the joint laws of motion of the mean and the variance of the distribution of the firms equity ratio.

4.5 Endogenous Dynamics

4.5.1 The Benchmark Case: a Representative Agent

Lets start from a convenient special case. If agents are identical - i.e. if the Representative Agent Hypothesis holds true - the law of motion of the equity ratio of the *representative firm* is:

$$a_t = \Gamma_1 a_{t-1} - \Gamma_2 a_{t-1}^2 - \Gamma_0 \quad (4.18)$$

(18) is the zero-variance case of (16). In this case, of course, equation (17) must be ignored by construction. (18) is a quadratic map, topologically conjugated to the logistic map. The dynamical properties of the logistic map are well known and we will not recall them here (for a comprehensive survey see Day, 1994). The average growth rate will be:

$$g_t = \frac{\lambda(a_{t-1} - 2r)}{2r - \lambda a_{t-1}}$$

There are configurations of the parameters such that the dynamics are chaotic, i.e. the equity ratio oscillates apparently at random around the steady state. In this case the economy follows a path of *endogenously determined fluctuating growth*.

4.5.2 The General Case: Heterogeneous Agents

If agents are heterogeneous, the evolution over time of the mean and the variance of the equity ratio is obtained by the iteration of a two-dimensional map $T : (a, V) \rightarrow (a', V')$ given by:

$$T : a' = \Gamma_1 a - \Gamma_2 a^2 - \Gamma_2 V - \Gamma_0 \quad V' = \Gamma_2^2(\theta - 1)V^2 - 2\Gamma_1\Gamma_2\mu^3 + (2\Gamma_2 a - \Gamma_1)^2 V + 4\Gamma_2^2\mu^3 a$$
$$(4.19)$$

---

[9] For instance, if the distribution were normal, $\mu^3 = 0$; $\theta = 3$.

where the symbol $'$ denotes the unit time advancement operator. As shown above, the coefficients $\Gamma_0, \Gamma_1,, \Gamma_2$ depend on the interest rate r , the stock adjustment parameter λ and the capital/output ratio σ . All in all, thew set of relevant parameters consists of 5 elements:$r, \lambda, \sigma, \mu^3, \theta$. The last two elements show up only in the second equation of (19).

T is a non-invertible map of the (a, V) plane, that is, starting from some initial conditions (a_0, V_0) , the iteration of (19) uniquely defines the trajectory $(a_t, V_t) = T^t(a_0, V_0) \forall t$ whereas the backward iteration of (19) is not uniquely defined.

The study of the dynamical properties of map T is not an easy task[10]. As usual, we begin from the determination of the fixed points. Imposing the steady state conditions $a = a'$, $V = V'$ into (19) we obtain:

$$-\Gamma_0 + (\Gamma_1 - 1)a - \Gamma_2 a^2 - \Gamma_2 V = 0$$
$$\Gamma_2^2(\theta - 1)V^2 - 2\Gamma_1\Gamma_2\mu^3 + \left[(2\Gamma_2 a - \Gamma_1)^2 - 1\right]V + 4\Gamma_2^2\mu^3 a = 0$$

Substituting the expression for V obtained from the first equation into the second one, we obtain a 4-th degree equation in a.

Symmetric Distribution

We shall consider first the case $\mu^3 = 0$. In this case, the distribution is symmetric. The system (19) specializes to:

$$T_0 : a' = -\Gamma_0 + \Gamma_1 a - \Gamma_2 a^2 - \Gamma_2 V \qquad V' = \Gamma_2^2(\theta - 1)V^2 + (2\Gamma_2 a - \Gamma_1)^2 V \quad (4.20)$$

Notice that $V = 0$ implies $V' = 0$. If we represent the initial conditions on the (a, V) plane, this property means that the coordinate axis $V = 0$ (i.e. the a-axis) is trapping. In other words, starting from any initial condition on the a-axis, say $(a_0, 0)$ i.e. a situation in which the Representative Agent Hypothesis holds true the dynamics are confined to the same axis for each t, governed by the restriction of the map T_0 to that axis. This restriction is the quadratic map obtained from (20) imposing V = 0, which coincides with equation (18).

One of the main consequences of the invariant line for the dynamics of T0 is the coexistence of *disjoint attracting* sets. An example is given in figure 1 in which we characterize each initial condition on the (a, V) plane by the dynamical properties of the trajectory associated with it for a given configuration of parameters.

There are four fixed points: P^*, Q^*, R^* and S^*, two of which (namely P^* and Q^*) are situated on the a-axis, i.e. they are characterized by zero variance. For each fixed point we can define an average growth rate:

$$g^* = g(a^*) + \frac{g_{11}}{2}V^*$$

---

[10] For an in-depth analysis of the properties of a map such as T, see Agliari, Delli Gatti, Gallegati, Gardini (1998).

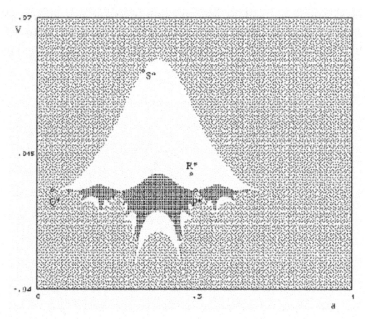

Fig. 4.1. Basins of attraction $\lambda = 0.5, r = 0.11, \mu^3 = 0, \theta = 1.8, \sigma = 3$

P^* and Q^* are repelling nodes. On the a-axis, we can detect a 2-cycle attracting node which we will denote

$$A_2 = \{(a_1, 0), (a_2, 0)\}$$

Also the fixed point R^* is attracting.

The white points belong to the basin of attraction of R^*, the dark grey points denote the basin of attraction of the 2-cycle A2 while the light grey points generate divergent trajectories. Depending on the position of the starting point, therefore, the initial distribution can

• converge to a stationary distribution characterized by an average equity ratio and a positive variance given by the coordinates of point $R^* = \{a_R^*, V_R^*\}$; in this case the average growth rate converges to a steady growth path given by $g_R^* = g(a_R^*) + \frac{g_{11}}{2} V_R^*$
• converge to a 2-cycle A_2 of a degenerate distribution (a representative agent) characterized by equity ratios a_1 or a_2 and zero variance; in this case the average growth rate oscillates between $g(a_1)$ and $g(a_2)$;
• diverge. We will characterize this situation as a '' financial crisis

Of course, different configurations of the parameters can yield different values of the fixed points and/or different configurations of the basins of attractions so that different dynamic trajectories can be associated with the same initial conditions and/or new dynamic patterns can emerge. For instance, figure 2 displays the dynamic properties of points in the (a, V) plane given the same configuration of parameters as in fig. 1 but for a small increase of λ .

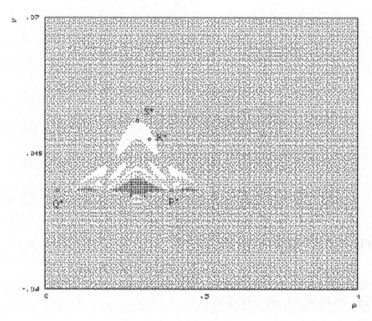

Fig. 4.2. Basins of attraction $\lambda = 0.595, r = 0.11, \mu^3 = 0, \theta = 1.8, \sigma = 3$

Once again there are four fixed points: P^* , Q^* , R^* and S^* , two of which are situated on the a-axis, but their levels are different from the ones reported in figure 1. Moreover, the point R^* is still an attracting node but its basin of attraction has shrunk dramatically. In other words both the values of the fixed points and the configurations of the basins of attractions are different. The implications are far reaching. If a given stochastic disturbance (an impulse), in fact, hits the stationary distribution in R^* , the distribution will go back to R^* in the case of figure 1 whereas it will diverge in the case of figure 2, provided the shock is neither '' too big nor '' too small , i.e. provided the distribution after the shock is still in the white region in the case of fig. 1 and is already in the grey region in the case of figure 2. An impulse therefore affects the nature of the dynamics of the equity ratio as well as the growth path.

If the stochastic disturbance modifies a parameter which is relevant in determining the dynamic properties of the map, the dynamics may be affected and shift back and forth from convergence to fixed points to endogenous cycles (Gallegati, 1993).

Figure 3 displays the dynamic properties of points in the (a, V) plane given the same configuration of parameters as in fig 1 but for a decrease of the parameter θ.

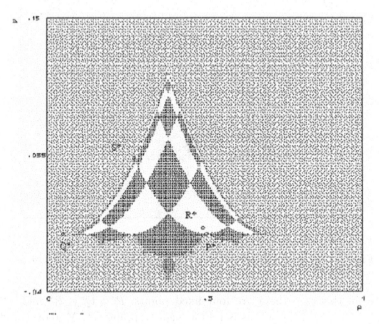

Fig. 4.3. Basins of attraction $\lambda = 0.5, r = 0.11, \mu^3 = 0, \theta = 1, \sigma = 3$

The fixed points P^* , Q^* , R^* and S^* have the same kind of local stability as in fig. 1 but globally different behavior.. In fact the configurations of the basins of attractions are different. By reducing r, we increase the nonlinear effect in the one-dimensional quadratic map (16), the attracting 2- cycle A_2 undergoes a flip bifurcation and an attracting 4-cycle appears: a new dynamic pattern emerges. As r is further reduced several bifurcations lead to chaotic attractors. In figure 4 we show a case in which there are three coexisting chaotic attractors: a 4-band chaotic attractor around the repelling fixed point R^* , a 2-band chaotic attractor near the a-axis and a 4-pieces chaotic attractor having one side on the a-axis. When the system is in the chaotic region, the dynamics will be complex. The system will display a process of fluctuating growth.

Asymmetric Distribution

If $\mu^3 \neq 0$ and '' small i.e.e; if the distribution is slightly asymmetric we can consider map T as a '' slight perturbation of map T_0. An example is given in figure 5 which is obtained with the same configuration of parameters as in fig. 1 but for an almost negligible increase of the parameter μ^3.

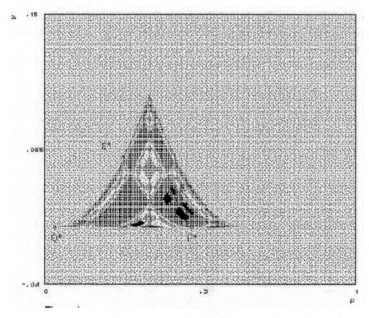

Fig. 4.4. Basins of attraction $\lambda = 0.5, r = 0.8599, \mu^3 = 0, \theta = 1, \sigma = 3$

Also in this case there are four fixed points: P^* , Q^* , R^* and S^* . In this case, however, the a-axis is no longer a trapping set. Moreover a closed invariant attracting *orbit* emerges. The fixed point R^* is repelling, while the other fixed points are situated along the frontier which separates the basin of attraction of R^* from the points having divergent trajectories. The dynamic behavior of T as the parameters are varied is similar to that of the symmetric case. In our framework, we can generate fluctuations of the Slutsky-Frisch type by forcing an exogenous shock upon the system in the vicinity of a stable steady state such as point R^* above. For instance, following Greenwald and Stiglitz (1993) we can think of a "nominal shock as a sudden change of the equity base of each and every firm due to a monetary innovation which brings about a price surprise (an unanticipated change of the price level), such that each and every firm perceives a lighter debt burden in real terms and a higher equity base[11]. The impact of a shock depends on the degree of heterogeneity of firms: the greater the variance, the larger the propagation mechanism. As a consequence, fluctuations associated to the same impulse will differ if the variance is different. Heterogeneity amplifies the effect of a nominal shock and increases its persistence which is positively correlated to the variance (see also Bernanke et al., 1999; Fisher, 1996).

Shocks to a relevant parameter may also change the dynamical behavior of the system if the value of the parameter is sufficiently close to the bifurcation

---

[11] In other words, the price surprise generates a Fisher effect.

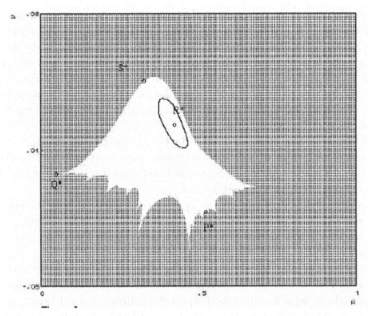

Fig. 4.5. Basins of attraction $\lambda = 0.5, r = 0.102, \mu^3 = 0, \theta = 1.8, \sigma = 3$

value. In such a case, the very nature of fluctuations changes, and a period of self-sustained oscillations may alternate with monotonic convergence or other dynamic patterns.

4.6 Conclusions

In this paper we have built a simple model of production and investment decision of financially constrained firms in an uncertain environment. In order to analyze the interaction between real and financial variables, we focus on the degree of financial robustness, proxied by the ratio of corporate net worth to the stock of capital, that is the equity ratio: the higher the equity ratio, the lower the degree of financial fragility of the firm. Production and investment are negatively related to the degree of financial fragility through bankruptcy risk.

We take seriously the role of heterogeneity in determining aggregate behavior and therefore do not extend individual behavior to the aggregate as in the representative agent framework. We keep track instead of the evolution over time of the distribution of agents according to the degree of financial fragility by studying the laws of motion of the first two moments of the distribution, i.e. the average equity ratio and the variance. These two laws of motion are described by a two-dimensional non-linear map which yield multiple '' equilibria (steady states), with different dynamical properties depending upon the chosen configuration of parameters. The map generates a wide range of dynamic patterns: convergence to a stationary distribution, periodic or aperiodic cycles, chaotic dynamics, divergence. To each dynamic pattern of the distribution corresponds a dynamic pattern of aggregate production and investment (capital accumulation). Therefore, the framework can generate convergence to a steady growth path, endogenous regular or irregular business cycles and growth which we label fluctuating growth for short divergent trajectories, which we label '' financial crises" .

A fluctuation can be generated also by an exogenous stochastic shock forced upon the system when it is in a steady state position. In particular, we have shown that the higher the degree of heterogeneity, the larger the effects of a shock and the longer its persistence.

Heterogeneity is very important in business cycle analysis for empirical as well as theoretical reasons (Kirman, 1992). When firms are heterogeneous, knowledge about the distribution of firms is crucial in order to understand the response of the system to aggregate and idiosyncratic shocks.

A step further should be made in the direction of explicitly modelling interaction. Two ways can be envisaged.. Social random interaction modifies the approach to equilibrium based upon deterministic equilibrium analysis and poses puzzling questions about the standard microfoundation procedure. In the case of strategic interaction, the individual laws of motion have to be modified in order to take into account other agents behavior.

A second line of inquiry which should be pursued in our framework is the analysis of the impact of policy interventions. Chaotic dynamics can be "controlled" by means of appropriate changes of the governing parameters, as shown by an extensive literature on chaos control. In our framework, the natural candidate to play the role of policy instrument to be "fine tuned" in order to stabilize the cycle is the rate of interest. We conjecture that interest rate changes can stabilize the cycle, preventing financial crises but also dampening fluctuations.

References

1. Agliari, A., Delli Gatti, D., Gallegati, M. and Gardini, L. (2000), "Global Dynamics in a Non-linear Model of the Equity Ratio", Journal of Chaos, Solitons and Fractals.
2. Altman, E.I. (1984), "A Further Empirical Investigation of the Bankruptcy Cost Question", Journal of Finance, 39:1067-1089.
3. Bernanke, B. and Gertler, M. (1989), "Agency Costs, Net Worth and Business Fluctuations", American Economic Review, 79:14-31.
4. Bernanke, B. and Gertler, M. (1990), "Financial Fragility and Economic Performance", Quarterly Journal of Economics, 105:87-114.
5. Bernanke, B., Gertler, M. and Gilchrist, S.G. (1998), "The Financial Accelerator and the flight to Quality", Finance and Economics Discussion Papers, Federal Reserve Board, n. 94-18.

6. Bernanke, B., Gertler, M. and Gilchrist, S.G. (1999), "The Financial Accelerator in a Quantitative Business Cycle Approach", in J. Taylor and M. Woodfors (eds), Handbook of Microeconomics, Amsterdam, Elsevier.
7. Caballero, R. J. (1992), "A Fallacy of Composition" , American Economic Review, 82, 5: 1279-92.
8. Caballero R., E.Engel and J.Haltinwanger (1997), " Aggregate Employment Dynamics: Building from Microeconomic Evidence" , American Economic Review, 87: 115-37.
9. Cochrane, J.H. (1994), "Shocks", NBER working paper, n.4698.
10. Day, R. (1994), Complex Economic Dynamics, Cambridge (Mass.), MIT Press.
11. Fazzari, S., Hubbard, G. and Petersen, B. (1988), "Financing Constraints and Corporate Investment", Brookings Papers on Economic Activity, I: 141-206.
12. Fisher, J. (1996), "Credit Market Imperfections and the Heterogeneous Response of Firms to Monetary Shocks", Federal Reserve Bank of Chicago, working paper 96-23.
13. Gallegati, M. (1993), "Capital Accumulation, Growth and Cycles", Banca Nazionale del Lavoro Quarterly Review.
14. Gertler M., and R.G.Hubbard (1988), "Financial Factors in Business Fluctuations", Federal Reserve Bank of Kansas City, Financial Market Volatility.
15. Gilson, S.C. (1990), "Bankruptcy, Boards, Banks and Blockholder: Evidence on Changes in corporate Ownership and Control when Firms Default", Journal of Financial Economics, 27:355-388.
16. Gordon, R.H., Malkiel, B.G. (1981), "Corporation Finance" in Aaron H.A. and Pechman J.A. (eds.), How Taxes Affect Economic Behaviour, Washington D.C., Brookings Institution.
17. Greenwald, B.C. and J.E. Stiglitz (1986) "Externalities in Economies with Imperfect Information and Incomplete Markets", Quarterly Journal of Economics, 101:229-64.
18. Greenwald, B.C., Stiglitz, J.E. (1988), "Imperfect Information, Finance Constraints and Business Fluctuations", in Kohn, M. and Tsiang, S.C. (eds.), Finance Constraints, Expectations and Macroeconomics, Oxford, Oxford University Press.
19. Greenwald, B.C., Stiglitz, J.E. (1990), "Macroeconomic Models with Equity and Credit Rationing", in Hubbard, R. G. (ed.), Financial Markets and Financial Crises, Chicago, Chicago University Press.
20. Greenwald, B.C., Stiglitz, J.E. (1993), "Financial Market Imperfections and Business Cycles", Quarterly Journal of Economics, 108:77-114.
21. Greenwald, B.C., Stiglitz, J.E., Weiss, A. (1984), "Informational Imperfections in the Capital Markets and Macroeconomic Fluctuations", American Economic Review, 74:194-200.
22. Kaplan, S.N., and Reishus, D. (1990), "Outside Directorship and Corporate Performance", Journal of Financial Economics, 27:389-410.
23. Kirman, A.P. (1992) "Whom or What Does the Representative Individual Represent?", Journal of Economic Perspectives, 6:117-36.
24. Kiyotaki, N. and J.Moore (1997), "Credit Cycles", Journal of Political Economy, 105: 211-48.
25. Lippi, M. and M.Forni (1996), Aggregation and Microundations of Macrodynamics, Oxford University Press, Oxford.
26. Mankiw, N.G. and D. Romer (eds) (1991) New Keynesian Economics, 2 vols, Cambridge, Massachusetts: MIT Press.

27. Minsky, H.P. (1982), Can "It" Happen Again? Essays on Instability and Finance, Armonk N.Y., M.E. Sharpe.
28. Myers, S.C., Majluf, N.S. (1984), "Corporate Financing and Investment Decisions when Firms Have Information that Investors Do Not Have", Journal of Financial Economics, 13:187-221.
29. Stanca L., D.Delli Gatti and M.Gallegati (1999), "Financial fragility, heterogeneous agents, and aggregate fluctuations: evidence from a panel of U.S. firms", Applied Financial Economics.
30. Stiglitz, J.E. (1992), "Methodological Issues and the New Keynesian Economics", in A.Vercelli and N. Dimitri (eds) Macroeconomics: A Survey of Research Strategies, Oxford: Oxford University Press, 38-86.
31. Stoker, T.M. (1993) "Empirical Approaches to the Problem of Aggregation over Individuals", Journal of Economic Literature, 21: 1827-74.
32. White, M.J. (1989), "The Corporate Bankruptcy Decision", Journal of Economic Perspectives, 3:129-151.
33. Zarnowitz V. (1998), "Has the Business Cycle been Abolished?", NBER working paper, n.6367.

5

Firms Interaction and Technological Paradigms*

Rainer Andergassen[1], Franco Nardini[2], and Massimo Ricottilli[3]

[1] Department of Economics, University of Bologna Italy
   anderga@economia.unibo.it
[2] Department of Mathematics for Social Sciences,University of Bologna Italy
   nardini@dm.unibo.it
[3] Department of Economics, University of Bologna Piazza Scaravilli 2, 40126,
   Bologna, Italy ricottilli@economia.unibo.it

Summary. This paper deals with the aggregate effects of small, exogenous but idiosyncratic technological shocks on locally interacting firms. Its main purpose is to model a situation in which technological paradigms emerge through endogenous propagation and diffusion of information leading to an aggregate pattern. We develop a theoretical framework in which large technological correlations emerge due to localised interaction of single firms. The paper states some simple results on spill-over dynamics determined by firms trying to improve their current technology and thus generating new information through investment in R&D and through localised technological search. The first part shows that different growth regimes can arise from the general framework of interaction that we propose. The second part shows that an interesting regime characterised both by long run innovation growth and endogenous short run fluctuations emerges spontaneously.

5.1 Introduction

The source of volatility of aggregates is one of the most debated and open questions of economic analysis in many decades. The conventional response is that in order to generate a business cycle aggregate shocks are needed since independent and idiosyncratic ones cancel out in the aggregate due to the law of large numbers (see e. g. Long Plosser, 1983, Dupor, 1999). If this were the case, system wide phenomena such as innovation waves and technological convergence would never occur since technological innovation is the foremost example of an idiosyncratic occurrence typically affecting the economic system at the individual firm level.

---

* The authors would like to thank participants at the 3rd Latin America - European Workshop on "New Tools of Qualitative Analysis of Economic Dynamics", October 2002, CIMAT, Guanajuato, Mexico for their comments.

Recent contributions have challenged the above mentioned conventional wisdom. Horvath (1998, 2000) has shown that sector specific shocks may cause aggregate fluctuations only when those involved are important input suppliers, thus recognising that firms do affect each other at least through an economys input-output structure. Contributions dealing with self organised criticality have provided a very interesting insight on the impact of exogenous events on systems populated by heterogeneous elements capable of interacting. Bak et al. (1993), Arenas et al. (2002), and Andergassen Nardini (2002) have proposed multi-firm, multi-sector models in which the effect of small independent shocks hitting single units do not cancel out in the aggregate owing to significant non-linear, strongly localised interactions between different members of the economy.

In this paper, we wish to inquire on the impact upon an economic system of idiosyncratic shocks which affect individual firms. The latter are heterogeneous, differently capable in respect to available technology, are rationally bounded, learn and need, therefore, to collect information to lay out their economic plans, in particular in order to technically improve. It is a consequence of these real world assumptions that firms knowledge base can possibly widely diverge determining the degree and strength of interaction and, thus, firm specific neighbourhoods of comprehension. It is within such neighbourhoods that information travels.

In Section 2 we show that, for critical values of the interaction strength between firms, even extremely small exogenous and idiosyncratic shocks of a technological nature may generate long run positive growth of the aggregate technological state of the economy yet exhibiting wide, short run oscillations. In Section 3 we identify, by taking into account their different levels of entropy, the neighbourhood which is capable to provide the strongest interaction and by a mean field approximation (see Vespignani and Zapperi 1998) we show that information propagating through minimal entropy neighbourhoods may generate innovation waves initiated by the input of a single bit of information at the individual firm level.

A final section draws the conclusions.

5.2 Growth Regimes: a General Framework

We consider a very large economy populated by n firms, in the limit for $n \to \infty$. Each firm i is characterised by an index $\phi_i(t)$ of productivity determined by innovations introduced up to time t; ϕ_i is a stochastic process, which compounds endogenous and exogenous technological adjustments occurring as a direct consequence of gathering information on technological innovations. Endogenous ones are direct consequences of technological information spillovers generated inside the system, while exogenous adjustments are autonomously generated by firms R&D activity[4] . The latter depend on firms capability to exploit new scientific advances and their frequency clearly depends on a given countrys technical and scientific progress but is largely independent of the number of firms populating the economic system.

We, therefore, assume the economy to possess technological capabilities that are conducive to generate exogenous technological shocks or to exploit technolotgical opportunities that are aggregate in nature and which then benefit single firms as an equal proportion of the whole. There are, accordingly, no scale effects due to the mere number of firms as argued by recent litterature of industrial economics.

We assume that they occur according to ϕ_i, i = 1, 2 ... n, the latter being random variables measuring the productivity rate of increase in the time unit which are assumed as i.i.d. with mean $\frac{\mu}{n}$ and variance σ^2. μ is the average and aggregate exogenous rate of increase generated by firms R&D activity in the unit time.

We assume that each firm i interacts with firms $j \in N_i$, where N_i is the neighbourhood of firm i. Neighbourhoods are so defined as being strictly independent in generating random transmission of information: there is, therefore, no overlapping. This assumption can also be taken as a first order approximation, higher order correlations being neglected, of more complex interactive structures. In what follows we will take the neighbourhood structure as given, while in Section 3 we will show how these neighbourhoods can suitably be defined. Because of bounded rationality, we assume that firms observe only a limited number of other firms, i.e.$|N_i| << n$. Within such a neighbourhood firm i receives a spillover from firm j when firm j adjusts. We define $\varepsilon_{i,j}$ the strength of interaction between neighbouring firms i and j. As a consequence, $\varepsilon_{i,j}\delta_j$ is a random variable indicating the productivity rate of increase achieved by firm i in the time unit induced by adjustments of neighbouring firm j. Given these assumptions, we obtain the following difference equation for the stochastic process $\phi_i(t)$ leading to technology upgrading

$$\frac{\Delta\phi_i(t)}{\phi_i} = \delta_i + \sum_{j \in N_i} \varepsilon_{i,j} \frac{\Delta\phi_j(t)}{\phi_j} \qquad (5.1)$$

According to (1) we have that each firm is able, at each point in time, to improve technologically either because of exogenous forces or because of endogenous spillovers from neighbouring firms. Notice that since $\mu > 0$ there is a positive drift in the exogenous driving force.

The technological state of any given economy can be rendered by several empirical measures. For instance, the overall technological prowess of the firms aggregate can be expressed by the overall number of engineers or by that of skilled workers or even by the number of technologically advanced blueprints.

---

[4] It is to be stressed that neither a single information spillover nor mere R&D activity necessarily generate a technological upgrading, as will be shown in section 3.

Assuming, therefore, additivity across firms, the aggregate technological state is represented by $Y = \sum_{i=1}^{n} \phi_i$. We begin our enquiry by studying the aggregate innovation growth rate of our countably infinite economy $\gamma j = \lim_{n \to \infty} \frac{\Delta Y}{Y}$ given by

$$\gamma_j = \lim_{n \to \infty} \sum_{i=1}^{n} \eta_i \frac{\Delta \phi i}{\phi i} \tag{5.2}$$

where $\gamma_j = \frac{\phi}{\sum_{j=1}^{n} \phi_j}$

Considering that in the long run firms that lag behind others in productivity growth are likely to be weeded out of the market, it is reasonable to assume that the weight of each firm is the same and equal to $\frac{1}{n}$.

In order to study the short and long run behaviour of technology growth in this economy we substitute (1) into (2) and calculate its average value and variance. Let us first consider a particular but important case.

Proposition 1. *Suppose $\varepsilon_{i,j} = 0$ for each $i, j = 1, \ldots, n$.*

$$E(\gamma_y) = \lim_{n \to \infty} \frac{\mu}{n} = 0$$

$$Var(\gamma_y) = \lim_{n \to \infty} \frac{\sigma^2}{n} = 0$$

Proof. Since the δ_i's are i.i.d. with mean $\frac{\mu}{n}$ and variance σ^2 and weights are uniformly equal to $\frac{1}{n}$, then given (1) and (2), the result follows straightforwardly.

Proposition 1 is a result of the law of large numbers. It states that, if there is no interaction between single firms, then, as their number diverges towards infinity, fluctuations average out. Further, the aggregate technology growth rate is vanishing small. If, however, the driving force of information depends on n and is such that $\lim_{n \to \infty} \frac{\mu}{n} > 0$, then the technological growth rate is entirely exogenous and in this case occurs without fluctuations. This is the standard neo-classical growth framework where the state of technology grows at the exogenously given parameter $\lim_{n \to \infty} \frac{\mu}{n}$.

Contrary to the standard neo-classical case, we are interested in technological growth taking place through endogenous diffusion of available information such that long ranged and highly volatile innovation waves are emergent features of the economy. Thus, in the next section we will study the case of non-negligible interaction, informational spillovers occurring between firms. It is interesting to remark how patterns of growth and fluctuations change when the interaction between single firms is allowed to increase.

We accordingly assume that $\varepsilon_{i,j} > 0$ for at least some i, j.

5.2.1 Strength of Spill-over Effects and Interaction: the Special Case of a Symmetric Economy

If no assumption is made on the neighbourhood structure of the economy and on the strengths of interaction between pairs of neighbouring firms, no general result can be stated about the expected growth rate. However it is clear that, due to bounded rationality and informational costs, each single firm can systematically interact only with a rather limited number of neighbours and, therefore, it will discard potential neighbours with comparatively low strength of interaction.

These considerations lead us to resort to a mean-field-type assumption and restrict our analysis to a uniform neighbourhood interacting economy which we call symmetric economy and define below.

Definition 1. *A symmetric economy is defined as follows:*

- *each firm i interacts with g other firms, where $2 \leq g$ finite: i. e. $|N_i| = g$ for each i.*
- *the strength of interaction determining the information spillover within each neighbourhood is uniform (and positive) throughout the economy: i.e. $\varepsilon_{i,j} = \varepsilon$ for each i, j.*

This mean field approach is perfectly natural, since firms with a systematically lower capability of interaction and technical innovation will soon fall much behind the top technological level and will systematically be weeded out by the market mechanism. Let us consider the case of $\varepsilon > 0$ and determine $E(\gamma_y)$ and $Var(\gamma_y)$.

Lemma 1. *Consider a symmetric economy as defined in Definition 2, where*

$$E(\gamma_y) = \lim_{n\to\infty} \frac{\mu}{n} \frac{1 - (\varepsilon g)^{m(n)+1}}{1 - \varepsilon g}$$

$$Var(\gamma_y) = \lim_{n\to\infty} \frac{\sigma^2}{n} \frac{1 - (\varepsilon^2 g)^{m(n)+1}}{1 - \varepsilon^2 g}$$

where

$$m(n) = ln\left(\frac{n}{y - n}\right)$$

Proof. Iterating (1) the long run technology growth rate and the variance are given by

$$E(\gamma_y) = \lim_{n\to\infty} \frac{\mu}{n}\left(1 + \sum_{j\in N_i} \varepsilon_{i,j}\left(1 + \sum_{k\in N_j} \varepsilon_{j,k}\left(1 + \sum_{j\in N_k} \varepsilon_{k,t}(1 + \sum \ldots)\right)\right)\right)$$

$$Var(\gamma_y) = \lim_{n\to\infty} \frac{\sigma^2}{n}\left(1 + \sum_{j\in N_i} \varepsilon_{i,j}^2\left(1 + \sum_{k\in N_j} \varepsilon_{j,k}^2\left(1 + \sum_{j\in N_k} \varepsilon_{k,t}^2(1 + \sum \ldots)\right)\right)\right)$$

Since we have assumed no overlapping between neighbourhoods, the number of iterations is $m(n)$. Using Definition 2 and Lemma 3 we can now characterise different growth regimes by the changing the relevant parameters: the strenght of interaction ε and the strength of the exogenous aggregate thrust μ. The next Proposition studies the different possibilities.

Proposition 2. *Let α and β be arbitrary finite constants. Consider a symmetric economy where $\varepsilon > 0$. The following growth regimes can be defined*

1. *if $0 < \varepsilon < \frac{1}{n}$, $\mu > 0$ and both independent of n, then*

$$E(\gamma_y) = 0$$
$$Var(\gamma_y) = 0$$

2. *if $\frac{1}{g} \le \varepsilon < \frac{1}{\sqrt{g}}$, $\mu > 0$ and both independent of n, then*

$$E(\gamma_y) = \infty$$
$$Var(\gamma_y) = 0$$

3. *if $\varepsilon = \sqrt{\frac{1}{g} - \frac{\alpha}{ng}}$ while $\mu > 0$ and independent of n then*

$$E(\gamma_y) = \infty$$
$$Var(\gamma_y) = \frac{\sigma^2}{\alpha} \frac{e^{\alpha-1}}{e^{\alpha}}$$

4. *if $\frac{1}{g} \le \varepsilon < \frac{1}{\sqrt{g}}$, while $\mu = n\beta(g\varepsilon)^{-m(n)}$ then*

$$E(\gamma_y) = \frac{g\varepsilon}{g\varepsilon - 1}\beta$$
$$Var(\gamma_y) = 0$$

5. *if $\varepsilon = \sqrt{\frac{1}{g} - \frac{\alpha}{ng}}$ and $\mu = n\beta(g\varepsilon)^{-m(n)}$ then*

$$E(\gamma_y) = \frac{g\varepsilon}{g\varepsilon - 1}\beta$$
$$Var(\gamma_y) = \frac{\sigma^2}{\alpha} \frac{e^{\alpha-1}}{e^{\alpha}}$$

6. *if $\varepsilon > \frac{1}{\sqrt{g}}$ while $\mu = n\beta(g\varepsilon)^{-m(n)}$ then*

$$E(\gamma_y) = \frac{g\varepsilon}{g\varepsilon - 1}\beta$$
$$Var(\gamma_y) = \infty$$

7. *if $\varepsilon > \frac{1}{\sqrt{g}}$ while $\mu > 0$ both independent of n then*

$$E(\gamma_y) = \infty$$
$$Var(\gamma_y) = \infty$$

mean \ variance	zero	finite	infinite
zero	case 1		–
finite	case 4	case 5	case 6
infinite	case 2	case 3	case 7

Fig. 5.1.

Proof. Consider first case 5. We substitute $\varepsilon + \sum \frac{1}{g} - \frac{\alpha}{ng}$ into the expression for the long run growth rate stated in Corollary 3 and obtain in this way

$$E(\gamma_y) = \frac{\mu}{n} g^{\frac{n}{2}} \frac{1 - \sqrt{g}(1 - \frac{\alpha}{n})^{\frac{n+1}{2}}}{1 - \sqrt{g}(1 - \frac{\alpha}{n})^{\frac{n+1}{2}}} \tag{5.3}$$

Since $(1 - \frac{\alpha}{n})^{\frac{n+1}{2}} \to_{n\to\infty} e^{\frac{-\alpha}{2}}$ and setting $\mu = n\beta g^{\frac{-\alpha}{2}}$ we obtain the result stated in the proposition. Consider now the variance of the growth rate.

Substituting $\varepsilon - \sqrt{\frac{1}{g} - \frac{\alpha}{ng}}$ into the expression for the long run growth rate stated in Lemma 3 and obtain in this way

$$Var(\gamma_y) = \frac{\sigma^2}{n} \frac{1 - (1 - \frac{\alpha}{n})^{n+1}}{1 - (1 - \frac{\alpha}{n})} \tag{5.4}$$

Since $1 - (1 - \frac{\alpha}{n})^{n+1} \to_{n\to\infty} e^{\frac{-\alpha}{2}}$ we obtain the result stated in the proposition.

Cases 1., 2., 3., 4., 6. and 7. are consequences of (4) and (5). As shown in Proposition 4, if the economy is driven only by small idiosyncratic information shocks and if each firm interacts locally with other g neighbouring firms, the strength of the spill-over effects being constant across firms, then we can identify seven growth regimes, as shown in Table 1.

- Case 1: negligibly small long run growth without short run fluctuations.

- Case 2: infinite long run growth rate but without short run fluctuations.
- Case 3: infinite long run growth rate with large short run fluctuations.
- Case 4: positive, finite long run growth but without short run fluctuations.
- Case 5: positive, finite long run growth rate with large short run fluctuations.
- Case 6: positive, finite long run growth rate with infinite fluctuations.
- Case 7: infinite long run growth with infinite fluctuations.

In the first growth regime, as the number of firms becomes ever larger ($n \to \infty$), the same aggregate and exogenous, μ, becomes negligible for each firm: the probability of making an adjustment tends to be nil. Moreover, the strength of the interaction is very small, it is lower than the probability of each firm contacting one of its equally reachable neighbours, $\frac{1}{g}$. In these circumstances, information being endogenously passed on is very scant and become lost in the process, each firm retaining what meagre information it gets. For the same reason the variance is also zero. No technological paradigm can, therefore, emerge.

The second growth regime considers the case in which, although the average exogenously induced adjustment tends to become very small, the strength of firm to firm interaction becomes sufficiently large to set off a process of information transmission such that, no matter how small, it is ever amplified. For very large economies, the resulting long run growth rate becomes infinite. Nevertheless, interaction is not quite as strong as to generate positive fluctuations. It can easily be checked from $Var(\gamma_y)$ as a function of ε that $\varepsilon = \frac{1}{\sqrt{g}}$ is the threshold before which fluctuations are dampened to zero but positive past it.

The third growth regime considers the case of ε becoming larger with n. As the latter becomes ever larger ($n \to \infty$), the said threshold is reached and the variance becomes positive. Hence, interaction is now as strong as to generate an infinite technological growth rate through strong endogenous diffusion but large aggregate fluctuations emerge.

The fourth growth regime deals with the case in which exogenous adjustments are a function of both the size of the economy and the strength of interaction, the former becoming smaller the larger are the latter. This case highlights the situation of an economy in which size and relation between firms render the original adjustments of less significance, all the more so the greater are these two characteristics, pre-eminence being acquired by endogenous transmission. Notice that the average long run growth rate is much larger than the exogenous one which is vanishingly small as the size of the economy is allowed to become very large. Assuming ε to be below the $\frac{1}{\sqrt{g}}$ threshold, fluctuations are zero but the long run rate of technological growth is finite and positive, the negligible size of the exogenous adjustments being exactly offset by the strength of the interaction. Technological paradigms, created through the endogenous diffusion of information and involving eventually the whole economy, possibly emerge.

Case five has a similar profile but the strength of the interaction is exactly the critical threshold above which fluctuations become positive while the long run growth rate is finite. This case marks a critical state since it separates non-exploding, absorbing dynamics from irregular ones with positive variance (regimes 2. and 3. respectively).

Growth regime six shows that as ε goes past the critical threshold, the average exogenously induced adjustment at single firm level $\frac{\mu}{n}$ vanishing with size, fluctuations become infinitely large on account of very strong interaction while growth regime seven generates infinite growth if μ is positive and independent of size with equally infinite fluctuations.

There is no a-priori reason to hold that technological progress should necessarily occur in the shape of any of the cases discussed above. It is, indeed, possible to state that the actual regime depends crucially on how the parameters entering Lemma 3 are structurally tuned. The history of economic development in large systems after the industrial revolution may suggest that the most likely case is growth regime five, featuring positive finite technological growth together with finite fluctuations, and possibly, but less likely, growth regime six in which positive growth is coupled to very large fluctuations. Many developing economies, on the other hand, may well be categorised by growth regime one in which both average growth and fluctuations are nil.

5.3 Emergent Technological Paradigms

In this section we proceed to describe single firms and their interaction capability. We consider an economy, populated by n firms, n very large. Each firm can either receive an exogenous bit of information or an endogenous one. In the former case, new information is borne directly by the firm, through successful investment in R&D, while in the latter case information is obtained by continuously observing a limited number of "neighbouring" firms. The first step is to describe in detail the neighbourhood structure.

5.3.1 The Interaction Structure

Since the capability of understanding and processing information coming from a different firm and a more advanced technological context depends on the common knowledge basis, the transmission of such information depends on the strength of this shared knowledge, i.e. the potential intensity of their interaction, and the probability of actually passing on relevant information. Let this strength be measured by $\varepsilon_{i,j} \in [0,1]$ for any two different firms.

The measures of cognitive distance, or proximity, thus defined and empirically observable through a statistical procedure, allow, in turn, a rigorous definition of the cognitive neighbourhoods through which innovative information can pass through. We are going to assume symmetry between firms i, j. There are, therefore, $\frac{n(n+1)}{2}$ couplings which compose the set.

Definition 2. $S = \{\varepsilon_{i,j} | i, j = 1 \ldots n\}$, *where* $|S| = \frac{n(n+1)}{2}$

Also in this case we wish to deal with symmetric economies (Definition 2). The set of all possible neighbours out of the total number n of firms in the whole economy can be defined as follows:

Definition 3. *The set of all possible neighbourhoods of a firm i is defined as*

$$\Gamma^i = \{\gamma_{i,j}\}_{j=1}^{|\Gamma^i|} \tag{5.5}$$

where $\gamma_{i,j} \subset \{1, \ldots n\} / \{i\}$ *and where* $|\Gamma_{j=1}^i| = g$ *for each* $j = 1, 2, \ldots |\Gamma_{j=1}^i|$ *and where* $|\Gamma_{j=1}^i| = \frac{(n-1)!}{g!(n-g-1)!}$, *this for each* $i = 1, 2, \ldots n$. *The collection of all sets of possible neighbourhoods for each firm defines the space of neighbourhoods*

$$\Gamma = \{\Gamma^i \mid i = 1, \ldots n\} \tag{5.6}$$

These definitions provide a map of cognitive neighbourhoods for each firm and for the entire set of firms. The set of neighbours in each Γ^i are of varying informative capability for the firm on account of the cognitive heterogeneity of its members. It follows that a ranking of these neighbourhoods can be compiled on the grounds of how enabling they are from the point of view of their informative content, given the combination of probabilities $\epsilon_{i,j}$. A convenient measure of such informative content and of the ease with which information percolates through to let the firm learn and cumulate knowledge for innovation is Shannons entropy measure (Klir and Folger, 1988). This entropy measure is:

$$M(\gamma_{i,j}) = - \sum_{k \in \gamma_{i,j}} \epsilon_{i,k} log_2(\epsilon_i, k) \tag{5.7}$$

Given all $M(\gamma_{i,j}) \in [0, \infty]$ for all $\gamma_{i,j} \in \Gamma^i$, it is possible to compute the minimum, thus identifying the firms neighbourhood which is most capable of carrying information and which best enables it to learn.

Definition 4. N_i *is the neighbourhood which is most likely to provide significant innovative information and which is, therefore, the cognitively relevant neighbourhood for the diffusion of innovative technologies:*

$$N_i = arg\ min_{\gamma_{i,j} \in \Gamma^i} M(\gamma_{i,j})$$

This definition allows us to identify a firms relevant neighbours. The probabilities $\epsilon_{i,j}$, measuring cognitive relationships, depend in each neighbourhood on the number of firms which are nested therein. This proposition follows directly from the very definition of neighbourhood as the locus of dense inter-firm externalities. Economic history and studies in the Marshallian tradition have provided plenty of evidence for this fact. In particular, literature on industrial districts indicates that firms tend to cluster according to a predictable pattern often determined by agglomeration economies based on shared knowledge and know how. The greater is the number of firms in any given cluster, the greater is the cognitive correlation and the greater the probability that information spread across the cluster. This is due to the fact that higher firm density fosters denser linkages and greater mutual understanding and awareness of eachothers technological status and capabilities easying the exchange of information.

Conjecture 1. The probability that information be passed on is

$$\epsilon_{i,j} = \epsilon = 1 - \frac{k}{n} \tag{5.8}$$

k is a parameter indicating the critical threshold at which no information can be exchanged. A large k implies hurdles to informational interaction possibly due to high heterogeneity in skills, knowledge and competence requiring, therefore, larger density for information to spread.

5.3.2 Spill-over Dynamics

We assume that the introduction of an innovation requires the accumulation of informational bits which is a process each firm has to complete if it wishes to do so. We consider a symmetric economy in which ϵ is the average strength of interaction[5] and g the number of informational bits each firm has to accumulate. In other words, we assume that more are the informational bits each firm has to gather, more are the neighbouring firms it continuously observes. We characterise each firm according to the number of bits of information it has accumulated and we label the possible states (elements of the state space) as follows: $\Omega = \{0, 1, 2, ..., c, a\}$, where the cardinality of the state space is $g + 1$. Thus, $x_i(t) \in \Omega$, for each $i = 1, ..., n$, where $x_i(t)$ indicates the state which characterises firm i at time t. 0 indicates that the firm has just upgraded its technology, and it has no new information; 1 indicates that the firm has accumulated one bit of information and so on. Finally, c indicates the state where it needs just one more bit of information such that technology upgrading becomes viable and a indicates the active state where the firm upgrades its technology level. We assume an average aggregate exogenous inflow of information h. We are interested in the case where technological paradigms are emergent features and, symmetric to the previous part of these notes, this will occur only if the time scale of the exogenous information inflow is very slow compared to diffusion dynamics, i.e. $h \to 0$. The dynamics are as follows: given that a firm is in a state u, if it receives a bit of information (either an exogenous or an endogenous one) it switches to state $u + 1$. If state $u + 1 < a$ nothing happens until the next bit of information arrives. On the other hand,

---

[5] This is consistent with the mean field approach used to describe the diffusion of information.

if $u + 1 = a$, then it upgrades its technology, and transfers in this way a bit of information with probability ϵ to g neighbouring firms. Notice that since each firm always observes the same neighbourhood, nothing happens until a new innovation is introduced. Only once technology upgrading occurs, will information accumulated by the innovating firm be freed to some degree. Thus, as long as the accumulation of information continues, acquired knowledge remains tacit and cannot help other firms introduce new innovations.

Vespignani and Zapperi (1998) show that the mean-field approximation to the interaction between single firms well approximates stationary state dynamics. Thus, we cluster firms according to their state. We call $\rho_u(t)$ the function of time indicating the average density of firms being in state u, where $u = 1, 2 \ldots c, a$ More formally, $\rho_u(t) = \frac{\sum_{i=0}^{n} I_{(x_i(t)=u)}}{n}$

We accordingly characterise the time evolution of average densities by specifying their transitional rates. Dropping t, t he state space dynamics are described by the following master equations

$$\rho_a = -\rho_a + \rho_c(\mu + g\varepsilon\rho_a)$$
$$\rho_c = -\rho_c(\mu + g\varepsilon\rho_a) + \rho_{c-1}(\mu + g\varepsilon\rho_a)$$
$$\ldots$$
$$\rho_{c1} = -\rho_1(\mu + g\varepsilon\rho_a) + \rho_0(\mu + g\varepsilon\rho_a)$$
$$\rho_O = \rho_O(\mu + g\varepsilon\rho_a) + \rho_a$$

The first term of the first differential equation in (6) indicates the outflow of firms from the active technology upgrading state: once a firm introduces a new technology it switches immediately to state 0 and starts the information collection process from the beginning. The second expression indicates the inflow of firms into the active state: firms being in state c, receiving either an exogenous bit of information with probability h or an endogenous one with probability $g\varepsilon\rho_a$, switch to the active state.

Thus, the probability defining the transitional rates is $h + \varepsilon a\rho_a$. The other expressions can be interpreted in a similar way. Since ρ_u, for $u = 1, 2, \ldots, c, a$ are average densities, the following normalisation condition has to be satisfied:

$$\rho_0 + \rho_1 + \cdots + \rho_c + \rho_a = 1$$

It can be shown that the stationary state is asymptotically stable (see Vespignani and Zapperi, 1998). We are interested in the stationary average number of firms introducing a new technology, given that a single, small idiosyncratic exogenous informational bit hits the economy.

Calling x the variable generated by process (6) indicating the number of firms introducing a new technology, $x = \sum_{i=1}^{n} I_{x_i=a}$ we state and prove the following proposition.

Proposition 3. . *The stationary average number of firms innovating, given that a single, small idiosyncratic informational bit hits the economy, is given by*

$$E(x) = \frac{1}{g(1-\varepsilon)}$$

Proof. Since the ρ_us can also be interpreted as the probability thata firm be in state u, let χ_i be the probability that a firm i is active ($x_i = a$) conditional to the arrival of one exogenous bit of information on the population n of firms:

$$\chi_i = \rho_c \Big(\frac{1}{n} + \varepsilon \sum_{j \in N_i} \chi_j \Big)$$

The average number of firms introducing an innovation is, therefore the sum over all χ_i:

$$E(x) = \sum_{i=1}^{n} \chi_i = \rho_c \sum_{i=1}^{n} \Big(\frac{1}{n} + \varepsilon \sum_{j \in N_i} \chi_j \Big)$$

Since $\sum_{i=1}^{n} (\sum_{j \in N_i} \chi_j) = g E(x)$, it is

$$E(x) = \rho_c + \rho_c \varepsilon g E(x)$$

and given that $\rho_c = \frac{1}{g}$ we obtain

$$E(x) = \frac{1}{(1-\varepsilon)g}$$

Notice that as long as $\varepsilon < 1$, the average number of firms introducing an innovation remains finite, even though the number of firms diverges towards infinity. This implies that the average number of firms introducing an innovation is vanishingly small compared to the total number of firms existing in the economy. In the latter case the correlation among the single firms remains small and as a consequence fluctuations average out in the process of aggregation.

On the other hand, if $\varepsilon \rightarrow 1$, then the average number of firms introducing an innovation diverges towards infinity. In this case it is no longer obvious that the average number of firms introducing a new innovation is vanishingly small compared to the total number of firms existing in the economy.

When a new technology is introduced productivity, sooner or later, normally increases. In the next section we investigate the consequences of innovation spill-overs on aggregate productivity growth.

5.3.3 Innovation and Productivity Growth

Let it be assumed that the rate of increase of productivity π is dependent on g, the informative content implied by the innovation process. The greater is the latter, the higher is the likely productivity increase. Hence, we postulate that $\pi = \pi(g)$ and such that $\pi' > 0$, $\pi'' < 0$. If N_i is the number of innovations introduced by firm i, the aggregate state of technology is given by

$Y = \sum_{i=1}^{n} e^{\pi(g)N_i}$ and $dN_i = I_{(x_i=a)}dt$, it follows from Proposition 9 given the assumptions made that the long run growth rate of technical progress is

$$E(\gamma_y) = \frac{\pi(g)}{n} \frac{1}{g(1-\varepsilon)}$$

We observe that if $\varepsilon < 1$, then as the number of firms diverges towards infinity the long run growth rate becomes negligibly small. In this case the innovation avalanches are too small, and as a consequence their impact on the long run growth rate is negligible. Thus, no innovation waves and technological paradigms can emerge.

On the other hand by using the Marshallian conjecture as stated in Conjecture 8 we obtain the following value for the long run growth rate

$$E(\gamma_y) = \frac{\pi(g)}{gk}$$

In this case the long run growth rate remains finite, even though the exogenous driving force is negligibly small. Spillover effects are such that innovation waves and technological paradigms are emergent features of the economy and growth in this case occurs through large fluctuations. Observe that the long run growth rate depends on structural parameters of the economy. In particular, the larger is the number of informational bits each firm has to accumulate such that the introduction of the innovation becomes viable, the lower is the long run growth rate of the economy.

5.4 Conclusions

This paper shows that structural characteristics matter. More specifically, growth and fluctuations depend on how each firms neighbourhood is determined in terms of the number of firms which belong to it and of the interaction strength they are capable of. Transmission of exogenous, idiosyncratic shocks and the diffusion of innovation waves are a consequence of the crucial role they play. As the strength of interaction rises in respect to a neighbourhood population, very different behaviour is obtained, in the limit of large systems, from negligible technological growth with no fluctuations to a very large size in both magnitudes through all possible intermediate cases. The paper reviews seven different cases. The first being the one in which interaction is so feable in relation to neighbourhood size that no growth and no fluctuations arise. The seventh, on the other hand, being the case in which interaction is so strong that positive shocks are ever amplified producing theoretically infinite growth and fluctuations. The remaining cases being intermediate ones showing that even when exogenous shocks are small in respect to the economy size, interaction is still capable of generating positive growth with significant fluctuations.

Technological paradigms emerge in consequence of such interaction. The paper highlights this point by describing an emerging regime in terms of a dynamic process in which informational spill-overs, together with an exogenpus drive, gradually build up a critical amount of information which triggers firms technological upgrading. The aggregate outcome depends on the strength of the interaction dependent on the cognitive relationship between firms in a given neighbourhood and on its size. Studies in the Marshallian tradition support this view. This aggregate outcome is defined as the average number of innovating firms that have gradually completed the dynamic process determining an avalanche of innovations and therby a technological paradigm.

It is finally interesting to note that at a micro-level neighbourhood size is the result of an evolutionary process; at a macro-level it concurs to determine the long run productivity growth rate. the last part of the paper discusses an implied trade-off: while a large and information-wise rich neighbourhood increases technological growth, the effort of collecting such information weakens it. Whether there are positive or negative feedbacks such that the economy autonomously settles on a given technological state , a process described in recent literature as self organisation, or a suitable policy is necessary to drive the economy towards the highest and feasible aggregate growth rate is an entirely open question. Further research will investigate this interesting point.

References

1. Andergassen, R., Nardini, F., (2004): Endogenous innovation waves and economic growth. Forthcoming in Structural Change and Economic Dynamics.
2. Arenas, A., Daz-Guilera, A., Prez, C. J., Vega-Redondo, F., (2002): Selforganized criticality in evolutionary systems with local interaction. Journal of Economic Dynamics and Control 26, 2115-2142.
3. Bak, P., Chen, K., Scheinkman, J.,Woodford, M.,(1993): Aggregate fluctuations from independent sectoral shocks: self organized criticality in a model of production and inventory dynamics. Ricerche Economiche 47, 3-30.
4. Dupor, B., (1999): Aggregation and irrelevance in multi sector models. Journal of Monetary Economics 43, 391-409.
5. Horvath, M. T. K., (1998): Cyclicality and sectoral linkages: aggregate fluctuations from sectoral shocks. Review of Economic Dynancis 1, 781-808.
6. Horvath, M. T. K.,(2000): Sectoral shock and aggregate fluctuations. Journal of Monetary Economics 45, 69-106.
7. Long, J., Plosser, C., (1983): Real business cycles. Journal of Political Economy 91, 39-69.
8. Vespignani, A., Zapperi, S., (1998): How Self-organized criticality works: a unified mean field picture . Physical Review E 57, 6345-6362.

6

Can Catastrophe Theory Become a New Tool in Understanding Singular Economies?*

Elvio Accinelli[1] and Martín Puchet Anyul[2]

[1] U. de la República, Montevideo, Uruguay and UAM-X, México D.F.
    elvio@fing.edu.uy
[2] Fac. de Economía, UNAM. anyul@servidor.unam.mx

Summary. The aim of this paper is to show that economic systems must be characterized by their possible singularities rather than by their regularities. Changes in parameters of a regular economy imply only small changes in the optimal choice of the agents, i.e. the economic system is structurally stable, and they are consistent with one another minor changes. But in a singular economy, small changes in parameters affect the choice of the agents in a relevant way. The equilibria, after and before the changes, are radically different states, i.e. the economic system is structurally unstable. Catastrophe theory and Morse theory are used here to characterize singular economies. These are classical theories in mathematics but nevertheless, they are new tools to help understand the behavior of an economic system. Also the approach of Negishi is followed, and this allows us to consider in a unified way economies with finitely or infinitely many goods.

6.1 Introduction

The purpose of this paper is to analyze the behavior of an economic system when singularities appear. To make this analysis we introduce some issues of the mathematical singularities and catastrophe theories in the framework of the economic general equilibrium theory. These mathematical theories constitute new tools to help understand the structural instability of an economic system. There are not many publications relating these two kinds of theories. The main papers on the subject are [Balasko, Y. (1978)] and [Mas-Colell, A.; Nachbar, J.H. (1991)]. In [Mas-Colell, A. (1985)] and in [Balasko, Y. (1988)] there are some considerations on singularities, but the main objective of these works is the study of regularities rather than singularities.

The existence of singularities determine the structural instability of an economic system. To identify the characteristics of the singularities we introduce the Morse theory of functions for 2-agent economies and the catastrophe

---

* The authors are indebted with Ramón García Cobbian, Paula Vera-Cruz for helping to improve our English, and with two anonymous referees for useful comments and suggestions. All remaining errors are responsibility of the authors.

theory for more general cases. We also use the $k-jet$ of smooth functions and the Thom's transversality theorem to classify the singularities. By translating these characteristics to the excess utility functions we describe some aspects of the behavior of the singular economies.

From a topological, or measure theory, point of view the set of singular economies is a small set. However, when the system traverses this set, big changes occur. Without such singularities the economic system would be a structurally stable and boring mechanism. When the system only shows regularities -regular economies- small changes in endowments do not imply important changes in its behavior, while it would bring about big changes in the performance if it had singularities -singular economies-.

The approach of the general equilibrium theory based on the excess utility function, originally due to Negishi, allows us to consider economies without having to specify whether they are modelled in finite or infinite vectorial spaces. So this approach allows us to analyze, in a unified form economies with commodity spaces with and without contingent goods. Also, the Negishi approach allows us to avoid all consideration of the smoothness or existence of demand function [Araujo, A. (1987)]. In order to prevent the problem of the generic nonexistence of equilibrium in spaces as L^p raised in [Monteiro, P.K. (1994)] we will introduce the, relatively strong, hypothesis that the first order derivatives of the utility functions belong to their dual space. For simplicity we will restrict the commodity space for which the interior of the positive cone is non-empty, i.e., spaces as L^∞, l^∞; in general spaces of functions with supremum essentially bounded.

Section 2 describes the Negishi approach. Section 3 defines regular an singular economies. Section 4 introduces Morse functions to classify 2-agent economies and shows an example of a singular economy. The aim of section 5 is to focus on some perspectives that help classify singular economies in more general cases. This section defines the kjet and sets the Thom's transversality theorem. By means of this theorem some general conclusions about singularities are obtained. Finally, the fold and the simple cusp catastrophes are considered to classify singular 3-agent economies. The last section contains some tentative conclusions and perspectives of our work.

6.2 The Negishi Approach

The original version of the Negishi approach was stated in the framework of the $\mathbb{R}^n$; space, see [Negishi, T. (1960)]. The framework of our paper is a Banach Lattice.

Consider an n-agent economic system $E = \{X, \omega_i, u_i, i = 1, 2, \ldots, n\}$ where the trading space or commodity space X is contained in the positive cone X_+ of a Banach Lattice B. The price space is contained in B^* the dual of B; i.e. the set of linear continuous functional on B: By w_i we denote the initial endowments of the agent represented by $i = 1, 2, \ldots, n$ and by $w \in \Omega$ we denote the profile of the initial endowments $w = \{w_1, w_2, \ldots, w_n\}$.

The value of a bundle set $x \in X$ when prices are $p \in B^*$ is given by the real number px.

The endowment of certain agents may be zero in some goods, but not identically zero, that is $w_i \in X - \{0\}$. However the aggregate endowment $\sum_{i=1}^{n} w_i = W$ is strictly positive, i.e. $W > 0$. The consumers preferences are represented by strictly quasi-concave smooth and increasing utility functions. Increasing preferences imply that $x > y \Longrightarrow u_i(x) > u_i(y)$.

In order to use the Fréchet differentiable techniques we assume that the interior of X_+ is nonempty[3]. The utilities are at least 2-times Fréchet differentiable, (F-differentiable). So $u_i'(x) \in B^*$ and $u_i''(x)$ is a bounded symmetric bilinear form, identified with the quadratic form $(h, k) \to u_i''(x)hk$ It follows that $\| u_i''(x)hk \| \neq q \| h \| \| k \|$ and hence $\| u_i''(x)hk \| \neq q \| h \|$ Thus $u_i''(x)h \in B^*$ and so the linear operator $u_i''(x) : X \to B^*$ is continuous.

In order to assure that in equilibrium each agent will consume inside its own space of consumption we can assume that utilities satisfy one of the following two conditions, these are widely used assumptions in economics,

1. (Inada condition) $\lim u_j' = \infty$ if $x \to \delta(X_+)$, for each $j = 1, 2, \ldots i$ and for each utility function, by $\delta(X_+)$ we denote the frontier of the positive cone. This assumes that marginal utility is infinite for consumption at zero.

2. All strictly positive bundle sets are preferable to all bundle sets with at least one zero component in one state of the world.

An allocation x is a vector $(x_1, x_2, \ldots x_n) \in \Omega$ where Ω is the n-cartesian product of the consumption spaces, it is feasible if $\sum_{i=1}^{n} x_i = \sum_{i=1}^{n} w_i = W$

6.2.1 Welfare Theory and Existence of Equilibrium

As it is well known, for every Pareto optimal allocation $(\overline{x_1} \ldots \overline{x_n})$ there exists a vector of nonnegative welfare weights

$$(\lambda_1 \ldots \lambda_n) \in \Delta = \{\lambda \in \mathbb{R}^n : \sum_{i=1}^{n} \lambda_i, \lambda_i \geq 0, \forall i\},$$

such that $(\overline{x_1} \ldots \overline{x_n})$ solves the maximization program:

$$max \left\{ W_\lambda = \sum_{i=1}^{n} \lambda_i u_i(x_i), s.t. \sum_{i=1}^{n} x_i \leq \sum_{i=1}^{n} w_i, \ and \ x_i \in X; i = 1, 2, \ldots, n \right\}$$
$$(6.1)$$

Reciprocally, for any $\lambda \in \Delta$ the solution of that maximization problem is a Pareto optimal allocation. The solution of this problem is symbolized by $x(\lambda, \omega)$. For any $\lambda \in int[\Delta]$ the relative interior of Δ the excess utility function is, $e : \Delta x \Omega \to \mathbb{R}^n$, with coordinates given by:

---

[3] This can be generalized using the Gateaux derivative. See [Accinelli, E. (1995)].

$$e_i(\lambda, \omega) = u_i'(x_i(\lambda, \omega) [x_i(\lambda, \omega) - \omega_i] :$$

Here $u_i'(x_i(\lambda, \omega)) : X \to R$ is the F-differential of the utility $u_i(x_i(\lambda, \omega))$: For a fixed ω from now on we will use the notation $e_\omega(\lambda) = (e_{\omega_1}(\lambda), e_{\omega_1}(\lambda), e_{\omega_2}(\lambda), \ldots, e_{\omega_n}(\lambda))$.

Definition 1. *For fixed $\omega \in \Omega$ we define the set*

$$Eq(\omega) = \{\lambda \in int [\Delta] : e_\omega(\lambda) = 0\}$$

*it will be called the set of the **Equilibrium Social Weights**.*

[Accinelli, E. (1996)] shows that the equilibrium social weights is a non-empty set. Since we assume that $u'(z) \in B^*$ the equilibrium set is not an almost empty set, see [Monteiro, P.K. (1994)], so the following theorem holds:

Theorem 1. *Let $\lambda \in Eq(\omega)$ and let $x^*(\lambda)$ be a feasible allocation, solution of the maximization problem (1) and let $\gamma(\lambda)$ be the corresponding vector of Lagrange multipliers. Then, the pair $(x^*(\lambda), \gamma(\lambda))$ is a Walrasian equilibrium and reciprocally, if (p, x) is a Walrasian equilibrium then, there exists 2 Eq such that x maximize W restricted to the feasible allocations set, and p will be the corresponding vector of Lagrange multipliers i.e., $p = \gamma(\overline{\lambda})$.*

The proof can be seen in [Accinelli, E. (1996)].

6.3 Regularities and Singularities

We begin this section with some fundamental mathematical definitions that will be needed later on. The main reference is [Zeidler, E. (1993)].

Definition 2. *Let $f : U(x_0) \subseteq \mathbb{R}^n \to R^m$ be C^1: Then de derivative of f is the $n \times m$ matrix given by $f'(x) = [\partial f_i(x)/\partial x_j]$ its rows are $i = 1, 2, \ldots n$ and its columns are $j = 1, 2, \ldots m$. If $n \geq m$ and rank $f'(x_0) = m$ then f is a submersion at x_0.*

Definition 3. *Let $f : U(x_0) \subseteq \mathbb{R}^n \to \mathbb{R}^m$ be C^1.*

- *(a) A point $x \in \mathbb{R}^n$ is called a regular point of f if and only if f is a submersion at x. Otherwise x is called a singular point.*
- *(b) A point $y \in R_m$ is called a regular value of f if and only if the set $f^{-1}(y)$ is empty or consists only of regular points. Otherwise y is called a singular value, i.e., $f^{-1}(y)$ contains at least one singular point.*
- *(c) A singular value y is called non degenerate if the corank of $f'(x)$ verifies that corank $[f''(x)] \leq 1$ for all $x \in f^{-1}(y)$ but at least one $x \in f^{-1}(y)$ satisfies the equality.*

Let B be a Banach space, it follows from definition (3) that a map f : $U(x_0) \subset B \to R$ has a singular point at x_0 if and only if $f'(x_0) = 0$ here $f'(x_0)$ is the Frèchet derivative at x_0.

Let X and Y be B-spaces, and $f : U(x_0) \subseteq B \to Y$ a $C^k - submersion$ at x_0 with $f(x_0) = 0$, then there exists a local C^k diffeomorphism ϕ with $\phi(x_0) = 0$ and $\phi'(x_0) = I$ (where I is the unitary matrix), such that

$$f(\phi(y)) = f'(x_0)y. \tag{6.2}$$

For details see [Zeidler, E. (1993)].

6.3.1 Regular and Singular Economies

We now introduce the concept of regular and singular economies. We characterize each economy by the excess utility function. As it is well known (see for instance [Accinelli, E. (1996)]) the excess utility function $e\omega(.)$ is a homogeneous of degree zero function, and verifies the identity $\lambda e_\omega(.) = 0 \forall \lambda \in int\,[\Delta]$.

In an analogous way to the definitions given in [Debreu, G.], and in agreement with our characterization we introduce the following definitions:

Definition 4. *An economy characterized by the excess utility function, $e\omega(.)$ is **regular** if and only if zero is a regular value for $e\omega(.)$; Otherwise we say that the economy is singular. In addition if 0 is a degenerate singular value for $e\omega(.)$; the economy is a degenerate singular economy.*

So, if we characterize each economy by its excess utility function, we will obtain a preliminary classification of the economies in three main groups:

1. Regular economies.
2. Singular non degenerate economies.
3. Singular degenerate economies.

Consider the excess utility function $e(\lambda, \omega) = (e_1(\lambda, \omega), (e_2(\lambda, \omega), \dots (e_n(\lambda, \omega))$. If the pair $(\overline{\lambda}, \overline{\omega})$ $e(\overline{\lambda}, \overline{\omega}) = 0$, and if the economy with endowments given by $\overline{\omega}$ is regular, then by applying the Implicit Function Theorem to the equality $e(\lambda, \omega) = 0$, it follows that there exist neighborhoods $V_{\overline{\omega}}$ and $U_{\overline{\lambda}}$ of $\overline{\omega}$ and $\overline{\lambda}$ respectively and a function $\Lambda : V_{\overline{\omega}} \to U_{\overline{\lambda}}$ such that $e(\Lambda(\omega), \omega) = 0$ for all $\omega \in V_{\overline{\omega}}$. We can then say that regular economies have the same local behavior in each $\lambda \in Eq(\omega)$.

This means that if endowments w are perturbed, the behavior of the economy will not change. Thus, the economy is stable in the following sense: When endowment are slightly perturbed, after the perturbation the economy will exhibit the same qualitative behavior. The number of equilibrium does not change and the behavior is the same in a neighborhood of the original endowments.

Singular economies are not stable. This means that small perturbations in the fundamentals of the economy imply big changes in its behavior. This characterizes the behavior of the singular economies, i.e. in all neighborhood of w there exist economies with different number of equilibria. The aim of the catastrophe theory is to classify systems according to their behavior under perturbations.

A singular economy will stay *non-degenerate* if the corank of the jacobian of the excess utility function $\Im_\lambda e_w(\lambda)$ has corank at most one for all $\lambda \in Eq$ i.e.: $corank\left[\Im_\lambda e_w\right] \leq 1 \ \forall \lambda \in Eq(\omega)$ and with at least one $\overline{\lambda} \in Eq(\omega)$ such that $corank\left[\Im_\lambda e_w(\overline{\lambda})\right] = 1$: Otherwise we say that the singular economy is *degenerate*. From an economic point of view, the difference between non degenerate and degenerate economies is as follows: only degenerate economies can have indeterminacy equilibrium, i.e. the equilibria could be non locally unique.

6.4 Morse Theory and Economics

> *Too often in the physical sciences, the space of the states is postulated to be a linear space when the basic problem is essentially nonlinear, this confuses the mathematical development. Steve Smale.*

This section is one first approach to a characterization of the singularities in economics. Often the linear approach prevents the true knowledge. This is what emphasizes the phrase of S. Smale previously mentioned. The Morse theory is a good tool in this way. The properties of the excess utility functions allow us to introduce the Morse theory to classify 2-agent economies. Observe that the dependence among the coordinates of the excess utility function $\{e_{w_1}, \ldots, e_{w_n}\}$ and its homogeneity allow us to consider the excess utility function as a function $\overline{e}_w : R^{n-1} \to R^{n-1}$, So, in the case of 2-agent economies the excess utility function will be a real function $\overline{e}_w : R \to R$.

The main theorem about non-degenerate critical points of smooth real functions is the Morses lemma, which classifies all such point. Roughly speaking, the Morses lemma states that locally around a non-degenerate critical point, a smooth real function of n real variables can be transformed to a simple standard form by changing coordinates.

A smooth function $f : U \to R$ with $U \subset \mathbb{R}^n$ without degenerate critical points is called a Morse function. Let $C^\infty(U)$ be the set of smooth functions on U, supplied with the Whitney $C^\infty - topology$[4].

Then the set of Morse function on U is an open and dense subset of $C^\infty(U)$,[5] As it is well know, this topology is finer than the uniform convergence on compact sets in U of all derivatives. For details see [Golubistki, M.; Guillemin, V. (1973)].

---

[4] We say that a sequence of maps f_n converges to f in the Whitney $C^\infty(X, Y)$ topology iff, for every non-negative integer k, $j^k f_n$ converges uniformly to $j^k f$ where $j^k g$ denote the $k - jet$ of g; see section (5.1).

[5] Recall that the complement of an open and dense subset in a metric space M is a subset such that its closure has no interior points, i.e. is a rare set in M:

Consider the Taylor expansion of a real function $f : U \to R$ $U \subset \mathbb{R}^n$, $f \in C^\infty(U)$:

$$f(x_0 + h) = f(x_0) + f'(x_0)h + f''(x_0)h^2 = 2 + \ldots$$

Let $f'(x_0) = 0$: Then f is not a submersion at x_0: In this case we cannot construct the local form $f(\phi(y)) = f(x_0) + f(x_0)y$ in a neighborhood of zero. Thus the following question arises: Is there a *coordinate transformation* $x = \phi(y)$ with $\phi(0) = x_0$ for maps $f : U(x_0) \subseteq B \to R$ with $f'(x_0) = 0$ such that

$$f(\phi(y)) = f(x_0) + f''(x_0)y^2/2$$

is satisfied for all y in a neighborhood of zero? The affirmative answer to this question is the main content of the Morse lemma. Then f will behave like the quadratic part of the Taylor expansion in a neighborhood of x_0: Then, non degenerate singular 2-agent economies will have one of the following three possible different behaviors according to the second order Taylor expansion of an equivalent excess utility function. We say that two excess utility functions e_w^1 and e_w^2 are equivalent at λ_0 if there exist diffeomorphisms ϕ^1 and ϕ^1 such that

$$e_w^1(phi^1(\lambda_0)) = e_w''(\lambda_0)\lambda^2/2 = e_w^2(\phi^2(\lambda_0))\forall\lambda \in U(\lambda_0)$$

Observe that for all no degenerate singular equilibrium the following identity is verified:

$$e_w''(\lambda_0)\lambda^2/2 = \pm\lambda_1^2 + \lambda_2^2\forall\lambda \in U(\lambda_0) \tag{6.3}$$

Then non-degenerate singular economies have one of the 3 simple stereotypes, namely

$$+\lambda_1^2 + \lambda_2^2, +\lambda_1^2 - \lambda_2^2, \ -\lambda_1^2 - \lambda_2^2$$

This means that after a suitable choice of coordinates, e_w is locally equal to its own second Taylor polynomial at the origin. A question that naturally arises is whether functions with only non-degenerate critical points occur frequently. As the Morse lemma shows the set of Morse functions is a subset of functions which its complement is a rare subset of $C^\infty(U)$.

Other characteristics of the singular non degenerate economies that can be found, from the theory of the Morse functions, are the following:

1. Non-degenerate singular economies are isolated.
2. The set of singular non degenerate economies with only one non degenerate singular equilibrium $\lambda \in Eq(\omega)$ is residual.

For details on the theory of Morse functions see [Golubistki, M.; Guillemin, V. (1973)].

6.4.1 Example

In this section we consider an example of a 2-agent economy, for more details of this example see [Accinelli, E.; Piria, A. Puchet, M. (2003)]. In this example we show that for a fixed set of utility functions it is possible to construct singular economies, that is, we show the existence of endowments ω such that zero is a singular value for $e_\omega : \Delta \to R^2$: In this example, considering strictly positive endowments, equilibrium allocations are in the interior of $\mathbb{R}_+$.

Example 1 Consider the economy $E = \{u_{alpha,i}, \omega_i, R^2_{++}, i = 1.2\}$ whose utility functions are:

$$u_{\alpha,1} = x_{11} - \frac{1}{\alpha}x_{12}^{-\alpha}$$

$$u_{\alpha,1} = -\frac{1}{\alpha}x_{21}^{-\alpha} + x_{22}$$

and endowments are $w_1 = (w_{11}, w_{12})$; $w_2 = (w_{21}, w_{22})$ and for a fixed $W \in R^2_{++}$ the equality $W = w_1 + w_2$ is verified. Following the Negishi approach we begin solving the optimization problem:

$$max\left\{W_\lambda(x) = \lambda_1 u_1(x_1) + \lambda_2 u_2(x_2), s.t x \in R^4_+ : \sum_{i=1}^2 x_i \le \sum_{i=1}^2 \omega_i\right\}$$

Denoting $\lambda_1 = \lambda$ it follows that $\lambda_2 = 1 - \lambda$. We then write the excess utility function:

$$e_w(\lambda) = \begin{cases} \left(\frac{1-\lambda}{\lambda}\right)^{\frac{\alpha}{1+\alpha}} - \left(\frac{1-\lambda}{\lambda}\right)^{\frac{1}{1+\alpha}} - w_{12}\left(\frac{1-\lambda}{\lambda}\right) + w_{21} \\ \left(\frac{1-\lambda}{\lambda}\right)^{\frac{-\alpha}{1+\alpha}} - \left(\frac{1-\lambda}{\lambda}\right)^{\frac{-1}{1+\alpha}} - w_{21}\left(\frac{1-\lambda}{\lambda}\right)^{-1} + w_{12} \end{cases}$$

Solving $e_\omega(\cdot) = e'_\omega = 0$ we obtain the following surface C_F , the catastrophe surface, formed by the pairs (λ, ω) such that $\lambda \in Eq(\omega)$ and where 0 is a singular value for $e_\omega(\cdot)$:

$$C_F = \left\{(\lambda, \omega_{11}, \omega_{12}) \in V_W : \omega_{12} = \frac{\alpha}{1-\alpha}h^{\frac{1}{1+\alpha}} - \frac{1}{1-\alpha}h^{\frac{\alpha}{1+\alpha}}; h + \frac{\lambda}{1-\lambda}\right\}$$

Then, economies whose endowments verify

$$\omega_{12} = \frac{\alpha}{1-\alpha}h^{\frac{1}{1+\alpha}} - \frac{1}{1-\alpha}h^{\frac{\alpha}{1+\alpha}}$$

are singular. Solving $e(\lambda, \omega) = 0$ it is easy to see that in all neighborhood of this singular economies, i.e in a neighborhood of these w; there exist economies with one equilibrium and economies with three equilibria, that is economies whose excess utility functions have one or three zeros.

6.5 Beyond the Possibilities of the Morse Theory

The behavior of a smooth function in a neighborhood of one of its regular points can be characterized by the first term of its Taylors polynomial (the liner part) despite, as we saw in the previous section, something more is necessary when we are characterizing the behavior of a function in a neighborhood of a singular point. The possibilities of the Morse theory are limited to the analysis of economies with two agents. In more general cases more sophisticate theories are needed. The main concern of the Catastrophe theory is to classify critical points of the smooth functions. Taylor expansion plays a central role in achieving this classification. In this section we introduce the concept of $k - jet$ an then we consider the class $J^k e_\omega$ i.e., the set of $k - jets$ of the excess utility function, to show some properties of the singular economies.

6.5.1 Taylor Expansions and Jets

In this section we will present a general definition of $k - jet$ for $C^\infty(X)$ functions between X and Y two any finite smooth manifolds. However our interest is to analyze singularities of a particular type of functions, $e_\omega(\cdot)$ whose domain is $int[\Delta]$ and with values in $\mathbb{R}^n$,

Definition 5. *Let X be a finite manifold and let x be a point of X. Let $f : X \to \mathbb{R}^n$ be $C^k(X)$, $k \neq 1$. Consider the up to k Taylor expansion of the i-th coordinate of f at x*

$$T_{x^k} f(h) = f_i(x) + \sum_{r=1}^{k} f_i^{(r)}(x) h^r / r!; \; i = 1, 2, \ldots, n.$$

Then we define the $k - jet$ coordinate of f at the point x as

$$j^k f_i(x) = (x, f_i(x), f_i', \ldots, f_i^{(k)}(x)),$$

and let $j^k f(x) = (j^k f_1(x), j^k f_2(x), \ldots, j^k f_n(x))$ Let f and g be two $C^\infty(X)$ maps from $X \to \mathbb{R}^n$ these are called k-equivalent at a point $x \in X$ if and only if:

$$j^k f(x) = j^k g(x)$$

Definition 6. *By means of $J^k f(x)$ we denote the set of all $C^\infty(X)$ maps $g : X \to \mathbb{R}^n$ which are k-equivalent to the $C^\infty(X)$ map $f : X \to \mathbb{R}^n$ at the point $x \in X$: And let $J^k(X, \mathbb{R}^n)$ be the set of all possible $J^k f(x)$ where f and x vary. And element σ in $J^k(X, \mathbb{R}^n)$ is called a $k - jet$ from X to $\mathbb{R}^n$*

Let $X \subset \mathbb{R}^n$, $Y \subset R^m$ be smooth manifold. Then the following general theorem holds:

Theorem 2. *The set $J^k(X, Y)$ is a C^∞ manifold. See [Golubistki, M.; Guillemin, V. (1973)].*

Observe that there is a canonical bijection $T_{X,Y} \colon J^k(X, Y) \to X x Y x B_{m,n}^k$ given by $T_{X,Y}(\omega) = (x, y, T_X^k f_1(x), \ldots, T_n^k f_n(x))$ where $B_{m,n}^k = \bigoplus_{i=1}^{n} A_m^k$, and A_m^k is the vector space of polynomials in m-variables of degree $\geq k$. As A_m^k is a smooth manifold, then $B_{m,n}^k$ is a smooth manifold too.

For a mapping $f : X \to Y$ we can make the following rudimentary classification of singularities. We say that f has a singularity of type S_r at $x \in X$ if the jacobian of this function at $xf'(x)$, drops rank by r; i.e. if rank $f'(x) = min(dimX, dimY) - r$. Then $S_r = \{\omega \in J^1(X,Y); corank(\sigma) = r\}$. The set S_r is a submanifold of $J^1(X,Y)$ consisting of $1 - jets$ of corank r. Let us now define $S_r(f) \subset X$; the set of singularities of f where $corank\left[f'(x)\right] = r$: So, $S_r(f) = (J^1 f)^{-1}(S_r)$. The following theorem will allow us to know some characteristics of the set $S_r(f)$ of singularities of a function f in $C^\infty(X,Y)$.

Theorem 3. *(Thom Transversality Theorem) Let X and Y be finite-dimensional real C^∞ manifolds with countable bases, Also, let S be a submanifold of $J^k(X,Y)$ for some fixed $k = 0.1 \ldots$ Then the set*

$$\{f \in C^\infty(X,Y) : J^k f transversal to S\}$$

is residual[6] and dense in $C^\infty(X,Y)$ equipped with the C^∞ - Whitney topology[7].

See [Golubistki, M.; Guillemin, V. (1973)].

As generically the set of functions $f \in C^\infty$ is transversal to S_r; then without restricting the relative dimensions of X and Y we can say that generically $S_r(f)$ is a submanifold of X and with $codim\ S_r(f) = codim\ S_r = r^2 + er$ where $e =| dimX - dimY |$.

On the other hand, the transversality of $k - jets$ can effectively be used to describe the nondegeneracy of higher-order derivatives. For example, let $f : R \to R$ be CY^∞: Then for $k \geq 1$ it follows that $\left[j^k f\right]'(x)h = (h, f'(x)h, \ldots f^{k+1}(x)h)$ for all $h \in R$. Let $S = \{(\xi_1, \ldots \xi_{k+2} : \xi_{k+2} = 0\}$. This implies the following statement: From $f^k(x) = 0$ it follows $f^{k+1}(x) \neq 0$ if and only if $J^k f$ is transversal to S.

6.5.2 Singular Economies and Catastrophe Theory

We characterized the behavior of an economy by its excess utility function. Moreover, as we did in the case of a 2-agent economy, the behavior of an economy may be characterized by the kind of singularities that arise from the excess utility functions. Now our objective is to give a classification of

---

[6] Recall that a subset of $D \subset B$ is residual in B if it is the intersection of open and dense subsets of B

[7] Recall that a function f : X ! Y intersects a submanifold $S \subseteq Y$ transversely at x if either:

(a) $f(x) \notin S$ or
(b) $f(x) \in W$ and $\tau_{f(x)} Y = \tau f(x)S + f_x'(\tau_x X)$,

where by $\tau_x X$ we denote the tangent space to the manifold X at the point $x \in X$

the economies by its *normal form*. From a mathematical point of view the aim is to get a survey of all possible qualitative structures. To obtain this classification we consider the following two questions:

1. When does the Taylor expansion up to some order k

$$f(x) + f'(x) + \cdots + f^{(k)}(x)u^k/k!$$

provide enough information to understand the local behavior of a function f at x?
2. What are the local normal forms for parameter families of functions?

The catastrophe theory, which was initiated by René Thom in the 1960s tries to answer the questions presented above [Thom, R. (1962)]. These two questions are of main importance in economics, to see this it is enough to consider the excess utility function $e_\omega(\cdot) : int\,[\Delta] \to \mathbb{R}^n$,, as the function to be analyzed. We will classify the behavior of the economies in this way. Functions of several variables which zero is not a singular value are determined by the linear part of their Taylor expansion, see [Arnold, V.; Varchenko, A.; Goussein-Zade, S. (1982)]. Thus, regular economies are characterized by the linear part of their excess utility functions. As we have shown, in the case of functions without degenerate singular points, Morse theory gives an answer to these questions. So, we have obtained a complete classification of 2-agent economies when there are no degenerate singular equilibria. As a consequence of the Thom's theorem the following two conclusions follow:

(a) The set of singularities of an economy generically forms a manifold, because generically at ω in Ω the k-jet of the excess utility function $J^k e_\omega$; is transversal to every manifold S in $J^1(\Delta, \mathbb{R}^n)$:
(b) For $k = 1$ the transversality conditions mean precisely that f has only non-degenerate singular points. So, in economics, this means that the set of non degenerate economies is generic. The fact that if f has no singularities implies that $S_r(f) = 0$ means that regularity is a generic property, that is generically economics are regular.

Unfortunately, at present, many classification problems seem to be difficult. Functions are no determined by their Taylor polynomials. As an illustrative example consider the functions $f(x, y) = x^2$ and $g(x, y) = x^2 - y^2$: Their have the same Taylor polynomial for all k but for all diffeomorphism: $\phi = (\phi_1, \phi_2)$ it follows that; $f(\phi(0, y)) = (\phi_1 1(0, y))^2 \neq -y^2$:

6.5.3 3-agent Economies

Let $E = \{X, u_1, w_i, i = 1, 2, 3\}$ be an economic system with 3 agents and total resources W. The excess utility function is $e_\omega() : \Delta^2 \to R^3$ where $\Delta^2 = \left\{\lambda \in R^3 : \sum_{i=1}^3 \lambda_i = 1, \lambda_i > 0; i = 1; 2; 3\right\}$

Definition 7. *We will say that a mapping* $f : X \to Y$ *is one generic if* $J^1 f$ *is transversal to* S_r *for all* r:

From now on, we will assume that all maps are one-generic. Whitney proved the only singularities that can occur for maps between 2-manifolds are fold or simple cusp, see [Golubistki, M.; Guillemin, V. (1973)]. A direct application of this statement in economics is as follows: In 3-agent economies, $\bar{e}_w : \Delta^2 \to R^2$ then only $S^1(\bar{e}_w)$ singularities occur, that is singularities whose codimension in R^2 is 1, and $S^2(\bar{e}_w)$ does not occur since it would have to be of codimension 4. So, it is straightforward that in this case only one of the following two situations can occur: singularities are folds or cusp points. These statement characterizes the kind of singularities in 3-agent economies whose excess utility functions are 1-generic.

These situation are characterized by the following two equalities:

$$\begin{cases} (a) \ \tau_{\bar{\lambda}} S_1(\bar{e}_w \bigoplus Ker(\bar{e}_w')_{\bar{\lambda}} = \tau_{\bar{\lambda}} \Delta^2 \\ (b) \ \ \ \ \tau_{\bar{\lambda}} S_1(\bar{e}_w = Ker(\bar{e}_w')_{\bar{\lambda}} \end{cases} \tag{6.4}$$

Where by $\tau_x X$ we denote the tangent space at $x \in X$ of the X manifold, and $Ker(\bar{e}_w')_{\overline{lambda}}$ is the kernel of the linear transformation e_w':

- **A fold.** If (a) occurs then one can choose a system of coordinates (λ_1, λ_2) centered at $\bar{\lambda}$ and $(\mu 1, \mu_2)$ centered at $\bar{\lambda}$ such that $e_w(\bar{\lambda}) = e_w(\bar{\mu})$ and $(\lambda_1, \lambda_2) \to (\mu_1, \mu_2^2)$
- **A simple cusp.** If (b) occurs, one can find coordinates (λ_1, λ_2) centered at $\bar{\lambda}$ and (μ_1, μ_2) centered at $\bar{\mu}$ such that $\bar{e}_w(\bar{\lambda}) = \bar{e}_w(\bar{\mu})$ such that:

$$\begin{cases} e^*(\mu_1) = \lambda_1 \\ e^*(\mu_2) = \lambda_1 \lambda_2 + \lambda_2^3. \end{cases} \tag{6.5}$$

6.5.4 The General Case

Provided that the set of singularities S_r of $C^\infty(\Delta, R^{n-1})$ functions $f : \Delta \to R^{n-1}$ whose jacobian drops rank by r is a submanifold in $J^1(\Delta, R^{n-1})$ then from the equality $S_r(\bar{e}_w) = (J^1 \bar{e}_w)^{-1} S_r$ and the transversality Thom's theorem it follows that the set $S_r(\bar{e}_w)$ is generically a submanifold in Δ: Their codimensions are the same, then to generalize some results obtained for 3-agents economies seems to be feasible. It is also possible to relate the kind of singularities with the number of agents in the economy. This and other considerations on this topic are made in [Accinelli, E.; Puchet, M. (2001)]. The classification of singularities appears to be difficult, nevertheless catastrophe theory and the geometric concept of transversality provide a good tool to achieve this goal.

6.6 Conclusions

In a neighborhood of a regular economy the behavior of an economic system is the same, but in a neighborhood of a singular economy big changes occur, then the characteristics of this changes are strongly related with the kind of singular economy. So the regularity implies the stability of the system, and on the contrary the existence of a singularity implies its instability. The Thom's transversality Theorem shows that a smooth mapping is generically as nice as possible because the jacobian has almost always the maximum possible rank. So, economies will be generically regular, i.e the rank of the jacobian of the excess utility function is equal to $n - 1$: In cases where $n = 2$ (2-agent economies), the Morse theory shows that when the excess utility function is not of maximal rank, the singularities are generically of a particular type, i.e. equilibria locally unique and determined by the initial distribution of the endowments. This means that singular 2-agent economies have a particular well known form.

Since the dimension of R is 1, the only non-empty submanifolds of $J^1(\Delta, R)$ of the type S_r are S_0 or S_1, i.e an economy is singular if and only if $j^1 e_\omega(\lambda) \in S_1$: We also saw that singularities in 3-agent economies have a particular form: a fold or a cusp. When the jacobian $\Im_\lambda e_\omega(\lambda)$ does not have the maximal rank and e_ω is one generic, there exist only two types of singular economies. If $j^1 e_\omega(\lambda) \in S_1$ and S_1 satisfies (a) in (3) the singular economy is a fold, otherwise it is a cusp.

So, structural unstable performance and sudden changes of the economic system can occur if the economy is singular, then the knowledge of the form of singular economies may be a starting point to describe the economic crisis.

Certainly it is not easy to classify the singularities, mainly because at the moment a completely developed mathematical theory of the singularities does not exist. The mathematical development of this theory is a necessary condition to achieve a deeper understanding of the economic system.

Let us make one more comment on stability. As it is well known, from the Sonnenschein- Mantel-Debreu theorem, there is not a satisfactory dynamics for the General Equilibrium theory. See [Mantel, R. (1976)]. From this point of view, the stability (in the Lyapunov sense) of the walrasian equilibrium has no point of being. However the question about the structural stability of the set of social weights equilibria does make sense. Regular economies mean structural stability of the system, then small changes in fundamentals of the economic systems do not imply big changes in its equilibrated behavior. Nevertheless, if the economy is singular, then there is no structural stability. That is, after a small change in the fundamentals, the set of equilibria is completely different with regard to the original one.

The performance of the economic system in the neighborhood of a singular economy, after a small perturbation, is absolutely unknown. This impossibility of the knowledge of the evolution of the system is structural, and does not depend on the precision of the gathered data. Moreover, since all measures are only an approximation to reality, in a singular economy all statements based on measures of the parameters of the excess utility function become false. Hence the theory of singularities (in particular catastrophe theory) allows us to say something about the possible forms of these economies and the performance of the system. Beyond this paper it is important to consider the following open issues.

(a) The consequences of perturbation of the utility functions on the economic system. It is possible to think that small changes in tastes could produce big changes in the behavior of an economic system. In particular it is of interest to know the kind of utility functions that generate structural stability or instability.

(b) To consider cases of the economies whose excess utility functions are non smooth. In particular to look for the minimal degree in the derivative of the excess utility function necessary to use differentiable techniques to obtain conclusions about structural stability.

(c) The extension of the classification of the singularities for the cases where the number of agent is greater than three, when endowments and/or utilities change.

Let us now make a final comment. The structural stability is strongly related to the geometry of the equilibrium set. For this reason, perhaps the statement of Galileo Galilei might be true also in the economic world:

He who understands geometry may understand anything in this world.

References

1. Accinelli, E. (1995) Notes about uniqueness of equilibrium on economies with infinitely many goods. Informes de matemàtica. Serie F-081-May/95, Impa.
2. Accinelli, E. (1996) Existence and Uniqueness of the Equilibrium for Infinite Dimensional Economies. Revista de Estudios Econòmicos 21, 313- 326.
3. Accinelli, E.; Puchet, M. (2001) A characterization of Walrasian economies of infinity dimension., Documento de Trabajo 7/01, March.
4. Accinelli, E.; Piria, A.; Puchet, M. (2003) Tastes and singular economies. Revista Mexicana de Econom`a y Finanzas 2, 35- 48.
5. Araujo, A. (1987) The Non-Existence of Smooth Demand in General Banach spaces. Journal of Mathematical Economics 17, 1-11
6. Arnold, V.; Varchenko, A.; Goussein-Zade, S. (1982) Singularites des Applications Differentiables. Editions Mir.
7. Balasko, Y. (1978) Economie et Thèorie des Catastrophes. Mathematiques et sciences humaines. 74. Gauthier-Villars.
8. Balasko, Y. (1988) Foundations of the Theory of General Equilibrium. Academic Press.
9. Castrigiano, D.; Hayes, S. (1993) Catastrophe Theory. Adisson-Wesley.
10. Debreu, G. Economies with a Finite Set of Equilibria. Econometrica 38 387-392.

11. Golubistki, M.; Guillemin, V. (1973) Stable Mappings and Their Singularities. Springer Verlag.
12. Mantel, R. (1976) Homothetic preferences and Continuity of the excess demand function Journal of Economic Theory 12, 197- 201.
13. Mas-Colell, A. (1985) The Theory of General Equilibrium: A Differentiable Approach. Cambridge University Press
14. Mas-Colell, A.; Nachbar, J.H. (1991) On the Finiteness of the Number of Critical Equilibria with an Application to Randon Selections. Journal of Mathematical Economics 20, 397- 409.
15. Monteiro, P.K. (1994) Inada's Condition Implies Equilibrium Existence is Rare. Economic Letters 44, 99-102.
16. Negishi, T. (1960) Welfare Economics and Existence of an Equilibrium for a Competitive Economy. Metroeconomica 12, 92- 97.
17. Thom, R. (1962) Sur la Thèorie des Enveloppes.Journal de Mathématiques 41.
18. Zeidler, E. (1993) Non Linear Functional Analysis and its Applications Vol.1. Springer Verlag.

7

Pretopological Analysis on the Social Accounting Matrix for an Eighteen-Sector Economy: The Mexican Financial System*

Andrés Blancas[1] and Valentín Solís[2]

[1] Instituto de Investigaciones Económicas. Universidad Nacional Autónoma de México, UNAM. neria@servidor.unam.mx

[2] Facultad de Economía, UNAM.valentinsolis@mexico.com

Summary. This paper analyzes the structural relationships of the financial transactions represented in a Social Accounting Matrix (SAM) for the Mexican economy through a pretopological approach. Based on a simple binary relationship between incomes and expenditures of institutional accounts, the pretopology is used as a mathematical tool to get an insight into the economic structure represented by the SAM. Such an analysis can be useful to identify the set of relationships between several institutional accounts, ordered according to their influence or domination.

Key words: JEL: C00, C69, G20, N20 Key Words: Social accounting matrix, pretopology, pseudoclosure, minimal closed set.

7.1 Introduction

The broad acceptance of a SAM like a data system and a conceptual framework to capture the interdependence within a socioeconomic structure has stimulated several extensions in the economic analysis. This event can be traced in the development of the system of national accounts proposed by United Nations since 1968 and its latest revision in 1993; as well in the use of SAM data to built up economic models in a well-behaved way. For example, the computable general equilibrium (CGE) models, are based on data in the form of a SAM, of the sort proposed by [24] and elaborated by [19]. Such models are used for macroeconomic and economic policy analysis in the 1980s by [11], and in the 1990s by [26], and by [12]. From an accounting multiplier approach,[5] has stated that there might be a "structural financial vulnerability" of the economy due to the fact that the financial transactions represented

---

* A preliminary version of this paper was presented at the New Tools of Qualitative Analysis of Economic Dynamics Workshop, CIMAT-Guanajuato, México. September, 2002.

in a SAM data system are weakly linked. The development of techniques to evaluate the interdependence of the economy has been the main concern of [17] and [22], directing his research to unify the graph theoretic and topological tools in the context of an input-output analysis and social accounting modeling.

This paper analyzes the structural relationships of the financial transactions through a pretopological approach applied to a SAM for the Mexican economy. Based on a simple binary relationship between incomes and expenditures of institutional accounts, the pretopology is used as a mathematical tool to inquire into the economic structure represented by a SAM. Such a method maps the sequence of pseudoclosure sets that converge toward closed topological sets or elementary closed sets. This information is utilized to obtain the so-called minimal closed sets, which constitute kernels that are used to characterize interdependent homogenous parts of such structure.

Section 2 presents a SAM for an eighteen-sector economy and its relationship with the pretopological approach. Section 3 shows some pretopological concepts and a computational algorithm. Finally, section 4 points out some results derived from the pretopological analysis of a SAM framework.

7.2 The Social Accounting Matrix and the Pretopology Approach

A SAM is a double-entry bookkeeping table used to display national income and product, interindustry, flows of funds, and other combined sets of accounts. Table 1 shows a SAM for an eighteen-sector economy that displays the accounting relationships between the real and financial sides of the Mexican economy in 1990.

This SAM table is an extension of what [27] have yet developed, and it was elaborated with data from [2], [3], [9], [13], [14] and [21]. All entries of the matrix are set out in nominal terms and totals for corresponding rows and columns are the same. Note that SAM row-column identities hold only in current price terms. The rows of this SAM record incomes (receipts) and the column expenditures (payments).

This table for an eighteen-sector economy distinguishes four general accounts: 1.Production Activities, 2.Factors of Production, 3.Current Account of Institutions, and 4.Capital Account of Institutions. It contains 35 specific accounts; each row and column pair represents an account with incoming in the row and spending in the column.

The 'real side' of the economy is represented by the production activities, factor of production transactions, and current account transactions of domestic and foreign institutions, accounts 1 through 26. The real side generates saving flows and the available savings put restrictions on the changes in assets and liabilities that appear in the saving and flow of funds sections that represent the 'financial side' of the economy, accounts 27 through 34. These

	1	2	3	4	5	6	7	8	9	10	11	12	13	14	15	16	17	18
1 Agriculture, Foresty and Fishing	7,7	0,0	0,0	28,0	0,7	1,6	0,0	0,2	0,0	0,0	0,0	0,0	0,0	0,0	0,0	0,0	0,0	0,2
2 Mining except oil	0,1	2,9	0,1	0,0	0,0	0,0	0,0	0,5	0,7	3,4	0,3	0,3	2,1	0,0	0,0	0,0	0,0	0,0
3 Oil and gas extraction	0,0	0,0	0,0	0,0	0,0	0,0	0,0	4,9	0,0	0,0	0,0	0,0	0,0	2,5	0,0	0,0	0,0	0,0
4 Food, Beverage and Tobacco	2,3	0,0	0,0	11,7	0,4	0,2	0,0	0,0	0,0	0,0	0,0	0,0	0,0	0,0	0,0	0,0	0,0	0,4
5 Textil	0,2	0,0	0,0	0,4	3,9	0,1	0,0	0,2	0,0	0,0	0,1	0,0	0,0	0,0	0,3	0,1	0,0	0,6
6 Wood	0,4	0,0	0,0	0,6	0,2	3,2	0,7	0,4	0,3	0,1	0,4	0,1	2,1	0,0	1,6	0,1	0,2	0,7
7 Paper and Print	0,0	0,0	0,0	0,5	0,2	0,0	0,5	1,2	0,1	0,1	0,5	0,1	0,0	0,1	1,6	0,1	0,4	0,5
8 Chemistry substances and oil derivates	5,3	0,3	0,1	1,6	3,7	0,0	0,5	11,3	0,8	0,5	1,9	0,3	1,0	0,3	2,1	4,6	0,5	4,8
9 Non Metalic Minerals	0,2	0,1	0,0	0,8	0,0	0,0	0,0	0,3	1,2	0,1	0,5	0,0	6,2	0,0	0,0	0,0	0,2	1,1
10 Basic Metalic Industries	0,2	0,1	0,1	0,4	0,0	0,1	0,1	0,2	0,1	6,8	4,9	0,2	7,4	0,0	0,2	0,1	0,0	0,2
11 Metalic products and Machinery	0,5	0,1	0,0	1,2	0,1	0,2	0,0	0,3	0,2	0,6	3,4	0,1	1,7	0,1	0,7	1,2	0,1	1,9
12 Other Manufactures	0,1	0,0	0,0	0,0	0,0	0,0	0,0	0,0	0,0	0,0	0,0	0,0	0,0	0,0	0,0	0,0	0,1	0,1
13 Construction	0,0	0,0	0,0	0,0	0,0	0,0	0,0	0,0	0,0	0,0	0,0	0,0	0,0	0,0	0,0	0,0	0,0	0,0
14 Electricity, Gas and Water	0,7	0,4	0,0	1,0	0,3	0,4	0,1	3,1	1,0	0,9	0,6	0,0	0,3	1,2	2,2	0,3	0,7	0,6
15 Trade, Restaurants and Hotels	2,8	0,4	0,3	9,1	2,9	1,2	0,6	3,9	0,6	1,3	5,9	0,4	3,6	1,2	4,7	3,5	0,8	3,7
16 Transport and Communications	1,1	0,1	0,4	3,3	0,8	0,4	0,3	1,6	0,3	0,5	1,8	0,1	2,5	0,3	5,9	1,8	1,1	2,6
17 Financial Services and Real State	1,5	0,1	0,1	1,0	0,6	0,3	0,3	0,7	0,2	0,2	1,2	0,1	2,5	0,3	10,9	1,4	9,1	6,0
18 Communal and Social Services	0,5	0,3	0,4	1,7	0,4	0,3	0,2	1,5	0,5	0,4	1,7	0,0	2,1	0,4	13,6	4,2	6,0	7,0
19 Wages	8,0	1,9	1,5	6,8	4,2	1,7	1,0	7,8	2,4	2,4	9,5	0,8	16,9	2,6	22,2	11,7	10,2	59,7
20 Corporate Profits	21,0	6,1	7,5	24,5	7,7	5,9	3,2	16,3	6,9	6,7	15,7	2,3	7,4	4,4	58,8	21,5	31,7	26,7
21 Unicorporated Profits	26,0	0,2	0,2	4,6	1,4	1,1	0,6	3,1	1,3	1,2	2,9	0,4	2,8	1,6	53,8	19,6	29,1	24,4
22 Households																		
23 Private Firms																		
24 Public Firms																		
25 Central Government	-0,1	0,2	0,1	4,7	1,5	0,8	0,6	2,0	0,7	0,0	2,9	0,2	0,1	0,8	44,0	7,1	2,4	1,4
26 Foreign	0,8	0,5	0,9	7,5	2,0	1,8	0,7	9,3	0,7	2,4	19,7	0,9	2,0	1,0	0,8	3,7	0,5	1,2
27 Households																		
28 Private Firms																		
29 Public Firms																		
30 Central Goverments																		
31 Central Bank																		
32 Commercial Banks																		
33 Development Banks																		
34 Foreign																		
35 Unidetified Firms																		
Total	79,0	13,7	11,7	109,2	31,2	20,1	9,2	69,4	18,1	27,4	74,0	6,5	61,9	16,9	223,5	81,1	93,2	143,9

Table 7.1. Social Accounting Matrix for an Eighteen-Sector Economy, Mexico. 1990 (Billion of New Pesos)

	19	20	21	22	23	24	25	26	27	28	29	30	31	32	33	34	35	Total
1 Agriculture, Foresty and Fishing				33,6			0,2	3,3	0,0	2,4	0,9	0,0				0,2	0,0	79,0
2 Mining except oil				0,0				3,2	0,0	0,0	0,0	0,0				0,1	0,0	13,7
3 Oil and gas extraction								4,4	0,0	0,0	0,0	0,0				0,0	0,0	11,7
4 Food, Beverage and Tobacco				85,3			0,1	3,9	0,0	2,5	0,9	0,2				0,7	0,0	109,2
5 Textil				20,9			0,1	2,8	0,0	0,8	0,3	0,1				0,2	0,0	31,2
6 Wood				5,4			0,7	1,4	0,0	0,9	0,3	0,1				0,2	0,0	20,1
7 Paper and Print				2,4			0,4	0,1	0,0	0,3	0,1	0,1				0,1	0,0	9,2
8 Chemistry substances and oil derivates				16,7			0,8	7,2	0,0	2,0	0,7	0,1				0,5	0,0	69,4
9 Non Metalic Minerals				4,2			0,6	1,5	0,0	0,6	0,2	0,0				0,2	0,0	18,1
10 Basic Metalic Industries				1,0			0,0	3,4	0,0	1,1	0,4	0,1				0,3	0,0	27,4
11 Metalic products and Machinery				12,2			0,2	24,5	0,0	14,6	5,3	1,0				3,8	0,0	74,0
12 Other Manufactures				2,8			0,2	2,4	0,0	0,4	0,1	0,0				0,1	0,0	6,5
13 Construction							0,0	0,0	28,2	20,0	11,3	2,4				0,0	0,0	61,9
14 Electricity, Gas and Water				2,5			0,3	0,2	0,0	0,0	0,0	0,0				0,0	0,0	16,9
15 Trade, Restaurants and Hotels				125,8			0,5	33,6	0,1	10,5	3,7	0,8				1,5	0,0	223,5
16 Transport and Communications				46,1			1,0	6,9	0,0	1,6	0,6	0,1				0,2	0,0	81,1
17 Financial Services and Real State				55,7			0,9	0,0	0,0	0,0	0,0	0,0				0,0	0,0	93,2
18 Communal and Social Services				49,2			50,8	2,7	0,0	0,1	0,0	0,0				0,0	-0,1	143,9
19 Wages								1,6										173,0
20 Corporate Profits								12,5										274,4
21 Unicorporated Profits								8,5										186,9
22 Households	173,0		186,9		137,8	31,9	24,3										1,1	563,5
23 Private Firms		231,0					25,3										-3,8	252,5
24 Public Firms		43,4		19,2	27,1													62,6
25 Central Government				18,3	15,1	6,8		13,4	0,0	18,2	6,6	1,3	0,0			0,0	1,6	119,9
26 Foreign						3,6			0,0	0,0	0,0	0,0				9,8	0,6	138,5
27 Households				20,1									0,0	26,2	0,8	0,0	0,6	69,3
28 Private Firms				42,2	68,7	20,3			3,5	0,0	0,0	0,0	2,7	26,5	0,5	9,8	0,3	112,3
29 Public Firms							1,8		13,7	11,0	0,0	-7,3	0,0	-2,6	-1,7	-1,4	0,1	32,3
30 Central Goverments									0,0	0,0	0,0	-1,7	2,2	12,1	-1,1	-17,8		-2,7
31 Central Bank									6,6	22,9	0,1	0,0	0,0	1,3	0,2	25,5	0,1	31,9
32 Commercial Banks									24,5		0,7	0,0	-0,4	0,0	8,0	12,9	-0,1	67,9
33 Development Banks									-1,5		0,0	0,0	23,0	2,2	0,0	-20,8		6,5
34 Foreign								15,0	-5,8		0,0	0,0	4,4	2,2	-0,2	0,0	0,7	16,3
35 Unidetified Firms					3,8		-1,6	-0,6		-0,6	0,0	0,0	0,0	0,0	0,0	0,0		1,1
Total	173	274,4	186,9	563,5	248,7	62,6	121,5	139,1	168,3	112,8	32,3	-2,7	31,9	67,9	6,5	16,3	1,1	3196,4

Table 7.1 continued

groups or institutions will be labelled by their specific income and spending flows. Account 35 displays the residual errors and unidentified items.

In such general outline the macroeconomic causality runs from demand injections to leakage, under conditions of endogenous finance. The expenditure patterns of endogenous accounts do not modify prices that remain unchanged when income is modified, under the assumption of excess capacity and hence fix price throughout.

The approaches to identify several subsets of highly interdependent industrial sectors in input-output analysis, the discovery of various cliques and cohesive subgroups into which a social network can be divided, the partition of social accounting matrices into accounts showing the directions of the general interdependence, are all familiar to researchers of matrices of relational data. There are several theoretical models which identify for a network "clusters", "cliques", "cores", "blocks", and other related concepts.

For the case of a SAM, which is a rather consistent relational data matrix, the different theoretical approaches to explore those matrices may be classified into three broad categories: a) algebraic methods, b) the identification of sub-graphs in path and flow analysis methods, and c) multivariate statistical methods.

The algebraic approach emphasizes an analytical partition of several accounts within the SAM. The estimation of the matrix multipliers is obtained thorough partitioned inverse methods, expressing its results in additive or multiplicative fashion. The most popular partitions are those of [18], [10], [25], and [23], and its interpretation is carried out in terms of transfer, open-loop and closed-loop effects. These partitions are good for understanding the consolidated result of the interactions among the institutional actors, but do not describe the subtle interaction of the relationships amongst production activities, factors and households that is carried along a complex paths within a network.

In order to obtain detailed paths of the economic impulses, graph theoretic tools had been used. This approach relies on the mathematics developed to study the distinct patterns of network formation on several fields, like chemistry, electric circuits, social sciences and others. The application of graph theory to economics can be seen as part of the mainstream of the so called "social network analysis", which introduced mathematical concepts such as valued and non-valued graphs, directed or not directed graphs, multi-graphs, hypergraphs, stochastic graphs and many other extensions to social phenomena [28]. These studies attempt to investigate structural properties of a whole graph and also to reveal sub-graphs into which it can be divided. A sub-graph can be identified measuring the connectedness of its points, the intensity of their connection, an so on. It is a sub-graph that is maximal in relation to a defining characteristic.

In many ways, graph theory applied to input-output and SAM tables relaxed the metric properties of those matrices, using extensively the adjacency relationships between sectors or institutional actors, focusing in their relative positions within a network. The statistical approach to networks analysis can take graph theoretical measures of the "closeness" of points and can express these relations of closeness and distance in metric terms. This involves the use of "proximity data" to construct a metric configuration of points. Proximity measures of relational data, such as the value of sales between sectors, the amount of wages paid to laborers, the transfers to international economy and other economic observations registered in input-output, SAM, and other sociological relational data, are often converted into correlation coefficients. These coefficients are known to conform an Euclidean metric and also scaled to comparable dimension. Two points with similar patterns of connection in a graph will be highly correlated and so the proximity measure, called "similarity", for this pair will be close to one. The other type of proximity measure is "dissimilarity" and based on the complement to one of the correlation value.

The measures based on the properties of similarity or dissimilarity are of great tradition in social sciences, allowing to researchers to apply them to relational data, so principal components, factor analysis, dual scaling, multidimensional scaling, an other multivariable techniques had been used to find out "structural equivalence", "cores", "cliques", "clusters" and related concepts [20].

As can be seen, algebraic, graph and statistical methods applied to relational data are both complementary and alternative to explore a network. However, their conceptual leading motive is rather similar, so it might be possible that they have common principles. In this paper we apply a topological analysis of a Mexican financial SAM within a framework that shows that the transitive closure of a discrete and finite graph is the result of applying a closure operator that induces a pretopological space based in a binary relationship between incomes and expenditures. The same framework can be applied to networks that require of valued and non-valued graphs, as well as graphs that do have a metric, reducing several approaches to a particular case of a general approach to study structural properties of relational matrices. With the use of pretopological spaces, as a generalization of both finite simple graphs and topological spaces, it is possible to unify these two seemingly distinct uses of metrics. Inasmuch as the use of metrics in topology and graph theory at first glance appears to be unrelated, because graphs in general cannot be regarded as topological spaces. In what follows we apply a pseudoclosure operation.

7.3 Some Pretopological Concepts and an Algorithm

The notion of pretopology[3] is used to operate on a discrete system like a SAM, where E depicts the set of points to be classified, and each element of E represents each entry of the matrix.

7.3.1 Pretopological Space

Let E be a set of n elements, where n is finite and different from zero. The relationships amongst the elements of E constitute a family, which enables to state the notion of neighborhood of each element of E:

$$\forall x \in E, \forall i \in \{1, ..., k\}, R_i(x) = \{y \in E/xR_iy\} \tag{7.1}$$

where $R_i(x)$ is the set of elements having a relationship i between x and y. $R_i(x)$ is called the neighborhood of x and each one of the elements of $R_i(x)$ is called a neighbor of x. The only requirement asked for $R_i(x)$ is to be reflexive, i.e. $x \in R(x)$ if and only if xRx.

Definition 1. *Let $P(E)$ be the set of the parts of E and let δ be a mapping from $P(E)$ to $P(E)$. δ is said to be a pseudoclosure mapping on E if*
K1: $\delta(\emptyset) = \emptyset$,
K2: $\forall A \in P(E), A \subseteq \delta(A)$

Definition 2. *The pair (E, δ) satisfying K1 and K2 is a Pretopological Space.*

Definition 3. *A more limited Pretopological Space can be built up if the mapping also satisfies*
K3: $\forall A \in P(E), \forall B \in P(E), A \subset B \Rightarrow \delta(A) \subset \delta(B)$

Definition 4. *The above constructed pretopology becomes topological if and only if the closure is idempotent, i.e., if and only if:*
K4: $\delta(\delta(A)) = \delta(A)$

In this case, the neighborhood relationships is used to build a pretopological mapping $\delta(A)$ from $P(E)$ into itself as follows:

$$\forall A \subset P(E), \delta(A) = \{x \in E/\forall i \in \{1, ..., k\}, R_i(x) \cap A \neq \emptyset\} \tag{7.2}$$

---

[3] In [7] is introduced the notion of pretopology. However, the first rigorous development about this topic is due to [8], who stated the concepts of closure space, order closures and the notion of generalized order closures for a monotone ordered set. We use the language and notation created by the French researchers, particularly, the work edited by [4].

7.3.2 Closed Set

A subset A of E is closed if an only if $\delta(A) = A$. Due to the fact that the pseudoclosure operator δ is not idempotent, it is possible to develop an iterative mapping of a subset A of E to generate a successive spreading of the subset:

$$A \subseteq \delta(A) \subseteq \delta^2(A) \subseteq \delta^3(A) \subseteq \delta^4(A) \subseteq ... \tag{7.3}$$

And because E is finite, there is an integer $m < n$ where the successive mappings stop, so the closure of a subset A of E is computed by successive pseudoclosures of A. When this occurs, the pretopology has become topology.

7.3.3 Closed Set, Elementary Closed Set and Minimal Closed Set

From all possible closed subsets in a discrete and finite pretopological space, there are some of them of particular interest. Such closed sets are the singletons of the set E. [4], [15] and [16] have shown that for each one of them, there always exists a closed set obtained from the spreading of the pseudoclosure operator. These closed sets are called elementary closed sets.

Let C_e be the set of elementary closed sets, which is composed of closures C_x of singletons $\{x\} \in E$, then the relationship between elementary closed sets is based on the property derived from the fact that if $C_x \cap C_y \neq \emptyset$ then for all z in $C_x \cap C_y$, $C_z \subset C_x \cap C_y$. This property shows the relationships amongst elementary closed sets: Any two elementary closed subsets are disjoint, either one is included in the other one or their intersection contains elements whose elementary closure is included in this intersection. From all possible elementary closed sets there is a particular subset that is important for us: the minimal closed sets, defined by inclusion within the elementary ones.

What characterizes this approach in economic terms is that it does model a diffusion process of the myriad of economic transactions between production sectors and institutional agents registered in a SAM. The diffusion process departs from the so called minimal closed sets, which constitute kernels of the SAM framework; they characterize interdependent homogenous parts of the structure that channelize exogenous economic impulses. With such an approach we can model straightforward a complex structure; while other methods used to classify or analyze an structure are so indirect methods. The notion of closed sets resembles directly an observable economic phenomena. For instance, an almost closed set could be beheld in the links between the agricultural goods and food industries without significant reference to relationships to other economic sectors; this observed fact is caught easily with the pretopological approach when we get the closure of the diffusion process established in simple income and expenditure relationship within food and agricultural sectors.

7.3.4 A Three Stage Computational Structuring Algorithm

The structuring algorithm used in this paper relies on the behavior of this pseudoclosure mapping, and it was suggested by [4], and further improved by [15] and [16], and by [6].

In the first stage we obtain the elementary closed sets applying iteratively the pseudoclosure operator to each one of the singletons of the set. The operator is stopped when the adherence of the singletons becomes idempotent. Computationally, this can also be obtained by the inverse of the adjacent matrix expressed in Boolean terms.

The second stage works over the elementary closed sets in order to obtain, by inclusion, the minimal closed sets. These sets constitute kernels of the main structure of the SAM.

The previous results are structured in the third stage. It is based on a recursive approach to enlarge the kernels obtained in the second stage, superimposing the elementary sets over the minimal ones. The result is that the SAM data are classified in highly interdependent and homogeneous groups.

In order to obtain the structuring of the SAM, we got a Boolean representation of the multiplier matrix, filtered it out at certain sensible level, and applied the algorithm described above. The algorithm was applied to several adjacency matrices of D based on distinct critical values that filtered out the size of its multipliers. The SAM was normalized by rows, i.e., $D_{ij} = X_{ij}/X_i$, so its multipliers $(I - D)^{-1}$ are read as elasticities when we transpose it as in [1]. So, the basic binary relationship xRy is read as "x is influenced by y".

7.4 Some Results

Table 2 shows the construction of each elementary closure of one element set. For example, the only pseudoclosure of $\{1\}$ is $\{1\}$. Thus, $\{1\}$ is the closure of $\{1\}$; it is noted F_1. The first pseudoclosure of $\{32\}$ is $\{27, 28, 32, 33\}$, the second is $\{11, 13, 27, 28, 31, 32, 33\}$, the third is $\{9, 10, 11, 13, 27, 28, 31, 32, 33\}$, the fourth is $\{2, 9, 10, 11, 13, 27, 28, 31, 32, 33\}$ and the fifth is the same one. Thus, $\{2, 9, 10, 11, 13, 27, 28, 31, 32, 33\}$ is the closure of $\{32\}$, i.e., F_{32}.

X=i	a(x)	$a^2(x)$	$a^3(x)$
1	$\{1\} = F_1$		
2	$\{2\} = F_2$		
3	$\{3\} = F_3$		
4	$\{1,4\} = F_4$	$\{1,4\} = F_4$	
5	$\{5\} = F_5$		
6	$\{6\} = F_6$		
7	$\{7\} = F_7$		
8	$\{3,8\} = F_8$	$\{3,8\} = F_8$	
9	$\{9\} = F_9$		
10	$\{2,10\}$	$\{2,10\} = F_{10}$	
11	$\{11\} = F_{11}$		
12	$\{12\} = F_{12}$		
13	$\{9,10,13\}$	$\{2,9,10,13\}$	$\{2,9,10,13\} = F_{13}$
14	$\{3,14\}$	$\{3,14\} = F_{14}$	
15	$\{15,20,21\}$	$\{15,20,21,22,23\}$	$\{1,4,5,6,7,8,9,12,15,16,17,18,20,21,22,23,27,28\}$
16	$\{16\} = F_{16}$		
17	$\{17\} = F_{17}$		
18	$\{18,19\}$	$\{18,19,22\}$	
19	$\{19,22\}$	$\{1,4,5,6,7,8,9,12,15,16,17,18,19,22,27\}$	$\{1,4,5,6,7,8,9,12,15,16,17,18,19,22,27\}$
20	$\{20,23\}$	$\{20,22,23,28\}$	$\{1,3,4,5,6,7,8,9,11,12,13,15,16,17,18,19,20,21,22,27,31,32\}$
21	$\{21,22\}$	$\{1,4,5,6,7,8,9,12,15,16,17,18,21,22,27\}$	$\{1,4,5,6,7,8,9,11,12,13,15,16,17,18,19,20,21,22,23,27,28,32,33\}$
22	$\{1,4,5,6,7,8,9,12,15,16,17,18,19,22,27\}$	$\{1,3,4,5,6,7,8,9,12,13,15,16,17,18,19,20,21,22,27,31,33\}$	$\{1,4,5,6,7,8,9,10,12,13,15,16,17,18,19,20,21,22,27,31,32\}$
23	$\{22,23,28\}$	$\{1,3,4,5,6,7,8,9,12,13,15,16,17,19,20,21,22,23,27,28,31,32,33\}$	$\{1,3,4,5,6,7,8,9,10,11,12,13,15,16,17,18,20,21,22,27,31,32,33\}$
27	$\{13,27,31,32\}$	$\{9,10,13,27,28,31,32,33\}$	$\{2,9,10,11,13,27,28,31,32,33\}$
28	$\{11,13,28,32,33\}$	$\{9,10,11,13,27,28,32,33\}$	$\{2,9,10,11,13,27,28,31,32,33\}$
31	$\{31,33\}$	$\{31,33\} = F_{31}$	
32	$\{27,28,32,33\}$	$\{11,13,27,28,31,32,33\}$	$\{9,10,11,13,27,28,31,32,33\}$
33	$\{33\} = F_{33}$		

Table 7.2. Pseudoclosures of SingleTons: $a(x)..a^3(x)$

Note that some closures are equal, such as F_{27}, F_{28} and F_{32}. In fact, there are 20 different elementary closed subsets: $F_1, F_2, F_3, F_4, F_5, F_6, F_7, F_8, F_9, F_{10}$, $F_{11}, F_{12}, F_{13}, F_{14}, F_{15}, F_{16}, F_{17}, F_{27}, F_{31}, F_{33}$.

From these 20 elementary closed subsets, we can draw out the minimal closed subsets: $\{1\}$, $\{2\}$, $\{3\}$, $\{5\}$, $\{6\}$, $\{7\}$, $\{9\}$, $\{11\}$, $\{12\}$, $\{16\}$, $\{17\}$ and $\{33\}$. Then, we put the other elementary closed subsets in the figure 1. This

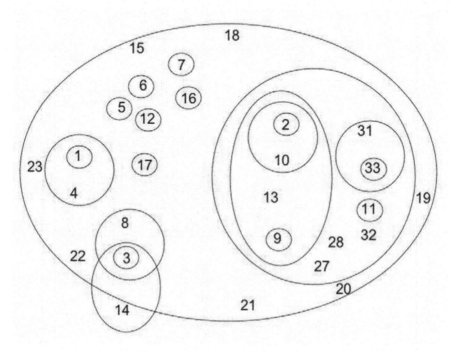

Fig. 7.1. Structure of Influence Transmission in the System

figure is a map of the structure of the influence transmission into the system. The map is built up from the minimal closed subsets contained in the different elementary closed subsets, from which we start to expand the map by adding up the subsets that contain such elements. Each of these minimal closed subsets is dependent of the other sets: each element has not influence on the other ones. Thus the same account is the closure of itself, which means that each one is influenced by itself but it does not influence to any other, although, they can be influenced by the exterior.

The pretopological interpretation of this map starts from the center or from each minimal closed subset toward outside of the map. For instance, the development bank $\{33\}$ is influenced by the central bank $\{31\}$, commercial banks $\{32\}$, private firms $\{28\}$ and households $\{27\}$. The metallic and machinery production sector $\{11\}$ is influenced by private firms, households and

x=i	$a^4(x)$	$a^5(x)$
1		
2		
3		
4		
5		
6		
7		
8		
9		
10		
11		
12		
13		
14		
15	{1,3,4,5,6,7,8,9,11,12,13,15,16,17,18,19,20,21,22,23,27, 28,31,32,33}	{1,3,4,5,6,7,8,9,10,11,12,13,15,16,17,18,19,20,21,22,23,27, 28,31,32,33}
16		
17		
18	{1,3,4,5,6,7,8,9,12,13,15,16,17,18,19,20,21,22, 27,31,32}	{1,3,4,5,6,7,8,9,10,12,13,15,16,17,18,19,20,21,22, 23,27,28,31,32,33}
19	{1,3,4,5,6,7,8,9,10,12,13,15,16,17,18,19,20,21,22,23, 27,28,31,32,33}	{1,2,3,4,5,6,7,8,9,10,11,12,13,15,16,17,18,19,20,21,22, 23,27,28,31,32,33}
20	{1,3,4,5,6,7,8,9,10,11,12,13,15,16,17,18,19,20,21,22,23 ,27,28,31,32,33}	{1,2,3,4,5,6,7,8,9,10,11,12,13,15,16,17,18,19,20,21,22, 23,27,28,31,32,33}
21	{1,3,4,5,6,7,8,9,10,12,13,15,16,17,18,19,20,21,22,23, 27,28,31,32,33}	{1,3,4,5,6,7,8,9,10,11,12,13,15,16,17,18,19,20,21,22 ,23,27,28,31,32,33}
22	{1,2,3,4,5,6,7,8,9,10,11,12,13,15,16,17,18,19,20,21,22,23, 27,28,31,32,33}	{1,2,3,4,5,6,7,8,9,10,11,12,13,15,16,17,18,19,20,21,22, 23,27,28,31,32,33} = F_{22}
23	{1,2,3,4,5,6,7,8,9,10,11,12,13,15,16,17,18,19,20,21,22,23, 27,28,31,32,33}	{1,2,3,4,5,6,7,8,9,10,11,12,13,15,16,17,18,19,20,21,22, 23,27,28,31,32,33} = F_{23}
27	{2,9,10,11,13,27,28,31,32,33} = F_{27}	
28	{2,9,10,11,13,27,28,31,32,33}	{2,9,10,11,13,27,28,31,32,33} = F_{28}
31		
32	{2,9,10,11,13,27,28,31,32,33}	{2,9,10,11,13,27,28,31,32,33} = F_{32}
33		

Table 7.3. Pseudoclosures of Singletons: $a^4(x)\dots a^5(X)$

x=i	$a^6(x)$	$a^7(x)$
1		
2		
3		
4		
5		
6		
7		
8		
9		
10		
11		
12		
13		
14		
15	{1,2,3,4,5,6,7,8,9,10,11,12,13,14,15,16,17,18,19,20,21,22,23,27,28,31,32,33}	{1,2,3,4,5,6,7,8,9,10,11,12,13,14,15,16,17,18,19,20,21,22,23,27,28,31,32,33} = F_{15}
16	{1,2,3,4,5,6,7,8,9,10,11,12,13,14,15,16,17,18,19,20,21,22,23,27,28,31,32,33}	
17	{1,2,3,4,5,6,7,8,9,10,11,12,13,14,15,16,17,18,19,20,21,22,23,27,28,31,32,33}	
18	{1,2,3,4,5,6,7,8,9,10,11,12,13,14,15,16,17,18,19,20,21,22,23,27,28,31,32,33}	{1,2,3,4,5,6,7,8,9,10,11,12,13,14,15,16,17,18,19,20,21,22,23,27,28,31,32,33} = F_{18}
19	{1,2,3,4,5,6,7,8,9,10,11,12,13,14,15,16,17,18,19,20,21,22,23,27,28,31,32,33} = F_{19}	
20	{1,2,3,4,5,6,7,8,9,10,11,12,13,14,15,16,17,18,19,20,21,22,23,27,28,31,32,33} = F_{20}	
21	{1,2,3,4,5,6,7,8,9,10,11,12,13,14,15,16,17,18,19,20,21,22,23,27,28,31,32,33} = F_{21}	
22		
23		
27		
28		
31		
32		
33		

Table 7.4. Pseudoclosures of Singletons: $a^6(x) \ldots a^7(X)$

commercial banks; the mining {2} is influenced by basic metallic {10} and construction {13} sectors; and non-metallic minerals {9} production sector is influenced by construction. Both mining and non-metallic minerals production sectors are influenced also by commercial banks, private firms and households.

Watching from outside toward inside of the figure 1, we can observe the trajectory that the demand exogenous injection follows. It goes initially to the income factors of production: wages {19}, corporate profits {20} and unincorporate profits {21}; then it goes toward the income of the private firms {23} and households {22}. These institutional units spend their current income in goods and services: utilities (electricity, gas and water) {14}, trade, hotels and restaurants {15}, communal and social services {18}, food, beverage and tabacco {4}, agricultural products {1}, chemistry products {8}, oil and gas {3}, textile {5}, wood {6}, paper and print {7}, transport and communications {16}. The difference between current income and current spending generates saving, which goes toward the financial services sector {17} in the form of financial investment or deposits. So, the exogenous injection goes through the financial flow of funds, where the transactions of the non-financial and financial institutions are determining the level of activity in the basic metallic sector {10}, mining sector {2}, construction {13} and non-metallic minerals {9}.

The influence of the exogenous injection travels from the current accounts to all the other accounts in both the real and financial sides of the economy. The path of the influence can be observed in both Figure 1 and Table 2. The demand exogenous injection represented by public and foreign spending influences the public services related with the public spending and then influences the level of income and the transactions of the non-financial and financial and financial institutions.

The moral of this pretopological analysis is that the financial system, represented by the banking sector and the financial services sector, have a limited influence in the primary and services activities, meanwhile it influenced the manufacturing production sectors. Call the attention the isolation of the financial services sector {17} and the development banks {33} since they are minimal closed subsets, which means that they do not influence to any other subset.

References

1. Auray J.P., Duru G. and Mougeot M.(1981). "Influence par les prix et influence per les quantitès dans un modele input-output." Economie apliqueè, vol. 1.
2. BANXICO, Banco de México.(1991). "Indicadores Económicos." Dirección de Investigación Económica. BANXICO,
3. BANXICO,Banco de México. "Sistema Nacional de Información Estructurada(SINIE)."
4. Belmandt Z.(1993). "Manuel de Prètopologie et ses applications." Hermes Publishing House.

5. Blancas Andrés. (2001). "Structural Vulnerability of the Mexican Financial System: A SAM multiplier approach." (Mimeo) presented at NAEFA meeting, New Orleans. 2001.

6. Bonnevay S., Lamure M.,Largeron-Leteno C., and Nicoloyannis N.(1999). "A pretopological approach for structuring data in non-metric spaces." Elsevier Preprint.

7. Brissaud Marcel. (1971). "Topologie et Prètopologie", Publications Econom ètriques, vol. IV no. 1.

8. Cech Edward. (1966). "Topological Spaces." John Wiley & Sons.

9. CIE, Consultoría Internacional Especializada. "MIP Matríz de Insumo Producto, 1990."

10. Defourny, J. and E. Thorbecke. (1984). "Structural Path Analysis and Multiplier Decomposition within a Social Accounting Matrix." Economic Journal, 94.

11. Devarajan, S., Lewis, D.J., and Robinson, S. (1986). "A Bibliography of Computable General Equilibrium (CGE) Models Applied to Development Countries." University of California Berkeley.

12. Hinojosa, O. R., and Yunez, N. A. (1999). "Economic Integration and Structural Adjustments in North America, Central America and the Caribbean A Computable General Equilibrium Approach." The North American Journal of Economics and Finance. Vol.10 No.1

13. INEGI, Instituto Nacional de Estadística Geografía e Informática.(1990). "Encuesta Nacional de Ingresos y Gastos de los Hogares, 1989."

14. INEGI, Instituto Nacional de Estadística Geografía e Informática. (1995). "Cuentas Nacionales de México."(CD)

15. Largeron C. and Bonnevay S.(1997). "Une mèthode de structuration par recherchè de Crmès minimaux-application á la modélisation de flux de migrations intervilles." Societè Francophone de Classification-Lyon, France.

16. Largeron C. and Bonnevay S. (1999). "Une approche pretopologique pour la structuration." Working Papers no.7, Centre de recherches èconomiques de luniversitè de Saint-Etienne.

17. Martínez, A. and Solís V. (1985). "Análisis Estructural e interdependecia sectorial: el caso de México." in Lifshitz E. and Zottele A. (editors) "Eslabonamientos productivos y mercados oligopólicos." Universidad Autonoma Metropolitana , México.

18. Pyatt, G and J.I. Round. (1979). "Accounting and fixed price multipliers in a social accounting matrix framework" . Economic Journal 89.

19. Pyatt G. and J. I. Round. (1985). "Social Accounting Matrices: A Basis for Planning." The World Bank, Washington D.C.

20. Scott, John. (2000). "Social Network Analysis: A Handbook". London: Sage Publications.

21. SHCP. Secretaria de Hacienda Pública.(1990). "Cuenta de la Hacienda Pública Federal." México.

22. Solís Valentín. (2001). "Pretopological spaces, closure operators and valued graphs." Mimeo, Facultad de Ciencias Económicas y Empresariales, Universidad de Castilla- La Mancha Spain.

23. Sonis M, G.J. D. Hewings and S. Sulistyowati. (1997). "Block structural path analysis: applications to structural changes in the Indonesian economy." Economic Systems Research 9, 265-280.

24. Stone, J. R. N. (1966). "The Social Accounts from a Consumer Point of View." Review of Income and Wealth. series 12, no.1.

25. Stone, J. R. N. (1985). "The Disaggregation of the Household Sector in the National Accounts." in Pyatt G. and J. I. Round. (1985). Social Accounting Matrices: A Basis for Planning. The World Bank, Washington D.C.

26. Taylor Lance. (1990). "Structural Computable General Equilibrium Models for Developing World." The MIT Press, Cambridge, Massachusetts, London, England.

27. Vos Rob. (1997). "External Finance and Structural Change." in Jansen Karel and Vos Ros (1997), External Finance and Adjustment Failure and Success in the Developing World. St. Martin's Press, Inc. and MacMillan Press LTD.

28. Wasserman, S. and K. Faust. (1994). "Social Network Analysis. Methods and Applications, Structural Analysis in the Social Sciences." Nueva York, Cambridge University Press.

8

Firm Creation as an Inductive Learning Process: A Neural Network Approach

Francesco Luna

International Monetary Fund
fluna@imf.org

Summary. I present a neural network model for the spontaneous emergence of enterprises following a *dynamic* approach to firm formation in the tradition started by Adam Smith and further pursued by Joseph Schumpeter. I suggest that the "natural propensity to truck and barter" is the observable behaviour of self-interested economic actors who are continually exposed to and confronted with an environment complexity, which transcends their limited cognitive and computational capabilities. In their learning process they build networks of relations with other agents. It is this interaction among heterogeneous agents that often leads to the formation of successful organizations that completely solve the original problem. Firms can be seen simultaneously as the result of the entrepreneurs induction process, but also as the essential instrument for the elaboration of a solution to the problem. Increasing returns to scale and market size find in this framework a very natural representation. The role of competition for the nurturing of efficiency, and the issue of protection of infant industries can be tackled by this model.

8.1 Introduction

One of the most fascinating issues which has engaged a large number of economistsand which perhaps is at the very foundation of the disciplineis the *paradox* of self-interested agents interacting with each other so as to create *unexpectedly* a coordinated system. Adam Smith employs the image of the "invisible hand" to specify the need for an explanation to such a "wonderful" phenomenon. It could be argued[1], that is precisely self-interest (with the connotation that Smith gives to the word), which can trigger an endogenous process of emergent phenomena leading to the "grand" harmony. However, if the "primitive drive" is self-interest, the visible *fumbling fingers*[2] will be perceived as a "propensity to truck and barter," leading to the creation of institutions.

---

[1] Luna (1996)

[2] This felicitous expression was coined by Professor Clower to indicate the observable working tools of some coordinating *invisible hand*

In this paper I want to follow up on this idea by showing how self-interest, represented by the need to deal successfully with a complex environment, can lead to the emergence of one fundamental institution of the market economy: the enterprise. The model I will propose is, as usual in economics, a metaphor, and an extremely simple one. Economic actors are described as finite automatamore precisely in terms of simple neural networksand their relations as the creation of a web of links.

The starting point is the role of the entrepreneur in the economy. Entrepreneurship can be identified as the ability to recognize the existence of some state of affairswhich may have passed unnoticedand to exploit such state to the entrepreneurs own advantage.

The model below proposes a framework of analysis (at the level of a simple metaphor) in which firms are created by innovating entrepreneurs as the result of a learning process triggered by a complex environment. A firm, an organization of heterogeneous agents, is the result of the inductive learning process and the way to deal successfully with the complex environment.

The use of finite automata in economics (especially in game theory) is not new. However, I somewhat depart from the usual approach. In fact, I employ a neural network framework to represent the typical economic agent. In such a way, the finite automata can endogenously change their own rules of behaviour according to their personal experience and learning process. At the same time, I use neural networks not much as a computational device (which has been exploited for some time now in various non-linear econometrics exercises), but as the way to model artificial economic actors.

I will propose the model in the following sections and employ neural networks to depict explicitly an inductive inference problem and the learning process of an agent endowed with limited cognitive and computational resources. The simulations I perform on this model (sections 2 and 3) show how the aggregation of individuals is "spontaneous" and how this combinatorial search often leads to the emergence of organizations that completely solve the original problem. Significantly, the architecture of the firm that emerges varies "randomly" and is rarely the simplest one that could, in principle, deal with the problem successfully.

In section 4 I argue that the individual characteristics of each member of the organization are exploited at best so that increasing returns to scale for the enterprise are clearly recorded. Finally, I find that introducing competition on the factor market imposes a degree of efficiency on the emerging firms; however, the results also suggest that competition may have a cost (for the system as a whole) in terms of a longer time required for the emergence of an enterprise.

8.2 A Typical Finite Automaton

The model I will propose is based on the interaction of *typical* artificial agents. By this I imply that a degree of heterogeneity is explicitly introduced and maintained among agents which share only a degree of similarity. The heterogeneity is essential to the results obtained. These results would not be captured by concentrating the attention on the *representative* agent (in terms of some average behaviour). Hence, such a concept is not relevant to the current exercise and I will describe the interaction of a set of typical agents each described by a particular finite automaton.

More precisely, each agent is described by a two-input single-layer neural network (pictured in figure 1). There is a considerable advantage in modelling the finite automaton in this way: each agent can modify its set of behavioral rules depending on its experience. In particular, the structure represented in figure 1 can simulate 12 different set of rules, that is, 12 different Boolean functions (given below). This artificial actor[3] is driven by self-interest. An ongoing learning process characterizes the actor's interaction with the environment. Clearly, the strategy I follow implements bounded rationality. Moreover, the very possibility of an actor facing a problem that is too complex to solve gives a credible justification for the observed interaction among agents. It is under the push of self-interest that actors, engaged in an inductive process of adaptation through trial and error, will manifest a "propensity to truck and barter and exchange one thing for another". The "squares" on the right are called "input neurodes"[4]. They simply receive the external stimuli as Boolean values: TRUE, FALSE. Assigning the value 1 to TRUE and 1 to FALSE, the set of instances that the net (our typical actor) can be confronted with is composed of four elements: {1, 1}, {1, -1}, {-1, 1}, {-1, -1}. The agent elaborates the observed input according to the function:

$$f(A, B) = 1 \ \ if(w_0 + w_A A + w_B B) > 0$$
$$f(A, B) = -1 \ \ if(w_0 + w_A A + w_B B) < 0$$

In other words, the net draws a separating hyperplane in the input space ($B = \frac{-w_0}{w_B} - \frac{-w_A}{w_B} A$) which classifies input instances as positive or negative.

---

[3] Another advantage of the following approach is that, by explicitly describing the actor in physical terms, I effectively set up a dichotomy between the problem and the problem-solver. In this way, I avoid some of the typical risks met by the usual modellization of the representative agent. It is not a rare occurrence to see representative agents coping with problems that are in general uncomputable, i.e. procedurally unsolvable. Nevertheless, if a nonconstructive proof can be given for the existence of a solution to such problems, the substantively rational agent is assumed to be able to find that solution.

[4] I follow Caudill and Butler (1990) and choose to use the term neurode to indicate the artificial equivalent of the neuron.

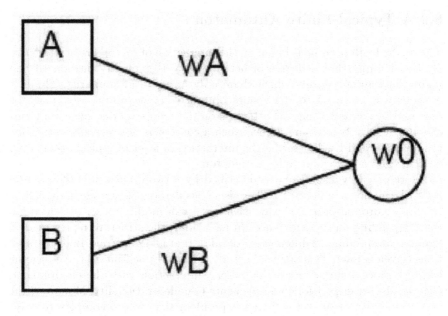

Fig. 8.1.

Such a simple structure captures the basic elements of an adaptation process undergone by any economic agent: perception, action and change promoted by trial and error. The objective is that of "reacting" correctly to the observed data. By a process of trial and error and weight adaptation[5], the net assumes in time different parametrizations and hence behaviour so that it may eventually simulate the function that actually generates the data. In other words, the agent has to induce[6] the "law" underneath the datagenerating process. It is easy to see that the automaton can assume at each point in time one of twelve possible qualitatively different states. They can be precisely characterized as the twelve linearly separable two-argument Boolean functions: functions whose positive instances can be separated from the negative ones

---

[5] The neural network learning algorithm adopted is the simple perceptron. Each time the networks makes a mistake, the weights are updated according to the formula $W = W + CE$, where E is the input received and C is the correct output expected. It is easy to see that a weight which was pushing in the right direction will be augmented, and vice versa in the opposite case.

[6] The reader familiar with COLT (Computational Learning Theory) will recognize my theoretical reference point as Gold's (1965, 1967) model of inductive inference. Following Gold's original study in order correctly to characterize the learning model, I have chosen the class of two-argument Boolean functions as the object of the inference process, an informant as the method for information presentation, and a tester as the naming relation.

by a straight line. The look-up table for these functions (plus for the XOR function discussed below) is the following:

Functions	Instances (1,1)	(1,-1)	(-1,1)	(-1,-1)
A	1	1	-1	-1
B	1	-1	1	-1
Not A	-1	-1	1	1
Not B	-1	1	-1	1
A ∧ B	1	-1	-1	-1
A ∧ Not B	-1	1	-1	-1
Not A ∧ B	-1	-1	1	-1
Not A ∧ Not B	-1	-1	-1	1
A ∨ B	1	1	1	-1
A ∨ Not B	1	1	-1	1
Not A ∨ B	1	-1	1	1
Not A ∨ Not B	-1	1	1	1
XOR	-1	1	1	-1

For example, the Boolean function $A \wedge B$ is TRUE when both inputs B and A are TRUE. Hence it will output 1 only on input $\{1, 1\}$. For any other instance it will output FALSE or -1.

Furthermore, it can be shown that a neural network (like the one depicted in figure 1) with random initial weights will reach an internal configurationafter a finite number of steps so as to simulate successfully any of the 12 simpler functions. Appendix 1 gives the Mathematica program for a neural network, which learns to simulate the Boolean function $A \wedge B$.

However, the class of two-argument Boolean functions also contains two members whichnot being linearly separablecannot be simulated by that architecture, hence they cannot be identifiedor learned. These much-too-complex elements[7] are the $Exclusive - OR$ function[8] and its complement, that is, the Boolean function whose output is True if both argumentsA and B in our caseare True or if both are False. I set the XOR function as the data generating process for the artificial economy. The following pictures show the **XOR** function in the instance space, one possible network to simulate it, and its graph in 2 and 3 dimensions.

With the following weights: W0 = 1, W1 = -1, W2 = -1, WA1 = 1, WA2 = -1, WB1 = -1, WB2 = 1, Wh1 = 1, Wh2 = 1 the above network simulates the XOR functions as shown below.

---

[7] Here obviously the complexity is relative to the agents capability.

[8] The XOR function, as recorded in the last row of the table, returns TRUE (represented by 1 in the table) whether either one of its input is TRUE, but not both.

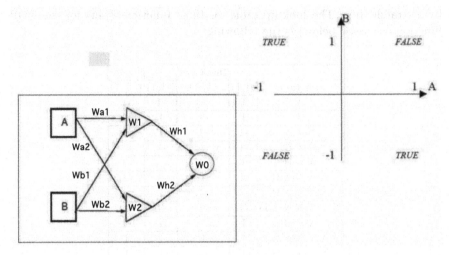

Fig. 8.2.

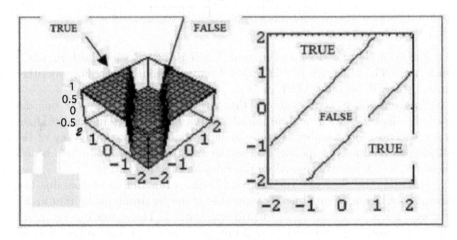

Fig. 8.3.

In the model, each actorwhich we can consider more similar to a producer than to a consumeris confronted with this (XOR-function) complex environment. It will obtain a positive payoff each time it replies correctly to a stimulus. Note that each automaton (as it simulates one of the twelve functions) outputs a correct signal on some subset of the instance set. For example, if the automaton is currently simulating the Boolean function $A \land not\,B$, it will be correct in its replies (giving the same classification the XOR function would give) on an average of 75% of stimuli with only the instance -1,1 wrongly and consistently classified as negative. Each agent observes exactly the same "time series" of random instances and is, initially at least, isolated in tackling the induction problem. The problem (and hence the series of data) can be interpreted as inherent to the production process itself–some chemical reaction that necessitates either one of two elements, but not both. Alternatively, it can be due to the nature of the demand the producer is trying to satisfy–for example, the producer has to decide whether to add a particular feature to the good she is trying to sell. A subset of customers appreciate the change, whereas others dislike it thoroughly. The producer has to infer, from two characteristics (A and B) that she observes in the customer population, whether a particular customer should be offered the gadget). In a cellular automata setting, the typical actor (T. A. in the figures below) is at the center of a (two-dimensional) neighborhood of identical automata as in the figure below.

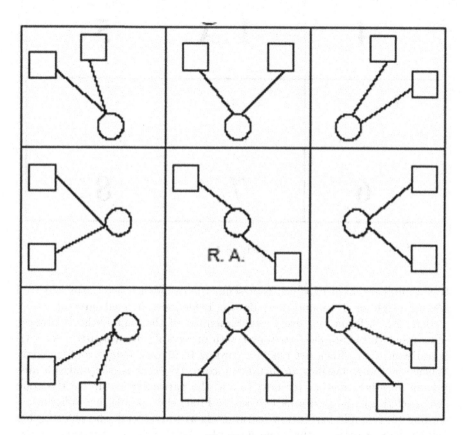

Fig. 8.4.

Figure 3 is a short-hand representation that I will use later when describing the evolution of the system.

The state of the typical automaton (Tania) is not only described by its current weight configuration (w0, wA, wB) which determines the intercept and the slope of the "separating hyperplane," but also by a vector of ra-

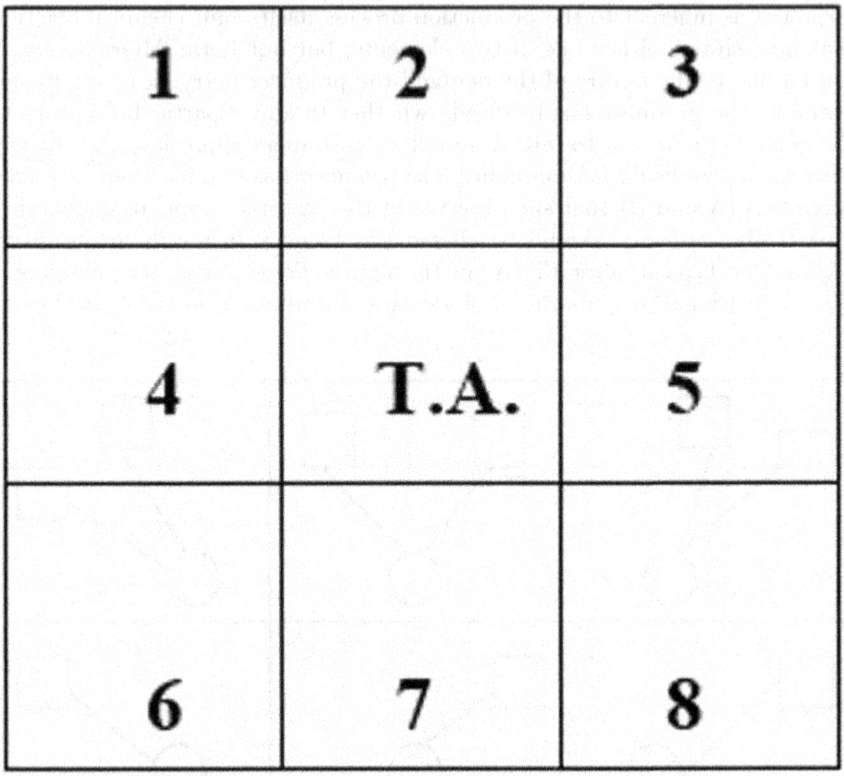

1	2	3
4	T.A.	5
6	7	8

Fig. 8.5.

tional numbers, each of which represents the confidence level attached to a specific neighbor. I assume the following behaviour: at fixed intervals (for t = 0, 1, 2,....) Tania and every other member of the neighborhood observe some input (for example, a customer with observable characteristics A, notB is to be satisfied). Each net reacts according to its own weight configuration. Whenever Tania classifies an instance incorrectly (that is, she makes a mistake in deciding whether (or not) to add the particular feature to the good she is offering), she performs two tasks. First, she updates her weights according to a simple perceptron learning algorithm; second, she observes her neighbors' responses to the same input. In case a neighbor has reacted correctly to the stimulus (i.e. it has satified that customer type), Tania increases her confidence in this neighbor. When this level is "sufficiently" high[9], Tania

---

[9] The process described is actually a threshold dynamics; however we adopt a somewhat different system relying on pseudo-random numbers. Hence, in our case, a confidence level (a number between 0 and 1) simply represents the probability that Tania will contact that particular neighbor.

becomes an entrepreneur. She convinces the neighbor to volunteer its input elaboration: in some sense, Tania is buying her neighbor's labor[10]. However, the emerging coordination cannot be attained simply through the market. In order for the "arrangement" to have a chance of success, it is necessary that the contacted neighbor interrupt its autonomous learning process. This is the essential feature that allows me to identify such phenomenon as the emergence of a firm: the entrepreneur needs to internalize the coordination of the different production factors. It is important to stress that the necessary coordination cannot be reached simply through repeated market exchanges (with each partner proceeding in its own search for a complete solution to the induction problem). This happens because, let me stress it once more, each actor (a single-cell net faced with a linearly non-separable set) will never settle down on any particular weight configuration. This means that, even though, by fluke, a successful configuration for the larger market-based interrelation is reached, its individual partners will persist in their personal attempt to find a successful configuration changing their own weights to the detriment of the occasional market-determined exchange[11]. Apparently, such a problem could be solved, at least in principle, contractually. To write such a contract the entrepreneur should take into consideration the whole set of contingencies. In our case these are not only all possible demand signals, but also each worker's "states of nature[12]" occasioned by the initial parameters and by the historical series of received signals (the XOR instances). Clearly, the cost of writing such a contract would quickly become unaffordable. Hence, under this perspective, it looks as though the model presented does not really suggest anything new with respect to the transaction-cost theory of firm formation.

---

[10] In this version of the model, there is no monetary exchange, and the employee simply supplies its work upon request. However, in Bruun and Luna 2000, the hiring process is modeled in such a way that the employee accepts an offer from the would-be entrepreneur only if it is greater or equal to the average of the past payoffs obtained working alone. On the other hand, a would-be entrepreneur will manage to "hire" a neighbor only in case its success history is longer than its neighbor's. Such mechanism satisfies one of the conditions recognized in the literature for the emergence of entrepreneurs: that of being better workers. The simulation results in that more complex version do not change qualitatively.

[11] This point in particular has raised some objections from a referee. It would seem that the very assumptions of the model prevent the system from settling down contradicting common human experience which ultimately concerns solving small, stationary problems. However, I ask the reader to look at the XOR problem from the eyes of the typical actor to realize that such a problem is not small. On the contrary, the creation of a firm (or of an institution for that matter), thanks to the complexity reduction it fosters, will precisely allow the agent to concentrate on the solution of small and stationary problems: the simulation of some particular linearly separable Boolean function in this metaphor. It is only the firm, which will solve the more complex problem.

[12] Minsky and Papert (1969) Perceptron Cycling theorem assures us that this set will be finite as well.

On the contrary, it is essential to stress that such a contract would itself transcend the computational and cognitive complexity of the entrepreneur; it very likely transcends the problem-complexity itself. A contract is simply not an option available to the agents in this scenario. The architecture describing the relation between the would-be entrepreneur and one trainee[13] is given below in Figure 6. For expository reasons we also depict in Figure 6 a different, but equivalent network.

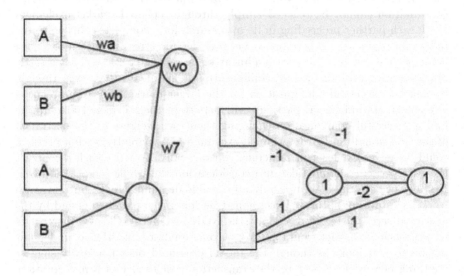

Fig. 8.6.

More precisely, Figure 6 pictures the relation between Tania and her neighbor "7" as perceived by Tania herself. She is combining the signals she directly receives from the environment (A and B) with the information obtained from the "hired" neighbor. Note that the weights characterizing neighbor "7" configuration are not known to Tania. Even more important is to stress that the learning process of this expanded organization will affect only Tania's (the entrepreneur's) original weight vector (w_A, w_B, w_0), plus the weight attached to the new link (w_7)–connecting Tania to its trainee. In other words, Tania's

---

[13] To avoid misunderstanding of this term, I want to stress that by trainee I do not mean a person being trained by an already established and successful organization. The phenomenon I am describing is, at this stage, the continual series of trial and errors that may, but not necessarily, lead to the emergence of a firm (i.e. a successful organization). Precisely because the individual we define as trainee partakes of this fluid set up, his status and function cannot be univocally defined either. The most effective way of labelling his condition is that of a tenure track. Only if and when the firm is established as such, will the trainee task be determined and his status become that of an employee.

induction procedure cannot modify the intimate configuration of its subordinate; it can only try to make the best out of its trainee's "acquired skills". The agents remain distinct from one another and contribute–each one according to its own role–to the construction of a higher level organization. Since Tania updates her confidence simultaneously over the whole neighborhood, it is possible that more than one neighbor be hired at any point. The set up proposed here is reminiscent of some neural network constructive algorithms such as the pyramid[14] algorithm. The fundamental difference is that in the pyramid case, once the net grows by adding a hidden neurode and appropriate links, any further learning will affect the whole architecture. Instead, here–as stressed above–only the original weights plus the newly established links are modified in the course of the adaptation process[15] pursued by the organization. Going back to Figure 6, it is evident that the two networks are structurally equivalent. Figure 6 represents the simplest neural net that correctly simulates the XOR function. The following picture shows how that particular net succeeds in simulating it.

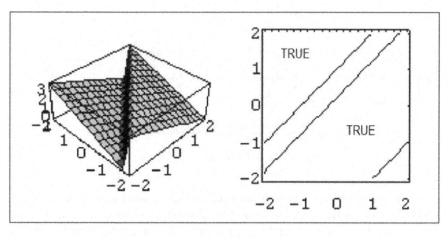

Fig. 8.7.

This implies that the architecture composed of Tania and one neighbor is potentially fit to induce correctly the XOR function. However, in order for that to happen the hired neighbor must play the same role as the hidden neurode in Figure 5. In that architecture the hidden neurode simulates the Boolean function $H(A, B) = B \land not A$ which is only one of the twelve[16] pos-

---

[14] Gallant (1993) is a good reference for these more advanced topics

[15] We may baptize this constructive algorithm the Acropolis algorithm (as opposed to pyramid) to stress that the newly acquired elements remain clearly distinct and "cooperate" with the leader as in Athens's democratic society.

[16] Since A and B are arbitrary and interchangeable "names", the favourable arrangement is actually one out of six. However, things are much less fortunate

sible arrangements. Unless the autonomous learning process of the neighbor is stopped in such a configuration, there is no possibility for Tania to coordinate successfully the organization. A similar reasoning can be repeated for larger "coalitions", hence, as found below during the actual simulations, the achievement of a successful organization is not at all a necessary event. The simulation will start with an initial configuration identical to Figure 2, where each agent is dealing with the problem on its own. However, as time goes on and the number of mistakes increases, Tania will start interacting, in the way described, with its neighbors. Graphically the emergence and evolution of such relations in one of the simulations performed can be represented by the following series of "slides" [17]

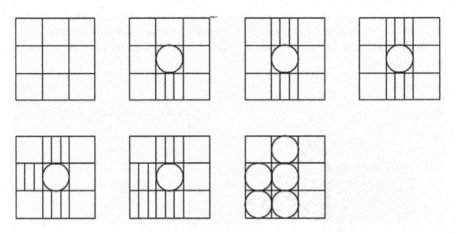

Fig. 8.8.

Each slide represents the state of Tanias neighborhood at a particular time. After the first period- -characterized as a training session of 30 instances– Tanias confidence in her neighbors is such that Tania decides to hire neighbor "7". At the end of the second period the organization has grown to include neighbor "2" without, however, accomplishing the induction task. In this instance six periods are sufficient to generate an organization–a firm in our metaphor–that completely solves the original problem. In particular, Tania has "hired" four neighbors managing to coordinate successfully their input

when simulating Boolean functions with neural nets. Since the choice of the initial weights is "random", in the set of signed integers -5,-4,...0,..5, there are 130 favourable cases out of 1331 possibilities: less than 10%.

[17] The program I wrote actually animates such slides. On a color monitor the effect is that of a cartoon. Tania (the would-be entrepreneur) is red and the trainees are blue. If the organization is successful, its members change color to signal the accomplished evolution. The program list of commands is available on request from the author.

elaboration. This is pictured in the last slide where I have represented the emerged firm as a collection of circles (the original Tania's crest). Employing once more Figure 2 the final structure can be represented as in Figure 9.

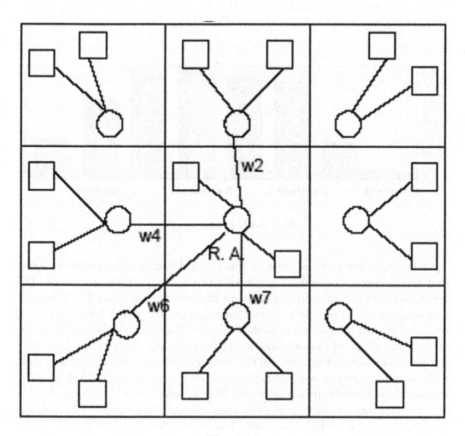

Fig. 8.9.

I ran 1000 simulations and the results are summarized in the following table The first row of Table 1, denoted by *struct*, records the physical structure of the

Struct	1	2	3	4	5	6	7	8
freq	3	47	145	156	172	167	116	52
ave tm	3.7	4.9	6.2	8.1	10	11	15	19

Table 8.1.

organization summarized by the number of members besides the entrepreneur. Clearly the architecture cannot be composed of more than eight elements (the complete neighborhood). In the second rowdenoted by freqI present the

number of runs that produced the corresponding structure[18]. The third row–
denoted by ave tm–records the average number of periods that were necessary
to accomplish the learning process or, equivalently, the average time necessary
for the emergence of an institution of that particular structure. The following
histogram gives an idea of the sample distribution.

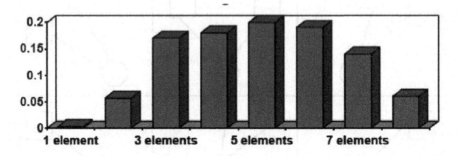

Fig. 8.10. Histogram 1

A few comments are in order at this point. First, it is important to stress
that the emergence of such an institution is not at all a necessary event. The
occurrence was recorded in 858 runs out of 1000. On the other hand, what is
undeniable is that the combinatorial optimization–the search for a solution–is
triggered by self-interest and justified by the complexity of the environment.
Second, the shape of the organization that eventually deals successfully with
the environment is not unique. On the contrary, in each case the search process
will continue as long as is necessary to construct some structure that "works".
I must admit that the program is such that in some cases the architecture of
an emerging organization will grow even though, in principle, the original
structure would have been successful had it had more time to learn. This is
clearly linked to the training session cardinality or, in other words, with the
size of the market[19]. However, rather than a limitation, I consider this feature
of the model very significant. It captures another phenomenon treated in the
literature, for example by Penrose (1959):

> if a firm deliberately or inadvertently expands its organization
> more rapidly than the individuals in the expanding organization can
> obtain the experience with each other and with the firm that is neces-
> sary for the effective operation of the group, the efficiency of the firm
> will suffer, even if optimum adjustments are made in the administra-
> tive structure.[20]

---

[18] For example, in 145 cases out of 1000, a firm with three workers was created.

[19] See section 5 below.

[20] Penrose (1959). Third edition 1995, p. 47.

Third, and linked to the previous point, it is not true that the simplest architecture–Tania and one employee–(which, moreover, is the one that on average takes less time to build) is the most prevalent one. Out of 1000 simulations performed, only three such cases were observed. It is easy to reply that the last result is due to the inherent cognitive limitation of the representative automaton and that a perfectly rational agent would have built his enterprise in the most efficient and simplest possible way. Hence, from the standpoint of the principle of simplicity and of Ockham's razor, the model does not seem to lead to plausible theoretical conclusions. The above observation follows in the steps of Simon's 1978 Ely lecture. In that paper, the parsimony principle is invoked in the comparison between optimizing and satisficing procedures.

On a first approximation, Ockham's razor seems to be in favor of "perfect" optimization since satisficing and bounded rationality require additional postulates to guarantee uniqueness of solutions. However, as stressed by Simon, "that argument only applies when we are trying to deduce unique equilibria, a task quite different from the one most institutional writers set for themselves". In other words, in a model like this, each actor seeks to solve the problem it is facing. We have seen (in Table 1) that the meta-procedure adopted to search and construct the solution (a procedure itself) can lead to various successful results. To restrict this set of solutions to the single most efficient one (in terms of components required) would force us to superimpose on our bounded rationality set up a number of extra constraints. Hence, uniqueness of the end product would probably be obtained at the cost of a complexity level higher than the one required for implementing perfect optimization. On the other hand, if the goal is "merely" that of constructing a solution that works, the search procedure I have set forward is much simpler (in terms of what is required of the entrepreneur) than perfect optimization[21].

8.3 The Model on a Torus

In the previous section, to describe the typical actor's behaviour I concentrated the analysis on a single neighborhood where only Tania had the chance of becoming an entrepreneur. The next obvious step is that of simulating a population of interacting automata each one potentially behaving as Tania. Some additional rules are required. In particular, a would-be entrepreneur cannot be hired as a trainee, whereas a trainee can collaborate with more than one would-be entrepreneur. Hence, in the complete model, we have to account for workers that are trainees for different organizations simultaneously,

---

[21] As an aside let me remind the reader that in general, that is, for problems for which the set of candidate solutions is not finite, the selection of a first best (required by the postulate of substantive rationality) would need–as a first step– the individuation of the solution set. From a recursive theoretical point of view this is shown to be uncomputable by Rice theorem. (See, for example, Rogers 1967). In such a case, finding a solution that works is all anybody can hope in.

or even deal with workers with two jobs. More substantially, I introduce a rule for the destruction of unsuccessful organizations that cannot improve. A would-be entrepreneur which has not reached complete learning in the current period and has no more neighbors that it can contact (perhaps because it is surrounded by other would-be entrepreneurs or double trainees that cannot be hired for the third time) is automatically demoted to the status of simple unemployed worker. Again the simulations are run on a color-monitor PC so that each particular state is immediately recognizable and the evolution of the system can be followed very easily. Because of calculation requirements, I limited the artificial population to 60 elements and in each simulation the population evolved for 15 periods only. I performed 100 simulations of this kind and the following histogram summarizes the results.

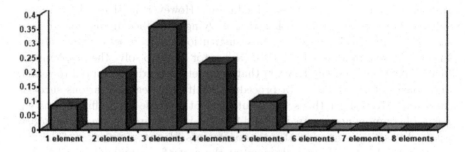

Fig. 8.11. Histogram 2

The above "distribution" was obtained from 605 successful cases (that is: 605 firms were created during the 100 simulations). It is essential to stress that the empirical distribution is substantially different from the one obtained for the single-Tania neighborhood case. The difference lies in the competition on the "factor market". Tania could proceed in her search on the whole neighborhood without any constraint, whereas, in this more sophisticated case, she implicitly competes with similar agents. Competition imposes a higher level of efficiency to the emergent organizations[22]. This proposition is not at all surprising and can be accepted without much argument. What is remarkable, however, is that such a finding is substantiated by the simulations of a model whose initial aim was to suggest an explanation for the emergence of a firm as a complex-problem solver. In pursuing the analysis on the role played by competition, I also performed a different series of simulations on a population of nine individuals so that the results could be compared with those obtained for

---

[22] Note that the entrepreneur-plus-one-worker structure now accounts for almost 10% of all successes, while it was recorded only 3 times out of 858 in the set up recorded in Table 1. Furthermore, the empirical mean and the mode of the distribution (summarized in Table 2) is 3 in presence of competition and 5 for the former "no-competition" case.

Tania's original neighborhood (summarized in Table 1). Besides the explicit introduction of competition, the difference with respect to that exercise lies in the mechanism for the destruction of "hopeless" organizations which has been introduced and described above. Hence, each experiment will proceed until a (successful) firm appears. I have run 800 simulations and, as for the original case, I have recorded frequency and average time to appearance for each possible architecture.

Struct	1	2	3	4	5	6	7	8
freq	56	171	280	192	88	11	2	0
ave tm	8.5	12.15	16.65	20.07	27.25	27.9	54	–

Table 8.2. Table 2

The size distribution is not very much different from the one depicted in figure 11, but the average time to the spontaneous emergence of the various structures differs significantly. In figure 12 I compare these average times obtained from a system characterized by the absence of competition (data from Table 1) and from the equivalent set up enriched with it (data from Table 2) respectively.

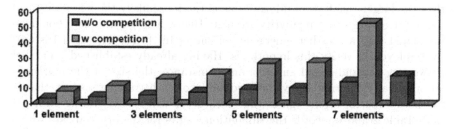

Fig. 8.12. Histogram 3

efficiency is a main concern. However, whenever priority is given to the very creation of some industry, the existence of a certain level of monopolistic power can be overlooked. The possibility of a technology transfer is not encompassed in this model. Even in that case the model may be relevant if we consider the resources and the time necessary to the building up of competent human capital.

8.4 Increasing Returns and the Size of the Market

I would like to offer now a particular interpretation of increasing returns to scale. Let us consider each neighbor as a production site with well-defined fixed capacity. A producer successfully delivers its output to the market every time it correctly responds to the demand signal it perceives. According to this interpretation, the model shows that it is only an "accomplished" firm that always correctly classifies demand signals. Thereforemaintaining the same productive capacity for each unit (constant local returns to scale)the emergence of a firm causes increased returns for the organization. Similarly, interpreting the problem as caused by technological sophistication, only a larger organization is able fully to exploit the productive capacity of each member. In both cases, the cause of the increasing returns to scale is the perception or recognition of the existence of a complex problem and the successful development of the induction process generated by it. In order to be successful, this process requires the exploitation of resources that no single actor commands on its own. Note that even Smith's source of increasing returns to scalethe division of laborcan be introduced in this framework. It is first necessary to identify a subset of operations in the production process; second, laborand possibly new physical equipmentof different qualities has to be coordinated. The result, as stressed above, is the creation of a new "trade". An essential element in Smiths analysis is also the size of the market. A large market will absorb the increased production justifying the original entrepreneurship. Obviously, from a different perspective, the size of the market can play another role. If, as I suggest, a firm emerges as the result of an induction process tending to the identification of some "regularity" hidden into the demand-signal flow, then the larger this flow of information is, the more likely the possibility will be of extracting some regularity from it. Hence a larger market should be translated into a speedier emergence of successful firms. The model is easily employed to verify this hypothesis. Having already established a parallel between a demand-signal and a training instance, the size of the market is automatically determined by the cardinality of each training session, that is by the number of input signals that each actor will receive during each period. Table 2 above records the simulation results for the original case where 20 instances compose each training session. So the initial size of the market is 20. Table 3 records the results of a "depressed" market with only 10 demand instances per period, whereas table 4 presents the results for a large market of 30 instances per period.

Table 3

Struct	1	2	3	4	5	6	7	8
freq	55	208	280	173	73	11		
ave tm	10.1	17.94	23.84	26.02	32.7	37.93	–	–

Table 4

Struct	1	2	3	4	5	6	7	8
freq	69	158	280	186	84	22	1	
ave tm	5.186	7.975	14.75	17.45	20.18	19	30	–

The hypothesis is certainly verified. There is a clear inverse relation between the training-session cardinality and the time to emergence. Another interesting feature results when looking at Tables 3, and 4. It is a well received tenet of the theory of competitive markets that sub-efficient firms will disappear in the long run and that they should be the first to go bankrupt during a recession. Symmetrically, it is reasonable to expect that during a boom sub-efficient enterprises may thrive. As summarized in the histogram below, the frequency distribution for the case of a "large" market appears to be slightly fatter in the tail corresponding to larger and, implicitly, more inefficient architectures. Again this observation requires a rigorous statistical test, but once more it suggests how rich in implications such a simple model can be.

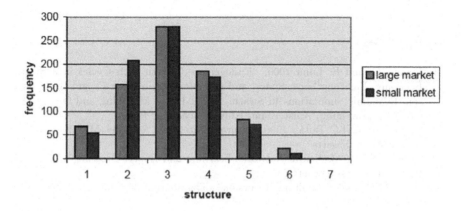

Fig. 8.13. Histogram 4

8.5 Conclusion

By explicitly describing economic organizations in their structural componentsa well defined neural network in the present caseI have suggested a "microfoundation" for the emergence of institutions (and firms in particular) as a viable solution to a complex induction problem. As indirect evidence of the interest this model may have on the theory of the firm, I want to stress that the same model also substantiated the positive role that competition (on the production-factor market in this case) plays for the efficiency of the emerging firms. Furthermore, a trade off between efficiency and the time necessary for the emergence of firms in a competitive environment has been tentatively recorded and needs deeper investigation. Last but not least, increasing returns to scale and the size of the market have been linked to the induction exercise leading to the formation of the firm.

This work can be extended in several directions. For example, I have implicitly modeled the demand side assuming that each producer would always receive a constant demand (the training session cardinality was the same for every cell). However, it may be important to model consumer behaviour explicitly by allowing some actors to move on the grid searching for consumption goods of a particular kind. Such exercise is performed in Bruun and Luna (2000) in a macroeconomic model of growth and cycles. Growth is endogenous as is triggered by the inherent complexity of the environment and the ongoing search of entrepreneurs. It would also be interesting to see whether a further increase in the environment's complexity may reverse the tendency towards aggregation in favor of imperfect, but more flexible smaller organizations. All of this would be based on the minimal hypothesis of individual economic actors continually faced with an induction problem.

References

1. Bruun, C. and F. Luna 2000, "Endogenous Growth with Cycles in a Swarm Economy: Fighting Time, Space, and Complexity," in Luna F. and B. Stefansson, Economic Simulations in Swarm. Agent-Based Modelling and Object Oriented Programming, Amsterdam, Kluwer Academic Press.
2. Caudill, M. and C. Butler 1990, Naturally Intelligent Systems, MIT Press, Cambridge Massachusetts
3. Gallant, S.I. 1993, Neural Networks Learning and Expert Systems, MIT Press, Cambridge Massachusetts.
4. Gold, M.E. 1965, "Limiting Recursion", The Journal of Symbolic Logic, 30, 1, 28-48.
5. Gold, M.E. 1967, "Language Identification in the Limit", Information and Control, 10, 447- 474.
6. Kasliwal, P. 1997, "A Cellular Automata Model of Schumpeterian Growth", paper prepared for the conference Computing in Economics and Finance, Stanford 30 June-2 July 1997.
7. Kihlstrom, R.E. and J.J. Laffont, 1979, "A General Equilibrium Entrepreneurial Theory of Firm Foundation Based on Risk Aversion", Journal of Political Economy, 87, 719- 748.
8. Laussel, D. and M. Le Breton, 1995, "A General Equilibrium Theory of Firm Formation Based on Individual Unobservable Skills", European Economic Review, 39, 1303- 1319.
9. Li, M. and P.M.B. Vitanyi, 1992, "Inductive Reasoning and Kolmogorov Complexity", Journal of Computer and System Sciences, 44, 343-384.
10. Luna, F. 1996, "From the History of Astronomy to the Wealth of Nations: Wonderful Wheels and Invisible Hands in Adam Smiths Major Works", Vaz, D. and K.Velupillai eds. Inflation, Institutions and Information, MacMillan Press, London.
11. Minsky, M. and S. Papert 1969, Perceptrons: An Introduction to Computational Geometry, MIT Press, Cambridge Massachusetts.
12. Penrose, E. 1959, The Theory of the Growth of the Firm, Oxford University Press, Oxford.

13. Roger, H. Jr 1967, Theory of Recursive Functions and Effective Computability, McGraw- Hill, New York N. Y.
14. Schumpeter, J.A. 1975, Capitalism, Socialism and Democracy, Harper and Row Publishers, New York.
15. Simon, H.A. 1978, "Rationality as Process and as Product of Thought", American Economic Association Papers and Proceedings, 68, 2, 1-16.
16. Smith, A. 1976, An Inquiry into the Nature and Causes of the Wealth of Nations, Chicago, University of Chicago Press.
17. Smith, A. 1980, The Principles Which Lead and Direct Philosophical Enquires; Illustrated by the History of Astronomy. Oxford University Press, Oxford.
18. Smith, A. 1983, Lectures on Rhetoric and Belles Lettres, Oxford University Press, Oxford.

Agent-Based Environments: A Review

Alessandro Perrone

University of Venice
alex@unive.it

Summary. Recent years have seen a proliferation of Multi-agent-based simulation (MABSS) models, in a growing range of domains, and using an increasing variety of software. In this article we compare some Agent Based environments anyone can download in Internet. The aim of this article is to discuss the general principles of each environment, and not to say which is the best or the worst one. The comparison is performed along several dimensions such as ease of learning, flexibility, available support, etc. It should also help the choice of a language by potential practitioners of agent-based economic. At the end of this article there's a personal proposition about these environments of the author.

Preface

Agent-based economic simulations are becoming part of the professions toolkit. Two recent special issues of the Journal of Economic Dynamics and Control and Computation collect several contributions of agent-based economics. They are the best evidence of the increased interest expressed by the profession.

As stressed by Testfatsion in her introduction to the JEDC special issue, agent-based simulations address at least four issues. First, heterogenous agents and interactions among them can be explicitly modeled in agent based simulations. Second, agents behaviour can change due to the interaction with the environment and other actors. Third, evolutionary processes can be implemented at the agent level, rather than dictated by population-level laws of motion. Finally, these models can lead to genuine emergence phenomena and provide a way out of the clockwork dynamics usually given by more traditional models.

Unfortunately, however, the spreading of this new tool is slowed down by the absence of a common language as underlined in Luna and Stefansson (2000) and Luna and Perrone (2002). This limitation prevents a cumulative learning process and finally the emergence of an agent-based school of thought.

This contribution reviews some of the most likely candidates to Esperanto of agent-based economic simulations. The comparison is performed along several dimensions such as ease of learning, flexibility, available support, etc. It should also help the choice of a language by potential practitioners of agent-based economics.

9.1 Introduction

In the sciences, especially in the study of complex systems, computer programs have come to play an important role as scientific equipment.

Computer simulations — experimental devices built in software — have taken a place as a companion to physical experimental devices.

Computer models provide many advantages over traditional experimental methods but the actual process of writing software is a complicated technical task with much room for errors.

Early in the development of a scientific field, scientists typically construct their own experimental equipment: grinding their own lenses, wiring-up their own particle detectors, even building their own computers. Researchers in new fields have to be adept engineers, machinists, and electricians in addition to being scientists.

Once a field begins to mature, collaborations between scientists and engineers lead to the development of standardized, reliable equipment (e.g., commercially produced microscopes or centrifuges), thereby allowing scientists to focus on research rather than on tool building.

The use of standardized scientific apparati is not only a convenience: it allows one to "divide through" by the common equipment, thereby aiding the production of repeatable, comparable research results.

Unfortunately, computer modeling frequently turns good scientists into bad programmers.

Most scientists, including social scientists, are not trained as software engineers. As a consequence, many home-grown computational experimental tools are (from a software engineering perspective) poorly designed.

The results gained from the use of such tools can be difficult to compare with other research data and difficult for others to reproduce because of the quirks and unknown design decisions in the specific software apparatus.

Furthermore, writing software is typically not a good use of a highly specialized scientist's time. In many cases, the same functional capacities are being rebuilt time and time again by different research groups, a tremendous duplication of effort.

A subtler problem with custom-built computer models is that the final software tends to be very specific, a dense tangle of code that is understandable only to the people who wrote it. Typical simulation software contains a large number of implicit assumptions, accidents of the way the particular code was written that have nothing to do with the actual model.

And with only low-level source code it is very difficult to understand the high-level design and essential components of the model itself. Such software is useful to the people who built it, but makes it difficult for other scientists to evaluate and reproduce results.

In order for computer modeling to mature there is the need for:

1. a supply of well-written tools useful for a wide range of simulations;
2. a set of shared models that allow the use and re-use of large parts of the software.

Luckily, on the web, there are a few toolkits which seem to achieve the previous goals. These tools were born either from a single programmer's ideas or from the effort of a large group of scientists.

9.2 Agent Based Modelling

Because of the way they are designed, Agent-Based models founded on object-oriented programming, are particularly suitable for software sense and recomposition.

One of the most important questions for Agent Based simulation users is: What is an Agent?

Is there a definition of Agent Based Modelling?

There are two possible ways to approach an Agent Based Model.

- **Model the Model.** ABM often offers certain system dynamic caused by simple interactions iterated across a number of agents. The rules may or may not be deduced from "nature". Some model dynamics might be observed and conclusions about the underlying properties and interactions betweenrules and states can be made. A classic example of this approachis the work of Reynolds on "Boids" wich turned out to closely mimic the flocking behaviour of birds
- **Mirror world.** It is however interesting to note that ABM also works in the opposite direction. Instead of trying to understand the emergent behaviour of interacting objects and deducing "natural laws" from such behaviour, we can model known interactions between natural agents". This approach attempts to reduce the amount of information that is necessary for describing the system and attempts to predict behaviour of the natural system

There is a crucial distinction to be made between modelling reality and "modeling the model". In the first case, the aim is to build a mirror world that can be used and examined. In the latter, the programmer is creating a conceptual model of a system and using modeling tools to realize that conceptualization.

An agent is an encapsulated computer system that is situated in some environment and that is capable of flexible, autonomous action in that environment in order to meet its design objectives. For clear understanding of the definitions, several points must be further elaborated . Agents are:

1. clearly identifiable problem-solving entities with well defined boundaries and interfaces.
2. situated (embedded) in a particular environment; they receive inputs related to the state of their environment and they act on the environment through effectors.
3. designed to fulfill a specific purpose; they have particular objectives (goals) to achieve.
4. autonomous; they have control both over their internal state and over their behaviour.
5. capable of exhibiting flexible problem solving behaviour in pursuit of their design objectives; they need to be both reactive (able to respond in a timely fashion to changes that occur in their environment) and active (able to act in anticipation of future goals).

9.3 Agent Based Tools

In this section we'll write a short description of a few popular free agent-based simulation tools which can be used to study complexity of social organisations. From the large numbers of packages, we chose Swarm, Ascape, Repast, and Starlogo simulation systems.

The aim is not to provide in-depth technical data, but to give an idea to those partners who have little or no experience of simulation software of whether the tool will be suited to their particular tasks and abilities. The goal is to give people an overall picture of the tool without digressing into too much detail.

9.3.1 Ascape

Ascape is a software framework for developing and analyzing agent-based models. In Ascape, agent objects exist within scapes; collections of agents such as arrays and lattices. These scapes are themselves agents, so that typical Ascape models are made up of "collections of collections" of agents. Scapes provide a context for agent interaction and sets of rules that govern agent behavior. Ascape manages graphical views and collection of statistics for scapes and provides mechanisms for controlling and altering parameters for scape models. According to Miles Parker, the Ascape designer, this toolkit "created to support the design, analysis and distribution of agent based models". Its principle design goals include abstraction and generalization of key agent modeling concepts, ease of use and configurability, best attainable performance, and deployment anywhere.

Ascape was developed primarily to support models of social and economic systems, which typically comprise agents with rules of behaviour interacting in networks (e.g. regular lattices, random graphs, soups), but the framework may be adabtable to other model types.

The main goal for the creators of Ascape was to develop a tool which could allow people with basic programming skills to build simple models from the existing ones which come with the package, and progress from there. Besides, the most complex models should be reasonably straightforward, and the code should remain easy to work with and understand.

Last but not least, as much functionality as possible should be provided "automatically", by some tools in Ascape.

The models built with Ascape are also very easy to understand for users, who have complete control of model parameters at runtime. Since the structure of Ascape allows to create graphs, no additional programming is required for that.

Here follow the design goals, provided by Miles Parker himself [JASSS vol.4 n.1]:

1. It should be expressive; it should be possible to define a complete model in the smallest possible description.
2. It should be generalized; it should be possible to express the same basic modeling ideas in one way and have them tested in many different environments and configurations.
3. It should be powerful; it should provide high-level user oriented tools that allow model interaction without programming, but it should also provide an environment capable of modeling diverse and complex systems.
4. Finally, to support these goals, it should be abstract; it should encapsulate modeling ideas and methodologies at the highest possible level.

A typical model environment in Ascape is shown in the figure 9.1

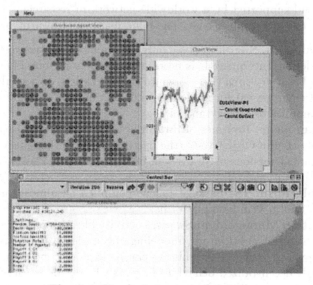

Fig. 9.1. Typical Ascape simulation

Ascape and Internet

Ascape is available at Brook Web Site[1]. The version which is actually available is a bit outdated, since the original programmer, Miles Parker moved from Brooking Institute. There's also a mailing list which help beginners to write simulations using Ascape tools. The Ascape Community is not as large as other Agent Based tools ones. Miles promised in the mailing list, a couple of months ago, to release the 3.0 version of Ascape within the end of the year (The Actual version is 1.9.1).

9.3.2 Swarm

Swarm was developed at the Santa Fe' Institute in New Mexico by a number of researchers including Chris Langton, Roger Burkhart, Manor Askenazi, Nelson Minar, Glen Ropella, Alex Lancaster and Marcus Daniels and is currently being supported by the Swarm Development Group (SDG), a nonprofit membership organization (http://www.swarm.org). In a paper written by the original Swarm Development team, the authors define Swarm as a "multi-agent platform for the simulation of complex adaptive systems". In practical terms, Swarm is a collection of software libraries: it is a discrete event simulator, meaning that time is broken-up into small time steps. Swarm is a programming library that facilitates both research and teaching in the field of complex systems and agent-based modelling.

The software is provided under the open-source terms of the the the GNU Greater Public License, meaning that anybody who receives an executable version of the program also has a legal right to receive the source code itself. It is used and supported by a worldwide network of researchers of all fields, from social to hard sciences ones. The main features of Swarm are:

- **General work.** Swarm imposes only minimal substantive restrictions on the modelling enterprise. It is necessary to represent individual agents by objects, but everything else is open to customization. As a result, Swarm is a useful environment for modelling in fields ranging from political science and economics to ecology and biology;
- **Multilanguage.** Routines of Swarm libraries can be accessed from programs written in Objective-C (the hystorical Swarm Language) and Java. Development versions of Swarm can be accessed from C++, COM Object and javascript, as well;
- **Multiplatform.** The package can be compiled on almost any Unix-like systems and in Windows platforms;
- **Simulation Management** Swarm's activity libraries provides scheduling tools to support "bottom-up" modeling of complex systems and emergent phenomena. A range of possibilities can be explored, including
  1. totally decentralized behavior of agents who decide when and if to act;

---

[1] http://www.brook.edu/es/dynamics/models/ascape/README.html

2. nested ordering of agent actions;
3. conventional cellular automata;

- **Graphical Input/Output Interaction.** Graphs and displays can monitor a simulation as it proceeds. Code can be designed to allow user interaction with simulation (e.g. create new agents, reposition agents, change their attributes);
- **High Quality Support Libraries.** State of the art library support for generation of random numbers and simulation of draws from statistical distributions;
- **Large Variety of example Applications.** The Swarm Development Group (SDG) provides a package of supported example programs as well as a Swarm Tutorial[2]. In the SDG ftp-site [3] there are other simulations that are developed from other authors: there are implementations of many classic agent-based models, such as Sugarscape, Conway's Game of life, Langton's ant.

Swarm is a software package mainly addressed to individual agents' interacting simulation by an object-oriented approach. Tipically, a Swarm based code is structured as a set of precoded individual agents interacting together following a common schedule, each of them characterized by its own set of internal state variables;[4] all that is called "Model Swarm". Moreover, it is also characterized by a suitable interface object, which is called "Observer Swarm", which manages both the simulation and the user input/output. A typical screen shot of a simulation developed using Swarm is shown in the figure 9.2

Swarm and Internet

The source code, a good tutorial and several commented examples are available at the Swarm Web Site [5]. There's also a mailing list that is the primary and best way to help users and developers to communicate to solve problems, checking bugs. Furthermore, Paul Johnson[6] has developed a FAQ where most common problems encountered by beginner and advanced Swarm users are addressed.

9.3.3 Starlogo

StarLogo is a language with a set of tools that was designed especially for students (and with children in mind) to make it possible for them to easily

---

[2] you can find all the material in SDG web site http://www.swarm.org
[3] ftp://ftp.swarm.org/pub/swarm
[4] In particular, an agent, in its turn, may be a set of predefined sub-agents, and so on.
[5] http://www.swarm.org
[6] http://lark.cc.ku.edu/ pauljohn

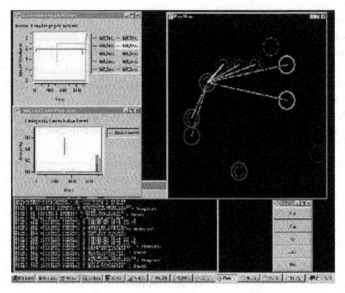

Fig. 9.2. Typical Swarm simulation

create simulations where emergent behaviour can be observed. This makes it possible to gain an intuitive understanding of a complex concept (emergent behaviours) through practical experience. Starlogo includes three types of objects:

1. **Turtles** are the objects that can move around in the world (like a virtual human in an VE)
2. **patches** is what the world consists of.
3. **Observers** are the interface towards the user and these can be used to create new turtles or monitor existing turtles and patches.

The world actually consists of a grid and it is in this grid that the patches and turtles exists. Every square in the grid contains one patch and some also contains one or more turtles. The turtles can interact with each other and with the patches but in both cases only if they are adjacent to each other. There is no perception past the adjacent squares. A Typical simulation model interface of Starlogo is show in the figure 9.3.

The figure show how StarLogo can be used to create a simulation of termites (the red/darker dots) picking up wood-chips (the yellow/brighter dots) and dropping them in piles. The rules guiding the termites are very simple and the pile collecting is an emergent behaviour.

Starlogo and Internet

Starlogo is one of the oldest Simulation toolkit and it has a wide variety of simulations of all fields written in these years and available in internet. The

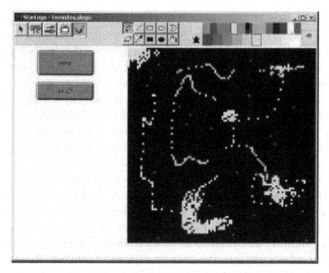

Fig. 9.3. Typical Starlogo simulation

official Web page[7] contains several tutorials for beginners and a list of all Starlogo commands along with a variety of examples (with full source code) anyone can download and study. Starlogo community has also various links over the world and an active mailing list with over 1000 active developer.

9.3.4 RePast

RePast (REcursive Porous Agent Simulation Toolkit) is an open-source enviroment for agent-based simulation using the Java programming language developed by Social Science Research Computing at the University of Chicago.

It provides an integrated library of classes for creating, running, displaying, and collecting data from an agent-based simulation.

According to Nick Collier Original RePast Paper "At its heart, RePast behaves as a discrete event simulator whose quantum unit of time is known as a tick. The tick exists only as a hook on which the execution of events can be hung, ordering the execution of the events relative to each other. For example, if event x is scheduled for tick 3, event y for tick 4, and event z for tick 5, then event y will execute after event x and before event z. If no events are scheduled for a certain tick, then it is as if that tick never occurred. Ticks are merely a way to order the execution of events relative to each other." Collier (2000) From the same paper "The typical RePast model contains a set of agents. These agents may or may not be homogenous, or perhaps these agents are arranged in a hierarchy (a firm and its employees, for example). Regardless of their composition, each agent has some behavior, the interactions of which

---

[7] http://education.mit.edu/starlogo/

a modeler is interested in exploring. In the early stage of the project, RePast wanted to be Modeled on Swarm Simulation toolkit, and so it can properly be termed a Swarm-little brother simulation framework." RePast was initially conceived of as a library of Java classes that would work together with and simplify the Swarm simulation framework. This initial conception was the result of University of Chicago researchers concerns with the complexity of both Swarm and Objective-C and our respect for the maturity and elegance of the Swarm API. This notion of RePast as an extension to Swarm was soon abandoned for a variety of reasons and made partially redundant with the release of Java Swarm (a Java layer running on top of the Swarm kernel and released by the Swarm Development Group). Prior to the release of Java Swarm, the original programmer at Chicago university had begun some exploration into developing an independent framework completely written in Java, but borrowing several of the key abstractions present in Swarm. Convinced of this frameworks viability and usefulness to University of Chicago researchers, the initial exploration grew into the current version of RePast. Today, RePast is used inside the University of Chicago and far beyond. Some features of rePast Package can be summarize in this way

Easy to learn it is fully written in Java language, so there's a short learning curve to use it. In Internet anyone can find a lot of exaustive tutorial of Java

Abstraction RePast abstracts most of the key elements of agent-based simulation and represents them as a Java class or classes. These classes cooperate to make a framework for creating agent-based simulations. They provide a ready-to-use class or classes for most of the common infrastructural abstractions of an agent-based simulation (e.g., scheduling, display, data collection, and so forth) and a variety of generic components for constructing representational elements. These generic components include such things as agent spaces (grids, torii, soups, etc.) and a few generic agent types.

Extendibility it is very easy to incorporate into RePast simulation external packages written in Java, for instance external graphical widgets (point graphs, 3 dimensional graphs) which are not programmed into the package kernel

Portability A simulation written in an environemt such as Windows or linux can be viewed without "re-compilation" into another platform

Good model Performance Even if the package contain a lot of classes and code, the simulation runs with a goood performances. RePast offers performance comparable to similar frameworks and will only get faster with the continual improvement of Java virtual machines. As a result of these goals, Repast is robust, extensible, and fairly easy to use, although the modeler must still learn Java.

Good support There's a growing community which use RePast for their work: there's a mailing list in which everyone with particular problem, can find the solution. Furthermore there's a frequent asked questions which help the beginner to solve initial problems.

Good documentation In these years people wrote a lot of good documented examples and a good initial tutorial for this environment

All components of RePast and the code libraries they are based on, is freely available under the GNU Library Public License. These components have been tested and implemented on a variety of different computer platforms and the code has been tested by numerous programmers. Also several of these programmers have published in their web site, a lot of models from different fields. A Typical simulation model interface of Repast is shown in the figure 9.4.

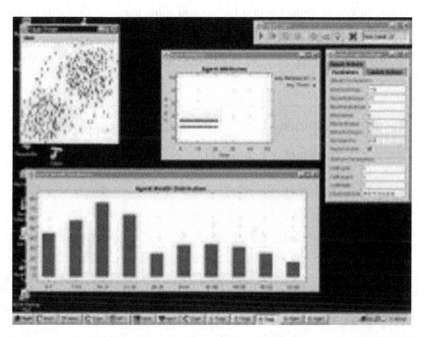

Fig. 9.4. Typical RePast simulation

RePast and Internet

The source code, a good tutorial and several commented examples are available at the RePast Web Site [8]. There's also a mailing list that is the primary and best way to repast users and developers to communicate to solve problems, checking bugs. Furthermore, Nick Collier has developed a FAQ where many the common problems a beginner RePast user can have, are solved.

---

[8] http://repast.sourceforge.net

9.3.5 Other tools

Browsing the net there are other tools written specifically for Agent Based simulation, such as Agentsheets[9], SDML[10], Mimose, Pangea, JAS but I chose not to present them because these tools are either commercial (such as agentsheet), not available for all platforms (Mimose), or it is in their first stage of development with a limited range of users,

9.4 Conclusion

In this review, I have attempted to give an general overview of some simulation packages. In this section, I would like to add some personal thoughts on each package. My comments are not intended as a criticism of any of the packages, but my aim in writing this article is to take into account the skills and abilities of the user and to suggest some simulation tools which could be of genuine use.

ABM	Learning Time	Presence on Internet	Compati- bility	Support	Installa- tion
Starlogo	Easy	Good	Good	Good	Easy
Ascape	Difficult	Not good	Good	Not good	Easy
RePast	Short	Good	Good	Good	Easy
Swarm	Difficult	Good	Good	Good	Difficult

Fig. 9.5. Table of comparison among ABM

Table 9.5 contains a comparison of different Agent Based Environments. Starlogo is a software tool-kit that was initially intended as an educational tool to introduce agent based modeling to high school students. The code is easy to read for a person who is not experienced with programming, and

---

[9] http://www.agentsheets.com/
[10] http://sdml.cfpm.org/

the documentation is written to facilitate easy entry into the process of programming, with a good number of example programs included in the basic download. These characteristics were extremely helpful for a large number people, in particular for those who had no experience in programming and allowed them to get started with any project with a minimum of time and effort (which has proven prohibitive with other systems). One of the most important factors in being able to implement any simulation was the ability to use existing code examples and programs as templates for my own. Starlogo seems flexible enough to allow easy addition and combination of code from different examples. The reference library of commands is fairly easy to use (again, from the point of view of a novice). Of the highest importance was the online support via the Starlogo mailing list, particularly since very often the actual developers were available to answer questions. The visual interface is another strong point; the graphic user interface is pretty much imbedded in the toolkit and requires very little adjustment to get something to look at. This is extremely important to a novice user, as it allows a very tangible feedback as to what the program is doing (as opposed to having to analyze a large series of numbers resulting from a batch run).

Some of the limitations of Starlogo as a research tool actually come from its strengths. There are topological limitations in so much as Starlogo pretty much assumes and enforces grid-space topography. This is useful for visualization of many projects, but may not be applicable to others. The plain-text nature of the code makes it easy for non-experienced programmers to get started, but apparently is not in conjunction with some programming conventions, making it difficult for experienced programmers to think in it. It is a fairly high-level programming language, in so much as certain basic processes are hidden that may be important in certain models (most notably scheduling). The imbedding of the graphic user interface (gui) precludes running a simulation in batch mode, and data acquisition/collection tools are limited. Both of these factors severely limit the scope of investigation offered by a Starlogo model.

I think that Starlogo is an excellent introduction to agent-based modeling, particularly to someone who is new to programming in general. It is easy to learn, and use, and programs can be written quickly. I think that it is also very useful as a prototype modeling system, a way to test out ideas, since a lot of the mechanics of the system are imbedded and a modeler can get right down to the essence of what the intended model should do. However, it is clearly limited in the type of systems it can model (essentially only grid-worlds) and its ability to handle a large number of runs and data.

Of the other packages reviewed, Swarm seems to be the most powerful and flexible and is a good tool for simulating social mechanisms. The main drawback with Swarm is its very steep learning curve. This is alleviated a little by the excellent support available (via the mailing lists) and a large collection of generally available simulation models. Whilst some of the other packages (e.g. RePast) aim to be specifically geared towards simulating social mechanisms, by providing built-in tools, there are no such facilities offered in Swarm. Thus, whilst Swarm would be able to implement such mechanisms it is likely to require some quite complicated programming.

Swarm's major downside (in my opinion, which is not necessarily the SDG opinion), is that we don't use garbage collection. Swarm has a well-tested multi-level discrete event scheduler, and relatively good software integration features (e.g. HDF5, R). Swarm also does not depend on any proprietary software, even in the case of Java. It is a complete tool, but it has the disavantage of not being easily installed on the Systems. While the other packages have their installer and anyone who has is somewhat familiar with computer can install it, Swarm has some particular "needed software" which have to be inside the computer. Swarm is released under the GNU General Public License (GPL)

Like Swarm, Ascape is a powerful tool with the advantage of being easier to learn than Swarm, though some Java programming experience is essential. Its current weakness stems from the lack of generally available simulation models. However, if the interest of potential usis in economic or market modelling then sufficient models exist and Ascape would be a very good tool. Similar comments could be made about RePast. The 'Swarm-like' approach adopted by Repast makes it a very powerful and flexible tool. In addition, since it uses a Java base (a common and powerful language) it should be relatively easy to find experienced programmers. However, since Swarm now also allows simulations to be developed in Java, this advantage has been somewhat eroded. In addition, Another drawback concerns the lack of generally available simulation examples. Ascape has strong features for doing lockstep/temporally-flat, CA-like simulations. It includes several fairly interesting/complex models. Ascape is freely usable (not sure about redistributable)

RePast has a nice looking control panel (it is derivated from Ascape environment). Repast is a respectable Java-based simulator implementation with interfaces highly similiar to many in Swarm. Repast has a more general scheduling engine than Ascape. It's a lot like Swarm's, but uses some different idioms. And of course, Repast and Ascape are 'pure java', which in principle means code portability is less of a problem, i.e. it is easier to make demos on the web, migrate code, and that sort of thing. Repast has a nice multilayer raster display, where you can turn different layers on and off. (In heatbugs, for example, you can see the bugs, the diffusion space, or both.) Repast has an 'advertising clause' type license

Concluding, in my opinion, for a novice programmer, the right tool is surely Starlogo, because the environment is very user-friendly, there are only a few commands to learn, and the result of the simulation is viewed immediately. So Starlogo environment is useful for "fast simulations"[11] and "easy simulation"

---

[11] I mean fast simulation all the simulation which do not need a lot of "time" to give results, even if those are "cpu intensive"

[12]. For an expert or an user who have to "build" different simulations, in which each agent is characterized by some variables and interact with other agents in the Space-world, the right tools can be one of the other. Anyone can build all kinds of simulation with Ascape, RePast or Swarm. Personally I prefer using Swarm Libraries to write my simulation because I am one of the old Swarm User, because it is the prototype of the environments and I know very well all the aspects of the package. Swarm has also the Visual Swarm Builder, a tool written from me, which permits to write visually simple simulations in Swarm using "drag & drop" mouse technique , Repast can be viewed in the same way as Swarm, with the advantage of the growing community and a large Web site in which there are a lot of commented code. Repast has also "Simbuilder" a rapid application development environment: the intention is that the user can visually construct a simulation out of component pieces and specify the behaviour of that simulation. Ascape is a good product, but, in my opinion, is less usable from the other two environments because when an user begins to write a simulation, he understand that the tools has been written from only a man, and the it's been written for "Prisoner's dilemma solving problem", in which there are only one main agent and it is very difficult to manage different agents in the space World. As already described on this paper, Swarm and RePast has examples in their web site of a large variety of fields: from economic to financial, to sociological, to antropological one.

References

1. Collier Nick, (2000) "RePast: An Extensible Framework for Agent Simulation"
2. Corazza M., Perrone A. (2001), "Simulating fractal financial markets", in F. Luna and A. Perrone (eds.), "Agent based methods in Economics and Finance: simulations in Swarm", Dordrecht and London, Kluwer Academic Press
3. Epstein J., Axtell R. (1996), "Growing artificial societies: social science from the bottom-up", Cambridge (Mass), MIT Press
4. Ferraris G. (1999), "Algoritmi genetici e *classifiers systems* in Swarm: due pacchetti per la realizzazione di modelli, due applicazioni economiche ed un. formichiere affamato", mimeo, Turin paper presented at IRES conference "Simulare con Agenti per l'analisi socieconomica" 28th October 1999
5. Foucart Louis "'A Small Multi Agent Systems Review" available at http://geneura.ugr.es/ louis/masReview.html
6. Holland, J. H., Reitman J. S. (1978), "Cognitive systems based on adaptive algorithms", in D.A. Waterman and F. Hayes-Roth (eds.), "Pattern-Directed Inference Systems", New York, Academic Press
7. Johnson P., Lancaster A. (1999),"Swarm user guide", mimeo
8. Le Baron B. (2000), "Agent based computational finance: suggested readings and early research", Journal of Economic Dynamics and Control, vol. 24, n. 5-7, pp. 679-702

---

[12] I mean "easy simulation" all the one characterized from agents running on a simple Cellular Automata world (ca-like)

9. Luna F., Stefansson B. (2000), "Economic simulation with Swarm: Agent based modelling and object oriented programming", Dordrecht and London, Kluwer Academic Press

10. Luna F., Perrone A. (2001), "Agent based methods in Economics and Finance: simulations in Swarm", Dordrecht and London, Kluwer Academic Press

11. Minar N., Burkhart R., Langton C., Askenazi M. (1996), "The SWARM simulation system: a tool-kit for building multi-agent simulations", working paper Santa Fe Institute

12. Parker, Miles (2000) "Ascape: Abstracting Complexity"

13. Parker, Miles (2001)"What is Ascape and why should you care" published in JASS and avaialble at http://jasss.soc.surrey.ac.uk/4/1/5.html

14. Parisi D. (2001), "Simulazioni: la realtà rifatta nel computer", Bologna, Il Mulino

15. Terna P. (2000)," Economic experiments with Swarm: a neural network approach to the self-development of consistency in agent behavior" in F. Luna and B. Stefansson (eds.), "Economic simulations in Swarm: agent based modelling and object oriented programming", Dordrecht and London, Kluwer Academic Press

16. Terna P. (2001), "Cognitive agents behaving in a simple stock market structure", in F. Luna and A. Perrone (eds.), "Agent based methods in Economics and Finance: simulations in Swarm", Dordrecht and London, Kluwer Academic Press

17. Tesfatsion L. (1996), "How economists can get alife", in B. Arthur, S. Durlauf and D. Lane (eds.), "The economy as an evolving complex system II", Santa Fe institute in the science of complexity, vol. XXVII, Addison-Wesley, Reading (MA), pp. 533-64

18. Tesfatsion L. (2001), "Introduction to the special issue on agent based computational Economics", Journal of Economic Dynamics and Control, vol. 25, n. 3-4, pp. 281-93

Econometrics and Time Series

10

Smooth Transition Models of Structural Change[*]

Bernhard Böhm

Institute of Econometrics, Operations Research and Systems Theory
University of Technology,
A-1040 Wien, Austria bboehm@e119ws1.tuwien.ac.at

10.1 Introduction

Monitoring the development of an economy requires a set of various techniques to define and measure the relevant variables and to specify and estimate important relationships among them. Despite the fact that many of the relationships are basically nonlinear it is customary to employ the linearity hypothesis for the parameters of most of the technical and behavioural economic equations. In many cases the simplicity and the approximation property deems sufficient. However, especially in the time series context the property of constancy of those linear parameters requires more attention pointing towards the need to improve specifications including nonlinear alternatives. Thus, after the early attempts to construct nonlinear models to explain business cycles and their abandonment in favour of the simpler linear dynamic models as proposed by Hicks and Samuelson which also could generate cyclical paths, interest in nonlinear (dynamic) models gains ground again. In particular the widespread and accelerating endowment with vast computing power and a few stimulating theoretical achievements facilitates consideration of this approach in model building.

This paper wants to contribute to recent developments in the specification of smooth transition regression (STR) models, a special class of nonlinear models that seems to offer widespread applicability. Some of the applications in the literature explore asymmetries of economic behaviour, like business cycle asymmetries (Anderson and Teräsvirta, 1992) or the relation between GNP and leading indicators (Granger, Teräsvirta, Anderson, 1993). Asymmetries and moving equilibria in unemployment rates have been investigated by Skalin and Teräsvirta (1998), the nonlinearity of Phillips curves by Eliasson (1999) and Heider (1999), applications to financial time series can be found in Mills (1999) and to money demand in Lütkepohl et al. (1997). Although estimation and testing procedures for such models are not straightforward

[*] Research assistance by M. Wagner and P.Pasching is gratefully acknowledged.

as the model selection process could become quite involved, empirical applications are on the increase. Particular emphasis receive models that do not involve many different variables but emphasise the autoregressive nature of time series. Smooth transition autoregressive (STAR) models are very reasonable alternatives when the nonlinear nature of time series plays an important role. Examples include applications to real exchange rates by Sarantis (1999) and Taylor et al. (2001) indicating their nonlinear dependence on the size of the deviation from purchasing power parity. STAR models are special cases of STR models when an influence of exogenous variables is not considered. Because of the increasing popularity of autoregressive models especially as vector autoregressive (VAR) models the exploration of modelling nonlinear single equations may constitute an interesting research objective. The extension to multivariate models has also begun to produce first theoretical results (Boswijk and Franses (1996), Anderson and Vahid (1998) and van Dijk (1999)) but very few applications exist.

The plan of this paper is to discuss first the basic model types and refer to possible extensions mentioned in the recent literature. The testing sequence required to arrive at well specified empirical models is briefly surveyed in order to facilitate interpretation of the empirical economic applications devoted to inflation and unemployment. German and British data are used to estimate basic relationships. Dynamic implications of these estimates are evaluated and followed by a conclusion on the relevance of smooth transition regression modelling.

10.2 Smooth Transition Models

10.2.1 Transition Functions

The development of smooth transition models can be seen as an attempt to extend the switching regression model to incorporate the transition between regimes in a continuous fashion. This idea was first developed by Bacon and Watts (1971), who gave their model the name 'smooth transition model'. Goldfeld and Quandt (1972) use a similar model. Assuming only for the moment x_t, y_t and z_t to be univariate time series the switching regression model can be written as

$$y_t = \alpha_1(1 - H(z_t)) + \alpha_2 H(z_t) + (\beta_1(1 - H(z_t)) + \beta_2 H(z_t))x_t +$$
$$+(1 - H(z_t))u_{1t} + H(z_t)u_{2t} \tag{10.1}$$

with the error u_{it} being $\mathrm{nid}(0, \sigma_i^2)$ and $H(z_t)$ the Heaviside function

$$H(z_t) = \begin{cases} 0 & z_t < c \\ 1 & z_t \geq c. \end{cases}$$

Because parameter estimation of the above model is complicated by the fact that $H(z_t)$ is not everywhere differentiable, the idea of Goldfeld and Quandt (1972) to approximate the Heaviside function by an everywhere differentiable function constituted the crucial step towards a smooth model. Their choice to employ the cumulative distribution function of a normal random variable with mean c and variance σ^2 (i.e. setting $\sigma_1^2 = \sigma_2^2$)

$$D(z_t) = \frac{1}{\sqrt{2\pi}\sigma} \int_{-\infty}^{z_t} \exp(-\frac{1}{2\sigma^2}(v-c)^2)dv \qquad (10.2)$$

was later (Maddala (1977)) replaced by the logistic function

$$D(z_t) = \frac{1}{1+\exp(\gamma_1 + \gamma_2 z_t)}. \qquad (10.3)$$

The use of the logistic function has been popularised by the work of Teräsvirta (1994, 1997, 1998) and Granger and Teräsvirta (1993). Earlier contributions to the literature are Tong (1983), Chan and Tong (1986), Luukkonen et al. (1988), and Tong (1990).

We follow the literature (e.g. [39]) in denoting a smooth transition regression model for the univariate time series y_t by

$$y_t = x_t\phi + (x_t\theta)G(\gamma, c, s_t) + u_t \qquad (10.4)$$

with $x_t = (1, \tilde{x}_t) = (1, x_{1t}, \ldots, x_{pt}) = (1, y_{t-1}, \ldots, y_{t-k}, z_{1t}, \ldots, z_{mt})$, and $p = k + m$. $\phi = (\phi_0, \ldots, \phi_p)'$, $\theta = (\theta_0, \ldots, \theta_p)'$ are the parameter vectors and $\{u_t\}$ is a sequence of iid errors. G is a bounded continuously differentiable function, which is normalised to lie in the interval $[0, 1]$. The variable s_t is the so called transition variable. This may be a stochastic variable, e.g. an element of the vector x_t or a linear combination of stochastic variables, but s_t may also be a deterministic variable like a time trend.

Among the most common choices for the transition function G we find

$$G_1(\gamma, c, s_t) = \frac{1}{1+\exp(-\gamma(s_t - c))}, \quad \gamma > 0 \qquad (10.5)$$

which leads to the logistic STR model of order one, abbreviated by LSTR1. The function G_1 is monotonically increasing in s_t. $\gamma > 0$ is an identifying restriction. The parameter γ indicates how fast the transition occurs from 0 to 1 as a function of s_t, c determines the location of the transition.

If we let $\gamma \to \infty$ the above model reduces to a 2-regime switching model. In this sense the STR models nest regime switching models as special (limiting) cases. An alternative non-monotonic transition behaviour can be captured by the LSTR2 model

$$G_2(\gamma, c, s_t) = \frac{1}{1+\exp(-\gamma(s_t - c_1)(s_t - c_2))}, \quad \gamma > 0, \quad c_1 \leq c_2. \qquad (10.6)$$

This formulation can be regarded as a generalisation of the exponential STR model (ESTR) (see Jansen and Teräsvirta (1996))

$$G_3(\gamma, c, s_t) = 1 - \exp(-\gamma(s_t - c)^2), \quad \gamma > 0. \tag{10.7}$$

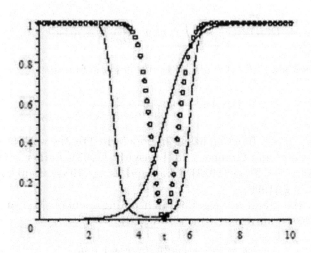

Fig. 10.1. LSTR1 Model (solid line) ($s_t = t, c = 5, \gamma = 2$)
LSTR2 Model (dashed line) ($s_t = t, \gamma = 2, c_1 = 3, c_2 = 6$)
ESTR Model (circles) ($s_t = t, c = 5, \gamma = 2$)

The STR approach allows for modelling a variety of different aspects, like e.g. the modelling of moving intercepts by including a term like $\mu G(\gamma, c, t)$ in the equation, where time is used as transition variable in any of the discussed transition functions. It is also possible to model changing seasonal patterns by incorporating terms like

$$\sum_{i=1}^{s-1} (\delta_{0i} + \delta_{1i} G_i(\gamma_i, c_i, t)) d_{it} \tag{10.8}$$

where d_{it} are seasonal dummies and s is the number of seasons. For an application of STR models including these features see e.g. Skalin and Teräsvirta (1998) and Franses, de Bruin and van Dijk (2000). In the following discussion we will ignore changing intercepts and seasons to keep notation tractable.

Rewriting the model in the form

$$y_t = x_t(\phi + \theta G(\gamma, c, s_t)) + u_t \tag{10.9}$$

one can directly see that STR models are locally linear in x_t. This parametrisation also relates to parameter constancy. If one takes $s_t = t$ the model can be interpreted as a model where the parameters are changing continuously over time. This might be a more realistic alternative to parameter constancy than an abrupt structural break. In the case of the rejection of parameter constancy against the alternative of an STR model, this parametric alternative can be estimated to provide further insights about where and in which way parameter constancy fails to hold.

In the process of specifying STR models one has to proceed in several steps. The first issue to be checked is whether it is necessary at all to leave the well understood class of linear statistical models. Thus, one has to test the hypothesis of linearity of the contemplated model against the alternative of nonlinearity in the form of a smooth transition regression model. In case the null of linearity is rejected an adequate specification of an STR model has to be found. This includes the decision about the functional form of the transition function G as well as the choice of the transition variable s_t and the dynamic specification of the nonlinear model. After this step the model is estimated using nonlinear least squares and must then be tested against several forms of misspecification (autoregressive errors, ARCH effects, normality of residuals) and in particular the absence of additive nonlinearity of the STR type and constancy of estimated parameters. The latter is usually tested against the alternative of smoothly changing parameters.

Because the model building procedure for STR models and their associated problems of statistical inference has been surveyed by Teräsvirta [39], we shall refer the reader to the literature for details and mention the specification and testing process only briefly to facilitate following the applications.

10.2.2 Specification and Testing

Empirical STR models then have the form $y_t = x_t(\phi + \theta G(\gamma, c, s_t)) + u_t$. The major advantage of using STR models is that any rejection of the constancy of parameters of the linear relationship is a rejection against a specific parametric alternative. If a rejection of constant parameters occurs, e.g. typically as the result of Chow tests or other instability tests, then the parameters of the alternative can be estimated. This helps in determining where and how parameter constancy of the linear model breaks down. In applications the transition function is usually renormalised to achieve that $G(\gamma, c, s_t) = 0$ for $\gamma = 0$. Still, there is an identification problem under the null hypothesis of linearity which is equivalent to γ being 0 in the above model. Under this hypothesis the parameters θ and c do not affect the value of the likelihood function, i.e. they are unidentified nuisance parameters. [2] This raises the problem of non-standard limiting distributions of the Likelihood ratio, Wald and Lagrange multiplier test statistics. Teräsvirta proposes to use a first order

---

[2] Such problems have first been discussed by Davies, [8] and [9]. See also [2] and [21].

Taylor series expansion of the transition function around $\gamma = 0$. The constant parameters of the approximation merge with the parameters of the model and permit the formulation of a null hypothesis for $\theta = 0$ against $\theta \neq 0$ and testing of the combined parameters by a Lagrange-multiplier (LM) test. The principle of this test is to evaluate the first order conditions derived from the likelihood optimisation problem at the estimated optimum of the likelihood function under the null hypothesis. If the restrictions are valid, the vector of first derivatives, the score, should be close to a vector of zeros. An important advantage of the LM testing principle is that no estimation of the model under the alternative is required.

The LM-test statistic (here for the case of testing an LSTR1 model asymptotically χ^2 distributed with $(p+1)$ degrees of freedom) is

$$\chi^2_{LM(p+1)} = \frac{1}{\tilde{\sigma}^2}(\sum_{t=1}^{T} \tilde{u}_t w_t)'(M_{11} - M_{10}M_{00}^{-1}M_{01})^{-1}(\sum_{t=1}^{T} \tilde{u}_t w_t) \tag{10.10}$$

with $M_{00} = \sum_{t=1}^{T} x_t x_t'$, $M_{01} = M_{10}' = \sum_{t=1}^{T} x_t w_t$, $M_{11} = \sum_{t=1}^{T} w_t w_t$, $w_t = x_t s_t$ and $\tilde{\sigma}^2 = \frac{1}{T}\sum \tilde{u}_t^2$, where $\tilde{u}_t$ are the residuals of the estimation under the null hypothesis. A multiplication of the type $x_t s_t$ means an element-wise multiplication of each column of x_t with s_t.

Applying a first order Taylor approximation to the LSTR1model one obtains

$$y_t = x_t \beta_0 + (x_t s_t)\beta_1 + u_t^* \tag{10.11}$$

where β_0 and β_1 are the parameter vectors of the transformed problem, which is linear in the parameters and $u_t^* = u_t$ plus a remainder term which is zero under the null. The null of linearity is equivalent to $H_0 : \beta_1 = 0$ in equation (10.11). This is now a standard testing situation with standard asymptotic behaviour of the test statistic. In case s_t is an element of x_t and a constant is included in the regression, only the non-constant regressors enter the computation of the LM test statistic (10.10). This stems from the fact that, using the notation $x_t = (1, \tilde{x}_t)$ again, the auxiliary model in this case is of the form

$$y_t = x_t \beta_0 + (\tilde{x}_t s_t)\beta_1 + u_t^*. \tag{10.12}$$

This can be easily seen by performing the Taylor approximation of the STR model and rearranging the regressors. Of course in this case the number of degrees of freedom is p instead of $p+1$.

There is one special case where the LM test has no power against the alternative. This case occurs when s_t is an element of x_t and $\theta = (\theta_0, 0 \ldots, 0)$, i.e. when the only nonlinear element is the constant. In this case β_1 in equation (10.11) is 0 also under the alternative. To remedy this problem Luukkonen, Saikkonen and Teräsvirta (1988) propose to use a third order Taylor approximation which leads to the auxiliary model of the form

$$y_t = x_t \beta_0 + (\tilde{x}_t s_t)\beta_1 + (\tilde{x}_t s_t^2)\beta_2 + (\tilde{x}_t s_t^3)\beta_3 + u_t^*. \tag{10.13}$$

The null of linearity is now given by $H_0 : \beta_1 = \beta_2 = \beta_3 = 0$ and w_t is now given by $(\tilde{x}_t s_t, \tilde{x}_t s_t^2, \tilde{x}_t s_t^3)$, the degrees of freedom are $3p$.[3] For a detailed discussion and the applicability to other functional forms of the transition function see [26]. One can also use an F-test which has higher power if the number of variables is large and the sample relatively small. Furthermore, any subset of parameters in the linear model may be tested for constancy using the same ideas as above. This is useful if one has to find out which parameters are not constant when the model has unstable parameters. Also other tests like Ramsey's RESET test would be appropriate to discover instabilities or non-linearities, however against a non-specific alternative. As linearity is tested against a specific STR alternative, any of the variables considered as candidates for the transition variable can be tested successively. A rejection of linearity implies therefore that the particular variable may be used as transition variable. However, performing several of these tests which are not independent implies that the true size of the multiple linearity test is not known. As a model specification tool in the case if more than one of the variables qualifies it has been recommended (cf. [19], [39], [45]) to select that one which yields the strongest rejection (in terms of probability values) of the linear hypothesis. This would correspond to the selection of that auxiliary model which minimises the residual variance, assuming it approximates the true nonlinear alternative to a certain degree of accuracy. Thus, the linearity test also provides a clue to the choice of the transition variable.

The decision which of the various types of transition functions to choose is again based on the auxiliary regression (10.13). Using a Taylor approximation of third order (considered appropriate for most practical purposes) one can start with testing $\beta_3 = 0$ (denoted F4 in the tables below) and continue under this condition to test for $\beta_2 = 0$ (F3), and finally under both previous conditions to test for $\beta_1 = 0$ (F2). Teräsvirta [39] suggests to select an LSTR2 or ESTR model if the second hypothesis is rejected strongest. Otherwise an LSTR1 model will do. Alternatively on may proceed with a computationally more intensive search strategy. For given values of the parameters in the transition function the estimation problem of the remaining model parameters is a linear one. A search grid could be used to find the set of transition parameters giving the best fit. This is also the criterion for the choice among models. LSTR1 and LSTR2 can be tested against each other. A choice between LSTR2 and ESTR can be based on a test for $c_1 = c_2$ within the LSTR2 model. Still better would be to perform nonlinear least squares estimation of the different models.

Having identified the transition variable and the functional form the parameter structure of the model must be specified. Using the estimated linear model as a starting point one re-estimates the parameters under the condition

---

[3] When performing this test the existence of the required moments has to be guaranteed. Which moments are required can be seen by looking at the information matrix.

of known parameters of the transition function, c and γ. These estimates provide starting values for the actual estimation which is carried out by nonlinear least squares. If normality of errors is assumed this is equivalent to Maximum Likelihood. Depending on the choice of the model different assumptions concerning the existence of higher moments and cross-moments of the regressors have to be made (cf. [39]).

The estimated STR model needs to be checked for the validity of the assumptions under which estimation has been carried out. Typically, normal errors, no autocorrelation of residuals, and parameter constancy are assumed. In addition to those assumptions any remaining non-linearities should also be checked. Absence of autocorrelation is tested by an LM test in its χ^2- or F- version. The LM-test statistic (10.10) as given above, with q degrees of freedom testing q-th order autocorrelation, is evaluated here by using the definitions $w_t = (\tilde{u}_{t-1}, ..., \tilde{u}_{t-q})$, the lagged residuals $\tilde{u}_{t-j}$ estimated under the null-hypothesis, and the vector $x_t = \tilde{z}_t$ representing here the gradient vector of the model with respect to the model parameters.

Constancy of parameters must be tested for an estimated STR model because it is a key assumption. To test the null that all parameters are constant against the alternative that the parameters are themselves moving over time according to some smooth transition model a LM test is used with

$$\hat{z}_t = \{(x_t^0 t), (x_t^1 G(\gamma, c, s_t)), \frac{\partial G}{\partial \gamma}(t), \frac{\partial G}{\partial c}(t)\} \qquad (10.14)$$

and

$$w_t = (x_t^0 t, x_t^0 t^2, x_t^0 t^3, x_t^1 t G(.), x_t^1 t^2 G(.), x_t^1 t^3 G(.)) \qquad (10.15)$$

where x_t^0 are those variables actually appearing in the linear part of the estimated STR model and x_t^1 are those that appear in the non-linear one (cf. [24]). It is a straightforward exercise to restrict parameter non-constancy to a subset of the parameters.

To complete the specification analysis of an STR model tests for ARCH effects and normality of the errors can be carried out as in linear models.

Finally, one has to analyse whether the estimated STR model is indeed capturing all nonlinearities present in the data taking as alternative additional nonlinearity of the STR type. Eitrheim and Teräsvirta (1996) define an additive STR model as

$$y_t = x_t \phi + (x_t \theta) G(\gamma_1, c_1, s_t) + (x_t \psi) H(\gamma_2, c_2, r_t) + u_t \qquad (10.16)$$

with H being a transition function normalised to $H(0, c_2, r_t) = 0$. r_t is the transition variable in the additional STR part of the model. If the estimated STR model represents a correct representation of the DGP, then H is equal to 0, and therefore also $\gamma_2 = 0$. In testing for additive nonlinearity an LM test for $H_0 : \gamma_2 = 0$ is set up. This is again a testing problem with nuisance parameters present under the alternative. The transition function H is developed in a third order Taylor series around the point $\gamma_2 = 0$, leading to the following auxiliary regression model

$$y_t = x_t\phi + (x_t\theta)G(\gamma_1, c_1, s_t) + (\tilde{x}_t r_t)\beta_1 + (\tilde{x}_t r_t^2)\beta_2 + (\tilde{x}_t r_t^3)\beta_3 + u_t^*. \quad (10.17)$$

The null of no additive STR nonlinearity is given by $H_0 : \beta_1 = \beta_2 = \beta_3 = 0$. Both, an F and a χ^2 version of the test can be formulated as above. The test statistic is computed as for the test of no error autocorrelation with $w_t = (\tilde{x}_t r_t, \tilde{x}_t r_t^2, \tilde{x}_t r_t^3)$ and $\hat{z}_t$ exactly as before. For the F version of the test the degrees of freedom are given by $3p$ and $T - 4p - 1$, for the χ^2 version the degree of freedom is p. If the null hypothesis is rejected another nonlinear component must be added whose properties are determined by following the modeling cycle as for the first nonlinear component.

10.3 Applications to Inflation and Unemployment

While for many economic applications a linear approximation might work sufficiently well, the Phillips curve ([31]) is one of those economic relationships where the actual form of the function has important implications for economic policy. Risk neutral policy makers will be averse to experiment with the unemployment – inflation trade-off when the Phillips curve is convex but not in the case of concavity. Thus, the question of linearity of the Phillips relation (in which case policy makers would be indifferent) as opposed to a non-linear curve and, in addition, the type of non-linearity, have been issues addressed in recent empirical work.

In most macroeconomic inflation theories the Phillips curve is taken as a component of the aggregate supply side in combination with Okun's law. Together with an equation representing the monetary system (e.g. the quantity equation in monetarist theory or a similar relationship) and an expectations hypothesis short and long run equilibrium growth, inflation and unemployment can be determined. In order to be able to identify and estimate the parameters of the Phillips curve we have to assume that the demand side contains variables which do not enter the Phillips curve. Since this is guaranteed even in the simplest models of the inflation process incorporating the growth of money and capacity output on the demand side, estimation of a Phillips curve may proceed representing one component of a short run equilibrium process.

Analytical works on inflation ([10], [15], [16], [18] among others) suggest the following empirical specification of a Phillips curve. The inflation rate of consumer prices is assumed to depend on the unemployment gap $(U_t - U^*)$ and the change in the unemployment rate (ΔU_t), the expected inflation rate, modelled by a backward looking component, a lag polynomial $a(L)$ in the inflation rate, and a forward looking component (π_t^f).[4] Supply effects will

---

[4] Forward looking price expectations are modelled here by the difference between the nominal and the real rate of interest according to Fisher's formula using the

basically come from the energy price side (π_t^o), tax policy (D_t) and random productivity shocks. In view of the modification of the original Phillips curve by Samuelson and Solow (1960) labour productivity growth (λ_t) could also be explicitly considered in the equation. This would leave the error term (ε_t) to take care of other sorts of supply shocks. Thus the model in its linear specification including several lagged effects looks as follows:

$$\pi_t = a(L)\pi_t + b(L)\pi_t^f + c(U_t - U^*) + d(\Delta U_t) + f(L)\pi_t^o + gD_t + h(L)\lambda_t + \varepsilon_t. \tag{10.18}$$

If the combined effects of backward and forward looking inflation expectations add up to unity, i.e. $a(L) + b(L) = 1$, we might call this an equilibrium of expectations. Then a natural rate of unemployment will be determined by

$$U^* = U_t + (d/c)(\Delta U_t) + 1/c(f(L)\pi_t^o + gD_t + h(L)\lambda_t + \varepsilon_t) \tag{10.19}$$

which reduces to a constant rate around an error term for unchanged energy prices, tax policy, and productivity growth once the unemployment rate is stabilised $(\Delta U_t = 0)$. Otherwise the natural rate will be a linear combination of the current and lagged unemployment rate, which implies that U_t approaches the steady state rate U^*.

In many studies (e.g. [22]) the unemployment gap is replaced by the output gap in view of Okun's law. This assumes a stable relationship between the unemployment gap and the output gap. Before we look at a Phillips curve for Germany we should first inspect the stability of Okun's law and base the choice of the gap variable on this outcome. This will provide a first attempt to enquire about a possible nonlinearity of this relation.

10.3.1 Okun's Law for West Germany

We shall assume the simple linear relationship between the unemployment gap, defined as the quartely difference of the West German unemployment rate, $DU4_t = U_t - U_{t-4}$, and the output gap, taken as the difference between the annual growth rate of quarterly GDP, y_t, and the rate of unobserved trend output, y^*. Okun's law then is given by the relation

$$DU4_t = U_t - U_{t-4} = -b(y_t - y^*) + e_t \tag{10.20}$$

with e_t the error term with usual properties. An estimate of unobservable trend output can be obtained from the estimate of b. We use an autoregressive specification for the linear model $A(L)DU_t = a + by_t + e_t$. Let $D(DU4_t)$ denote the first difference of $DU4_t$, the least squares estimates with 127 observations (1969:1 2000:3) are

---

long term government bond yield and a proxy variable for the real rate (cf. Böhm (2001)).

$$D(DU4_t) = - \underset{(0.0296)}{0.1705} \ DU4_{t-1} + \underset{(0.0571)}{0.5354} \ D(DU4_{t-1}) +$$

$$+ \underset{(0.0708)}{0.1516} \ D(DU4_{t-3}) - \underset{(0.0655)}{0.1489} \ D(DU4_{t-4}) - \underset{(0.0087)}{0.0470} \ y_t +$$

$$+ \underset{(0.0316)}{0.1505} - \underset{(0.1356)}{0.6502} \ D1 + \underset{(0.0992)}{0.5856} \ D2$$

$$(10.21)$$

$$R^2 = 0.668, \sigma = 0.183079, SSR = 3.9886, F = 34.144$$

The Dummies D1 and D2 reflect the effects of German unification. Without their inclusion residuals of this equation exhibit strong ARCH effects up to fourth order. Otherwise all diagnostic checks are passed. In order to test for linearity four variables qualify as candidates for the transition variable as can be seen from the following table (with p-values in brackets).

LM tests of linearity	t	$D(DU4(-1))$
F (linearity)	1.9288 (0.0213)	2.3146 (0.0045)
F4 (3rd against 4th)	1.3673 (0.2350)	2.2660 (0.0430)
F3 (2nd against 3rd)	1.6103 (0.1513)	2.3675 (0.0347)
F2 (linearity against 2nd)	2.6341 (0.0199)	1.8650 (0.0929)

LM tests of linearity	y_t	$DU4(-1)$
F (linearity)	2.0206 (0.0148)	1.8044 (0.0346)
F4 (3rd against 4th)	1.3217 (0.2544)	1.2407 (0.2920)
F3 (2nd against 3rd)	1.0192 (0.4169)	1.7486 (0.1168)
F2 (linearity against 2nd)	3.6332 (0.0025)	2.2776 (0.0411)

After several failed attempts due to non-converging parameter estimates the trend variable was chosen as transition variable in a LSTR1 transition function which produced the following equation:

$$D(DU4_t) = -\ 0.3232\ DU4_{t-1} +\ 0.5497\ D(DU4_{t-1}) +$$
$$(0.0514) \qquad\qquad (0.0541)$$

$$+\ 0.1683\ D(DU4_{t-3}) -\ 0.1244\ D(DU4_{t-4}) -\ 0.0686\ y_t +$$
$$(0.0665) \qquad\qquad (0.0628) \qquad\qquad (0.0095)$$

$$+\ 0.3572\ -\ 0.5922\ D1 +\ 0.6481\ D2 +$$
$$(0.0550)\quad (0.1289) \qquad (0.0940)$$

$$\left\{ -\ 0.1899\ +\ 0.1428\ DU4_{t-1} \right\} *$$
$$\quad\ (0.0431)\qquad (0.0475)$$

$$* \left\{ 1 + \exp\left[-414.2162\,(t\ -\ 34.0213\,)/\sigma_{(t)} \right] \right\}^{-1}$$
$$(0.1310) \qquad\quad (0.2574)$$

$$(10.22)$$

$$T = 127, R^2 = 0.719, S.E. = 0.1711, \hat{\sigma}_{nl}/\hat{\sigma}_{lin} = 0.873$$
$$F_{AR(1)} = 0.209(0.648), F_{AR(4)} = 0.168(0.954), F_{AR(8)} = 0.142(0.997)$$
$$LM_{ARCH(1)} = 1.689(0.196), LM_{ARCH(4)} = 0.553(0.697),$$
$$LM_{ARCH(8)} = 0.377(0.931)$$

Diagnostic checking gave no reason to reject these estimates which turn out to be close to a switching model due to the large estimate of γ. The switch occurs after the 34th quarter which corresponds to 1977:2. This implies that after this period the unobserved trend growth must have changed. Calculations show that the implied long run estimate of y^* before the break was approximately 5.2% and only 2.4% after the break. This compares well with the overall mean of GDP growth of 2.6%. We conclude that the unemployment gap would be the appropriate variable in a Phillips curve for Germany.

10.3.2 A Phillips Curve for West Germany

We now investigate whether non-linearities are also present in a proper specification of a Phillips curve for (West-) Germany. Quarterly data taken from OECD statistics from 1969:1 until 1998:4 are used. Starting from a linear specification containing inflation lag effects of up to 8th order, the usual oil price impact over about one year and a significant stimulus of the unemployment rate, the regression diagnostics indicates residual autocorrelation as well as significant ARCH effects and heteroscedasticity. Surprisingly the weak test for unspecified non-linearities by Ramsey does not suggest improvements achievable by non-linear specification. On the other hand the CUSUM of squares test hits the lower and upper bound, pointing towards structural breaks in the early eighties and nineties. Thus, the next step is to test whether such a linear specification is rejected in favour of a STR formulation. Table

1 reports a selection of the estimated F-values and corresponding p-values obtained from LM tests of linearity.

The minimal p-value is found for π_{t-1} rejecting linearity against 2nd-order nonlinearity. This suggests to employ this variable as transition variable and to select a logistic model (LSTR1) because second against third order is not rejected. After a few attempts to select the appropriate variables for the nonlinear part one ends up with the following estimation result:

$$
\begin{aligned}
\Delta\pi_t = \ & 0.7205 \ - \ 0.1803 \ \pi_{t-1} \ - \ 0.2536 \ \pi_{t-2} \ + \ 0.2241 \ \pi_{t-3} \ - \\
& (0.1905) \quad (0.0785) \qquad\quad (0.0820) \qquad\quad (0.0841) \\[4pt]
& - \ 0.2845 \ \pi_{t-4} \ + \ 0.2191 \ \pi_{t-5} \ + \ 0.2434 \ \pi_t^o \ - \ 0.1471 \ \pi_{t-1}^o \ - \\
& \quad (0.0852) \qquad\quad (0.0525) \qquad\quad (0.0259) \qquad (0.0274) \\[4pt]
& - \ 0.0355 \ U_t \ + \ (\ 0.2474 \ \pi_t^f \ - \ 0.1381 \ \pi_t^o \ + \ 0.1183 \ \pi_{t-1}^o \ - \\
& \quad (0.0201) \qquad\quad (0.0625) \qquad\quad (0.0302) \qquad\quad (0.0292) \\[4pt]
& - \ 0.2261 \ U_t \ - \ 0.0884 \ \lambda_t) \ * \\
& \quad (0.0474) \qquad\quad (0.0276) \\[8pt]
& * \left\{ 1 + \exp\left[- \ 12.3673 \ (\pi_{t-1} \ - \ 3.5590 \ )/\sigma_{\pi_{t-1}} \right] \right\}^{-1} \\
& \qquad\qquad (6.0555) \qquad\quad (0.2061)
\end{aligned}
$$

$$(10.23)$$

$$T = 114, R^2 = 0.789, S.E. = 0.4388, \hat{\sigma}_{nl}/\hat{\sigma}_{lin} = 0.663$$
$$F_{AR(1)} = 2.166(0.144), F_{AR(4)} = 2.068(0.091), F_{AR(8)} = 1.487(0.173)$$
$$LM_{ARCH(1)} = 0.018(0.893), LM_{ARCH(4)} = 1.381(0.245),$$
$$LM_{ARCH(8)} = 0.907(0.514)$$

LM-tests of no error autocorrelation as well as tests of no ARCH are satisfactory.[5] The test of normality of residuals according to the Lomnicki-Jarque-Bera statistic yields 0.202 (p-value 0.904). Constancy of parameters against nonlinearity of the LSTR1 or LSTR2 variant can also not be rejected.

---

[5] p-values in brackets.

Table 10.1. Tests of no additive non-linearity not ignoring linear restrictions

LM tests of linearity		trend	π_{t-1}	π_{t-2}	π_t^o
	df	F-stat (p-val)	F-stat (p-val)	F-stat (p-val)	F-stat (p-val)
F (linearity)	37,65	3.8571 (0.0000)	2.5258 (0.0005)	1.6520 (0.0381)	2.0235 (0.0064)
F4	12,65	5.1466 (0.0000)	1.9988 (0.0384)	0.4076 (0.9556)	0.8499 (0.6000)
F3	12,77	1.4469 (0.1635)	1.3838 (0.1920)	1.9132 (0.0455)	2.3747 (0.0116)
F2	12,89	2.5004 (0.0071)	3.4472 (0.0003)	2.8932 (0.0020)	2.6520 (0.0044)

The LM(j) tests of parameter constancy reject non-constancy in the form of lower orders, but a small subset indicates a slight tendency of nonmonotonic and nonsymmetric parameter non-constancy (cf. table 2).

Table 10.2. LM-tests of parameter constancy (p-values)

	(1)	(2)	(3)	(4)
LM(3)	0.027241	0.040366	0.608199	0.028400
LM(2)	0.129543	0.292729	0.599590	0.210163
LM(1)	0.862656	0.750098	0.541445	0.373334

(1) Parameters of intercept, $\pi_{t-1}, \pi_{t-2}, \pi_{t-3}, \pi_{t-4}, \pi_{t-5}$ are constant
(2) Parameters of $u_t, \pi_t^o, \pi_{t-1}^o$ in the linear part are constant
(3) Parameters of π_t^f, U_t in the non-linear part are constant
(4) Parameters of $\pi_t^o, \pi_{t-1}^o, \lambda_t$ in the non-linear part are constant

It is interesting to note that none of the lagged inflation rates enters the nonlinear part, except as transition variable itself. This implies that effects of backward looking inflation expectations are not varying much. On the other hand the forward looking component only enters the nonlinear part and thus is operative only in the regime defined by an excess of lagged inflation above a value of 3.5 percent. Taking into account only the linear part we detect a long run Phillips curve for Germany. The long run coefficient of the unemployment rate is approximately -0.05 in this case which means that one percentage point increase in the rate of unemployment will ceteris paribus lead to a fall in the inflation rate by 0.05 percentage points, i.e. very little. The estimated short run trade-off is even a bit smaller. Adding now for the nonlinear effect we observe two facts. First, the (negative) sum of the lagged inflation parameters is just matched by the opposite effect of forward looking expectations, implying $\Delta\pi_t = 0$ in the long run. Second, the combined effects of the energy price changes in the long run are around 0.076 and that of productivity growth -0.088. Taking the combined parameters of the unemployment rate (-0.262) yields an equation for the natural unemployment rate of

$$U^* = 2.77 + 0.292\pi^o - 0.328\lambda \qquad (10.24)$$

which, evaluated at average values for energy price changes and productivity growth yields approximately a value of 3.23. We may conclude that this long run rate lies approximately around a value of 3 percent with the energy price changes roughly offsetting the productivity gains. When plotting the transition function against time together with the unemployment rate, it is remarkable that the transition function becomes active in exactly those periods where the largest increases in unemployment are observed. Furthermore, the economic changes during the unification period seem to have been captured by the nonlinear model rather well, as there was no need to introduce specific dummy variables for this purpose.

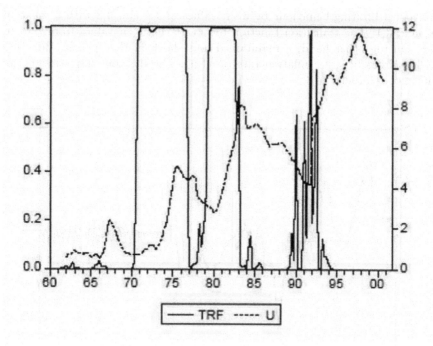

Fig. 10.2. Transition function and German unemployment rate plotted against time

This exercise shows how extreme the changes between different regimes can be. The Phillips curve changes its slope from very flat in either short or even long run in periods of relatively low inflation to a little steeper in the short run when inflation steps up to the extreme case of vanishing in the long run altogether.

A little more insight is provided by an inspection of the dynamic behaviour of the non-linear model. So far we have only looked at extreme values of the transition function. Inspecting the skeleton model obtained by stripping the fitted model of the innovations we are able to get information about the adjustment processes triggered by changes in the determining variables. We have four exogenous variables in the model $(\pi_t^f, \pi_t^o, U_t, \lambda_t)$. Fixing these variables for the simulation period 2001:1 to 2050:4 at values (1, 1, 10, 2.5) and simulating the model generates the base solution with a long run value of the inflation rate of 1.68. An additive unit shock in the first quarter of 2001 produces a damped oscillating reaction in the inflation rate, reaching the base level again after some five years. Because the shock is too small to influence the transition function the resulting effect derives basically from the linear part of the model. This situation changes, however, when the unit shock is maintained, corresponding to an inflation acceleration of one percent due to sources not captured in the model. The resulting inflation path then shows a cyclical limiting behaviour between values of 1.6 and 3.8 with a mean of around 3.1. The transition function too cycles between values from zero to 0.8. On the other hand, a maintained unit shock in the opposite direction yields a steady state inflation rate of −1.95. The dynamic adjustments are thus not symmetric.

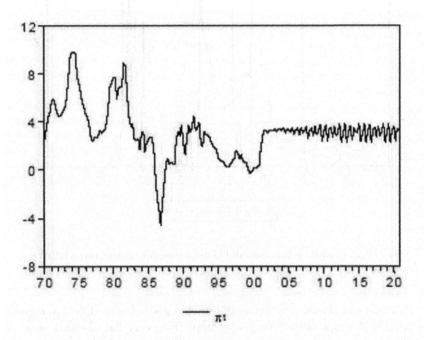

Fig. 10.3. Simulated inflation rate resulting from a maintained additive positive unit shock over the period (2001:1-2020:4)

A similar limit cycle obtains when the model is simulated with a maintained energy price inflation rate of around 10 percent. The bifurcation value of energy inflation seems to lie between 8 and 9 percent. Short energy price hikes (i.e. rates of 20% for one to four quarters) yield damped oscillations and more asymmetric responses the more the transition function becomes effective.

10.3.3 A Phillips Curve for the United Kingdom

Another application of a smooth transition regression is the British Phillips curve as given in equation (10.25). It features two transition functions and, thus, altogether four extreme regimes.

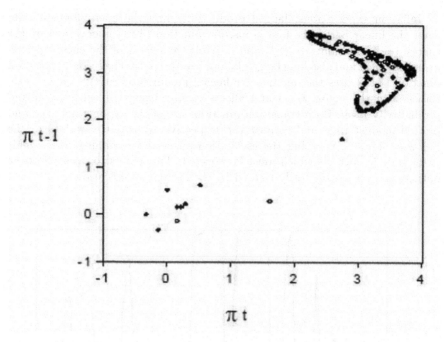

Fig. 10.4. Directed scatter diagram of the simulated inflation rate

$$\Delta\pi_t = \begin{array}{l} 1.5047 \\ (0.3758) \end{array} - \begin{array}{l} 0.7881 \\ (0.0731) \end{array} \pi_{t-4} - \begin{array}{l} 0.4106 \\ (0.0527) \end{array} \pi_{t-5} + \begin{array}{l} 0.3477 \\ (0.0681) \end{array} \pi_t^f -$$

$$\begin{array}{l} - \; 0.1889 \\ (0.0508) \end{array} \lambda_{t-4} - \begin{array}{l} 0.0930 \\ (0.0325) \end{array} U_t + [\begin{array}{l} 0.2622 \\ (0.0436) \end{array} \pi_t^o - \begin{array}{l} 0.1568 \\ (0.0486) \end{array} \pi_{t-1}^o -$$

$$\begin{array}{l} - \; 0.4980 \\ (0.1662) \end{array} U_t + \begin{array}{l} 0.4493 \\ (0.1700) \end{array} U_{t-4} - \begin{array}{l} 0.3607 \\ (0.0694) \end{array} \lambda_{t-8}] *$$

$$* \left\{ 1 - \exp\left[- \begin{array}{l} 0.2456 \\ (0.1029) \end{array} (\pi_t^o - \begin{array}{l} 1.4542 \\ (0.9841) \end{array})/\sigma_{\pi_t^o} \right]^2 \right\} +$$

$$[- \begin{array}{l} 0.3123 \\ (0.0477) \end{array} \pi_{t-1} + \begin{array}{l} 0.3539 \\ (0.0547) \end{array} \pi_{t-4} - \begin{array}{l} 0.3432 \\ (0.0866) \end{array} \lambda_t + \begin{array}{l} 0.2333 \\ (0.0784) \end{array} \lambda_{t-1}] *$$

$$* \left\{ 1 - \exp\left[- \begin{array}{l} 1.8272 \\ (0.7552) \end{array} (\lambda_{t-1} - \begin{array}{l} 1.5304 \\ (0.1620) \end{array})/\sigma_{\lambda_{t-1}} \right]^2 \right\}$$

$$(10.25)$$

$$T = 130, R^2 = 0.783, S.E. = 0.8454, \hat{\sigma}_{nl}/\hat{\sigma}_{lin} = 0.435$$

Inflation expectations, lagged productivity and the unemployment rate enter the linear part. The first transition function (TF1) is governed by the energy price inflation rate (π_t^o) and controls the effects of the energy prices, unemployment and also long lagged labour productivity. The second transition function (TF2) uses the one quarter lagged labour productivity growth rate (labelled in the figure λ_{t-1}) and affects mainly backward expectation and productivity itself. Both transition functions are of the exponential type and plotted against time and against the respective transition variables in the following figures. Assessing the estimation properties normality of residuals is the only hypothesis which must be rejected. This points towards effects of outliers not yet satisfactorily treated by the current specification.

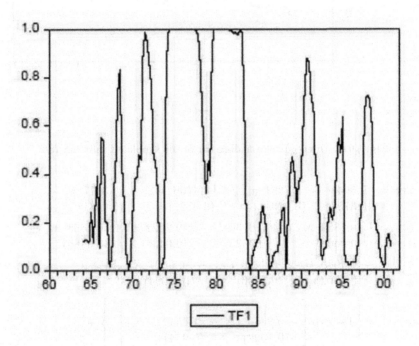

Fig. 10.5. UK, Transition function 1 against time

For the interpretation of the estimation result it is interesting to note that the estimated expectations parameters roughly imply a unit root of inflation in the long run under all regimes. There exists, thus, only a short run trade-off between inflation and unemployment. The sum of the lagged inflation coefficients plus the effect of forward expectations in the linear part is –0.0298, very close to zero in view of the variance of this sum. Adding the lagged inflation coefficients of the second non-linear part would yield 0.0118, implying an unstable model for a value of the transition function equal unity (which

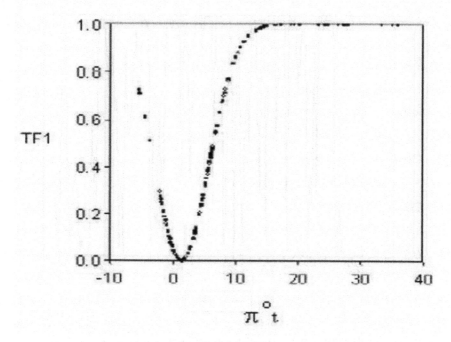

Fig. 10.6. UK, Transition function 1 against transition variable

occurs rarely). From the simulated impulse response function of the skeleton model the closeness to a unit root model is apparent. If forward inflation expectations are treated as exogenous at the historical average of 4.2% the impulse response function decays much faster.

Considering the long run implications by taking average values for the energy inflation rate and productivity growth the implied "natural" rates of unemployment (given the unit root in the inflation rate) vary between 10 and 20 percent for all combination of transition functions (i.e. four extreme regimes). The linear model implies a long run unemployment rate of 12.3% which varies with productivity growth with a long run multiplier of -2. Evaluated at sample means the full model implies a long run unemployment rate of 16.6%. While these numbers might not necessarily be accurate to the digit, they very well show what a range of variability of an implied natural rate one can get and how often these implied values might change as one can see from the number of transitions between states.

10.4 Conclusions

Even our small set of applications of smooth transition regression models to problems of unemployment and inflation provides sufficient evidence on the

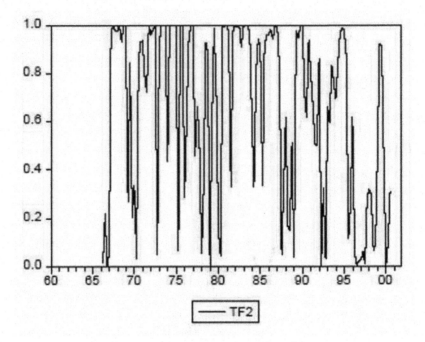

Fig. 10.7. UK, Transition function 2 against time

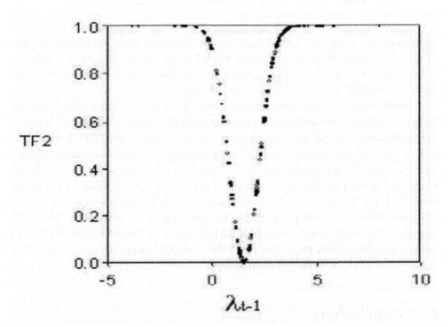

Fig. 10.8. UK, Transition function 2 against transition variable

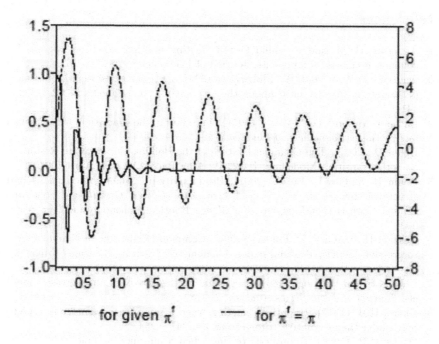

Fig. 10.9. UK, Impulse response functions of one unit shock for given forward expectations and for forward expectations equilibrium (2001:1 – 2050:4)

rich diversity of dynamic implications of such models. It is especially this evidence which has to be closely inspected in order to determine whether a particular specification makes sense in economic as well as in econometric terms. Due to the large number of possible nonlinear models which can be applied to a data set the simpler specifications capable to approximate fairly well relatively complex dynamic developments should be preferred. There are many events in the economy which are associated with changes in its structure.[6] Econometric models are trying to capture such changes with a wealth of different approaches, starting with the simplest dummy variable approach, and extending to very refined variable parameter models. Among the parametric approaches the smooth transition idea seems to be at the same time simple enough to yield to tractable estimation and testing procedures and sufficiently general to integrate abrupt switching behaviour and, however fast, parameter changes identifying variations in the underlying economic structure.

---

[6] The potential of STR models to represent structural change in a multi-sectoral framework has been indicated in Böhm and Punzo (2001).

References

1. Anderson H. M. and F. Vahid (1998), Testing multiple equation systems for common nonlinear components, Journal of Econometrics, 84, 1 - 36.
2. Andrews D.W.K. and W. Ploberger (1994), Optimal tests when a nuisance parameter is present only under the alternative, *Econometrica*, **62**, 1383 – 1414.
3. Bacon D.W. and D.G. Watts (1971), Estimating the transition between two intersecting straight lines, *Biometrika*, **58**, 525 – 534.
4. Böhm B. (2001), The Changing Nature of the Phillips Curve. Some Results in Smooth Transition Modelling, IDEE Working Paper, Vienna, April 2001
5. Böhm B. and L. F. Punzo (2001), Productivity-investment fluctuations and structural change, ch. 3 in L. F. Punzo (ed.) *Cycles, Growth and Structural Change. Theories and emprirical evidence*, Routledge, London and New York, 2001
6. Boswijk H. P. and P. H. Franses (1996), Common Persistence in Nonlinear Autorgressive Models, Working paper, Econometric Institute, Erasmus University Rotterdam
7. Chan K.S. and H. Tong (1986), On estimating thresholds in autoregressive models, *Journal of Time Series Analysis*, **7**, 178 – 190.
8. Davies R.B. (1977), Hypothesis testing when a nuisance parameter is present only under the alternative, *Biometrika*, **64**, 247 – 254.
9. Davies R.B. (1987), Hypothesis testing when a nuisance parameter is present only under the alternative, *Biometrika*, **74**, 33 – 43.
10. Debelle G. and J. Vickerey (1998), Is the Phillips curve a curve? Some Evidence and Implications for Australia, *Economic Record* , 74, 227, 384-98
11. Eliasson A.-C. (1999), Is the short-run Phillips curve nonlinear? Empirical Evidence for Australia, Sweden and the United States, Working Paper in Economics and Finance 330, Stockholm School of Economics, June 1999
12. Eitrheim O. and T. Teräsvirta (1996), Testing the adequacy of smooth transition autoregressive models, *Journal of Econometrics*, **75**, 59 – 75.
13. Engle R.F. (1984), Wald, Likelihood Ratio, and Lagrange Multiplier Tests in Econometrics, in Griliches Z. and M.D. Intriligator, eds. *Handbook of Econometrics*, Vol. 2, 775 – 826, Amsterdam: Elsevier.
14. Franses, P. H., P. de Bruin, D. van Dijk (2000), Seasonal Smooth Transition Autoregression, Econometric Institute Report 2000-06/A, Erasmus University, Rotterdam
15. Frisch H. (1983), *Theories of inflation*, Cambridge Surveys of Economic Literature, Cambridge University Press
16. Gali J. (1999), The Return of the Phillips Curve and other Recent Developments in Business Cycle Theory, Mimeo, Universitat Pompeu Fabra.
17. Goldfeld S.M. and R.E. Quandt (1972), Nonlinear methods in econometrics, Amsterdam: North-Holland.
18. Gordon, R. (1997), The time-varying NAIRU and its implications for economic policy, *Journal of Economic Perspectives*, 11 (1), 11-32
19. Granger C.W.J. and T. Teräsvirta (1993), Modelling nonlinear economic relationships, Oxford: Oxford University Press.
20. Haldane A. and D. Quah (1999), UK Phillips Curve and Monetary Policy, LSE Working Paper.

21. Hansen B.E. (1996), Inference when a nuisance parameter is not identified under the null, *Econometrica*, **64**, 413 – 430.
22. Heider M. (1999), La non-linéarité de la courbe de Phillips de court terme en France et en Allemagne, Research Paper, Université Montesquieu Bordeaux IV
23. Jansen E.S. and T. Teräsvirta (1996), Testing the parameter constancy and super exogeneity in econometric equations, *Oxford Bulletin of Economics and Statistics*, **58**, 735 – 763.
24. Lin C.-F. and T. Teräsvirta (1994), Testing the constancy of regression parameters against continuous structural change, *Journal of Econometrics*, **62**, 211 – 228.
25. Lütkepohl H., T. Teräsvirta and Wolters J. (1997), Investigating Stability and Linearity of a German M1 Money Demand Function, Stockholm School of Economics, Working Paper Series in Economics and Finance
26. Luukkonen R., P. Saikkonen and T. Teräsvirta (1988), Testing linearity against smooth transition autoregression, *Biometrika*, **75**, 491 – 499.
27. Maddala D.S. (1977), Econometrics, New York: McGraw-Hall.
28. Mankiw N. G., (2000), The Inexorable and Mysterious Tradeoff Between Inflation and Unemployment, Working Paper 7884, NBER, Sep. 2000
29. Mills T. C. (1999), The Econometric Modelling of Financial Time Series, Cambridge, Cambridge University Press
30. Neftçi S. (1984), Are economic time series asymmetric over the business cycle, *Journal of Political Economy*, **92**, 307 – 328.
31. Phillips A.W. (1958), The relation between unemployment and the rate of change of money wages in the United Kingdom, 1861 – 1957, *Economica*, **25**, 283 – 299.
32. Roberts J. M. (1995), New Keynesian Economics and the Phillips Curve, *Journal of Money, Credit, and Banking*, 27 (4), 975-984
33. Samuelson P.A. and R. M. Solow (1960), Analytical Aspects of Anti-Inflation Policy, *American Economic Review* 50 (Papers and Proceedings), 177-194
34. Sarantis N. (1999), Modeling non-linearities in real effective exchange rates, Journal of International Money and Finance, 18, 27-45
35. Skalin J. and T. Teräsvirta (1998), Modelling asymmetries and moving equilibria in unemployment rates, Stockholm School of Economics, Working Paper Series in Economics and Finance, No. 262.
36. Taylor M. P., Peel D. A., Sarno L. (2001), Nonlinear mean-reversion in exchange rates: towards a solution to the purchasing power parity puzzles, *International Economic Review*, Vol. 42, 1015-42
37. Teräsvirta T. (1994), Specification, estimation, and evaluation of smooth transition autoregressive models, *Journal of the American Statistical Association*, **89**, 208 – 218.
38. Teräsvirta T. (1997), Smooth Transition Models, in Heij Ch., H. Schumacher, B. Hanzon and K. Praagman, eds. *System Dynamics in Economic and Financial Models*, 109 – 123, Chichester: John Wiley and Sons.
39. Teräsvirta T. (1998), Modelling economic relationships with smooth transition regression, in A. Ullah and D.E.A. Giles (eds.) *Handbook of Applied Economic Statistics*, 507-552, New York: Marcel Dekker.
40. Teräsvirta T. and H.M. Anderson (1992), Characterizing nonlinearities in business cycles using smooth transition autoregressive models, *Journal of Applied Econometrics*, **7**, 119 – 136.

41. Teräsvirta T., D. Tjøstheim and C.W.J. Granger (1994), Aspects of modelling nonlinear time series, in Engle R.F. and D. McFadden, eds. *Handbook of Econometrics*, Vol. 4, 2919 – 2957, Amsterdam: Elsevier.
42. Tjøstheim D. (1994), Nonlinear time series analysis: A selective view, *Scandinavian Journal of Statistics*, **21**, 97 – 130.
43. Tong H. (1983), Threshold models in non-linear time series analysis, Lecture Notes in Statistics, No. 21, Springer, Heidelberg
44. Tong H. (1990), Non-linear time series. A dynamical systems approach, Oxford: Oxford University Press.
45. van Dijk D. (1999) Smooth Transition Models: Extensions and Outlier Robust Inference, Tinbergen Institute Research Series No. 200, Rotterdam
46. Wyplosz C. (2000), Do we know how low should inflation be?, First Central Banking Conference on "Why Price Stability?", European Central Bank, November 2000.

Fraction-of-Time Approach in Predicting Value-at-Risk*

Jacek Leśkow[1] and Antonio Napolitano[2]

[1] Department of Econometrics,
The Graduate School of Business WSB-NLU
NOWY SACZ, POLAND.
leskow@wsb-nlu.edu.pl

[2] Dipartimento di Ingegneria Elettronica e delle Telecomunicazioni,
Università di Napoli Federico II
NAPOLI, ITALY.
antonio.napolitano@unina.it

Summary. The aim of this work is to present a new method of Value-at-Risk calculation using the fraction-of-time probability approach used in signal processing (see e.g Leśkow and Napolitano (2001)). This method allows making statistical type inferences based only on a single observation of phenomenon in time. Such setup is very convenient for time series data in financial analysis, when an assumption of having multiple realization of time series is very seldom realized. Another advantage of this method is the possibility of using it without assumptions on the distributions of returns. The aim of the paper is to present the method as well as application to financial data sets.

11.1 Introduction

Value-at-Risk (VaR) has recently become a basic measure of risk in financial transactions. It quantifies a level of potential loss that can be incurred with a given probability level (Leśkow (2001)). Value-at-Risk has a important role in quantifying the necessary minimum of capital a financial institution needs to maintain given the history of transactions risk and the chosen probability level. Currently, there are many methods to predict the value at risk, most popular being the one proposed by Risk Metrics algorithm of the JP Morgan. This method is based on the assumption that the historical data concerning transactions come from the normal distribution with a changing variance. In such an approach, the Value-at-Risk calculation is reduced to calculating

---

* This paper was read at the 2nd. LA-EU Workshop "New tools of qualitative analysis of economic dynamics" in Cholula, Puebla, México, September 17th. and 18th, 2001

estimates of the parameter in an appropriately chosen model of variance dynamics, most popular being the GARCH(p,q) model.

The assumption of normality in the Risk Metrics model has received a lot of criticism from the research community. Recently, a more distribution-free method - CAViAR model - was proposed by Engle and Manganelli (1999). In such model, however, the calculation of VaR is quite complex and frequently invoke assumptions that are hard to prove in practice (Engle and Manganelli (1999)). One of such assumptions is the asymptotic independence of the series of returns. In technical language, φ-mixing or, more generally, ergodicity of the time series of returns. As a result, therefore, we have a choice between the Risk Metrics method: a popular model based on normality of returns, that is easy to verify and quite frequently verifies negatively or a general, CAViAR model that is quite difficult to verify for the concrete data sets.

The aim of this paper is to provide a new method, easy to verify and general that does not resort to normality distribution of the historical returns. In fact, the approach is non-stochastic so no stochastic distribution of the returns is required. Such new approach has been quite successfully used in signal processing and time series in the last decade. This approach is called *fraction-of-time probability* (FOT) analysis and has been presented e.g. in the monograph of Gardner (1994). The basic difference between the classical stochastic approach and the FOT approach is in defining measures corresponding to observations through time-averages rather than ensemble averages.

We think that the FOT approach has several advantages over the classical, stochastic approach and that these advantages are especially important in the context of econometric or financial data analysis. First main advantage of the FOT approach over the stochastic approach is its focus on a single realization of time series that is available in observations. When we analyze econometric data - like GNP, unemployment - or financial data, like prices of stocks or market indices - it is not realistic to assume that the single realization on hand can be repeated in the same time interval. The GNP of Poland in the years 1989 - 2002 or daily levels of Dow Jones in the year 2002 are observed obviously only once. In the classical stochastic approach, in order to circumvent this fundamental difficulty, one assumes that increasing the time interval of the observations is equivalent to having multiple realizations of the observed time series. However, such equivalence requires a strong assumption on the underlying data, this assumption being ergodicity or φ-mixing. In practice, those assumptions are impossible to verify. The FOT approach, on the other hand, does not postulate this type of a property. Using relatively simple analytical tools known for almost periodic functions we obtain the convergence as a simple corollary true for the class of functions on hand.

The other advantage of FOT approach over the stochastic approach is in its use of measure theoretic concepts. In the classical, stochastic approach, the convergence of the sequences of random variables can be analyzed in many different senses: almost sure, in probability, in L^p or in the weak sense. However, the statistical analysis really requires appropriate convergence of empirical distribution function based on the data. In the FOT approach we closely follow the concept of convergence of empirical distribution function without the inconvenience of different types of stochastic convergences. Therefore, the FOT approach is heavily based on measure theory and analysis but is not including the stochastic approach. We do not claim, that FOT analysis is always better than the classical, stochastic approach. If one already knows, that the time series in hand can be successfully modeled with the typical model - like ARMA or GARCH , then FOT models should not be considered as a alternative. On the other hand, in many econometric or financial data models, one is confronted with nonstationarities, structural breaks, changing volatilities - and then a nonstochastic approach can be viewed as a feasible alternative. In that sense, FOT approach is a new tool of analysis of dynamics of economic and financial data.

More formally, our approach here is that we have at our disposal a *single* realization of a time series $x(t)$ on the interval $[t, t + T]$, where $t > 0$. In such context, we will introduce the concept of the FOT probability . In Section 2, general concepts and definitions of the FOT probability framework are presented. Section 3 will be devoted to the main issue of the paper- that is to the calculation and prediction of Value-at-Risk (VaR). Section 4 contains application of the presented method to four data sets: series of daily returns of 6 months treasury bills, series of Dow Jones Industrial Average, values of stocks of the Polish copper conglomerate KGHM and the blue-chips Polish stock index Techwig.

11.2 Fraction-of-time Probability Background

Let $x(\cdot)$ be a Lebesgue measurable function defined on the real axis **R**.

Definition 2.1 The *empirical fraction-of-time* probability distribution function $F_T(t; \xi)$ of $x(u)$ observed on the time interval $[t, t + T]$ is defined as

$$F_T(t; \xi) = \frac{\text{meas} \{u \in [t, t+T] \ : \ x(u) \le \xi\}}{\text{meas} \{u \in [t, t+T]\}} = \frac{1}{T} \int_t^{t+T} \mathcal{U}(\xi - x(u)) du, \quad (11.1)$$

where meas$\{\cdot\}$ denotes the Lebesgue measure and

$$\mathcal{U}(t) = \begin{cases} 1 & t \ge 0, \\ 0 & \text{elsewhere.} \end{cases} \quad (11.2)$$

The function $F_T(t; \xi)$ represents the proportion of time where $x(u) \le \xi$ while $u \in [t, t + T]$.

It is easy to see that the expected value of the distribution F_T is given by

$$E\{x(u), u \in [t, t+T]\} = \frac{1}{T} \int_t^{t+T} x(u)du \qquad (11.3)$$

For the details of the above derivation see Leśkow and Napolitano (2002).

Definition 2.2. Assume that $\lim_{T \to \infty} F_T(t; \xi)$ exists. The (limit) *fraction-of-time* probability distribution function $F(t; \xi)$ is defined as

$$F(t; \xi) = \lim_{T \to \infty} F_T(t; \xi)$$

$$= \lim_{T \to \infty} \frac{1}{T} \int_t^{t+T} \mathcal{U}(\xi - x(u))du.$$

The above function can be interpreted as the proportion of time in which the values of $x(t)$, when t ranges in $\mathbf{R}$, are less or equal to ξ.

The limit in Definition 2.2 exists for a large class of functions that are called *relatively measurable* (see e.g. Urbanik (1958)). Well-known examples of relatively measurable functions are the *almost periodic functions* (Besicovitch(1932)) and stepwise functions of the type $x(t) = \sum_{k=0}^{\infty} a_k p(t - kT_s)$ where $p(t)$ is a rectangular function defined on the interval $[0, T_s]$ and the symbols a_k belong to a finite set $\{A_1, \ldots, A_M\}$ and are such that the FOT probability that $a_k = A_m$ exists for all $m \in \{1, \ldots, M\}$. For both types of functions the limit stated above exists. More information regarding the stepwise functions in the above sense is included in the preprint of Leśkow (2000).

Fact 2.3. Assume that the limit in Definition 2.2 exists. Then it does not depend on t.

Proof of this fact is based on work of Leśkow and Napolitano (2002) and will not be presented here.

In the following, it well be assumed $t = 0$ and, hence, whenever possible t will be dropped from the notation.

Remark. To understand better the role of the FOT distributions F_T and F consider first a Lebesgue measurable real function $x(t)$. For such function consider the set of observations $\{x(t) : 0 \le t \le T\}$ and define the measure P_T corresponding to those observations. The measure P_T is defined on $\mathbf{R}$ equipped with the σ-field generated by the sets $\{t \in \mathbf{R} : x(t) \le c; \ 0 \le t \le T, \ c \in \mathbf{R}\}$. The natural link between P_T and F_T is established via the equality

$$P_T\{x(t) \le c\} = F_T(c),$$

where $F_T(c)$ can be calculated from the Definition 2.1. Identical construction can be done to link the FOT distribution F and the measure P obtained as the limit of P_T for $T \to \infty$, provided that such limits exist.

11.3 Calculation and Prediction of VaR

Definition 3.1. Let F be an arbitrary distribution function. Then the α-quantile q_α of F is defined as :

$$q_\alpha = \inf\{s \in \mathbf{R} : F(s) \geq \alpha\},$$

where $0 < \alpha < 1$. Similarly, we define the inverse F^{-1} of a given distribution function F as $F^{-1}(v) = \inf\{s \in \mathbf{R} : F(s) \geq v\}$ for $v \in (0, 1)$.

The Definition 3.1 can be applied to both empirical FOT distribution and the FOT distribution. For the empirical FOT distribution F_T an α-quantile q_α^T is such a number that the proportion of time that $x(u) \leq q_\alpha^T$ while $u \in [0, T]$ is equal to or greater than α.

It is straightforward right now to see the method of Value-at-Risk estimation in this approach. For a finite segment of data $\{x(u); u \in [0, T]\}$ we may calculate the empirical FOT and then calculate q_α^T. This will be our estimate of Value-at-Risk for that data segment. Since for the observed time series $x(t)$ the limit $\lim_{T \to \infty} F_T(\xi)$ exists for each ξ then we are guaranteed that q_α^T converges to q_α, which, in turn, is the quantile of the limiting FOT. **This is the main difference between the FOT and the stochastic approaches.** In our approach, convergence is a simple analytical fact, valid for a large class of observed functions. In the stochastic approach, however, the convergence of empirical distribution is a main problem to be established using hard to prove asymptotic independence assumptions (see Engle, Manganelli (1999)).

Note the following

Definition 3.2. The FOT confidence interval $[L, U]$ at the confidence level α is defined as the interval between $L = \frac{\alpha}{2}$-quantile and $U = (1 - \frac{\alpha}{2})$-quantile of the FOT distribution F. Similarly, the empirical FOT confidence interval $[L_T, U_T]$ at the confidence level α is defined as the interval between $L_T = \frac{\alpha}{2}$-quantile and $U_T = (1 - \frac{\alpha}{2})$-quantile of the empirical FOT distribution F_T.

The consistency of the Value-at-Risk estimate is very easy to establish when the significance level α is a continuity point of the inverse F^{-1} of the limiting FOT distribution F (e.g. see Reiss (1989)). However, in practice, the observed financial time series may have discontinuities. Nevertheless, in such case there is also a way to construct a consistent procedure of Value-at-Risk estimation. The details of such method are described in Leśkow (2000). In the sequel, for brevity, we focus on VaR prediction for the continuous case.

We will now pass to the problem of prediction of Value-at-Risk. In general, the prediction for Value-at-Risk is a rather complicated task and in stochastic approach several models have been already suggested (see e.g. Engle and Manganelli (1999)).

Here we would like to present a method of prediction of VaR using the FOT probability approach. The general assumption is that we observe a function $x(u)$ on the interval $[0, T]$. This function is, in the language of stochastic analysis, the realization of a stochastic process. Then, the basic question is to provide predictors for the VaR given the observations until the time moment T.

The answer to the question above will be based entirely on the previously presented method of estimating VaR and empirical FOT distributions. The FOT approach has a lot of advantages over the classical, stochastic methods. Most importantly, no assumption on independence of observations or stationarity of distributions is necessary.

11.3.1 VaR Prediction

We start with the technical result, convenient while calculating the predicted value.

Fact 3.3. Assume that the time horizon δ is arbitrary and let $F_{T+\delta}$ be the empirical FOT distribution generated by x on the interval $[0, T + \delta]$. Then

$$F_{T+\delta}(\xi) = \frac{T}{T+\delta} F_T(\xi) + \frac{1}{T+\delta} \int_T^{T+\delta} \mathcal{U}(\xi - x(u)) du. \qquad (11.4)$$

Proof of this Fact is standard and will not be presented here.

It is easy to see that the second term of the right hand side of (11.4) is smaller than $\frac{\delta}{T+\delta}$. In another words,

$$F_{T+\delta}(\xi) \in [\frac{T}{T+\delta} F_T(\xi), \frac{T}{T+\delta}(F_T(\xi) + \frac{\delta}{T})].$$

This observation allows us to formulate the following

VaR prediction algorithm

Step 1. Given the observation of the continuous function x calculate the FOT F_T.

Step 2. Calculate

$$\hat{q}_\alpha^{T+\delta}(1) = \inf\{\xi : \frac{T}{T+\delta}F_T(\xi) + \frac{\delta}{T+\delta} \geq \alpha\}$$

and

$$\hat{q}_\alpha^{T+\delta}(2) = \inf\{\xi : \frac{T}{T+\delta}F_T(\xi) \geq \alpha\}.$$

It is easy to see that the above algorithm produces the interval of predictions of the form $[\hat{q}_\alpha^{T+\delta}(1), \hat{q}_\alpha^{T+\delta}(2)]$ and that the length of this interval, $\frac{\delta}{T+\delta}$ converges to zero while $T \to \infty$ for any fixed time horizon δ.

11.3.2 VaR Prediction in Nonstationary Case

The method of predicting Value-at-Risk presented above works well for such data sets, where there are no significant structural breaks, so all the data points can be given a similar weight in prediction algorithm. This equal weight can be seen at the very beginning of our theory, namely in the Definition 2.1, where the proportion of time is calculated with respect to the total length of the observation interval, that is T.

However, in financial and econometric data we frequently encounter rapid changes, structural brakes or changing volatilities. In such situations it is not reasonable to put equal weight on remote past data and recent data in prediction algorithms. Rather, one is interested in looking for some concept of window for the past data. Within the window, the data will have more importance, outside the selected window the data should be given less weight. In what follows, we will include such concept into the FOT approach via the following

Definition 3.4. The *weighted FOT distribution function* $F_{T,Z}(t;\xi)$ of $x(u)$ observed in the time interval $[t, t+T]$ is defined as

$$F_{T,Z}(t;\xi) = \frac{1}{\displaystyle\int_t^{t+T} w_Z(u - (t+T))du} \int_t^{t+T} \mathcal{U}(\xi - x(u))\, w_Z(u - (t+T))du\,.$$

(11.5)

In the definition above, the weight function w_Z is defined for $t \leq 0$, with the window width $Z > 0$, and such that

$$\lim_{Z \to \infty} w_Z(t) = c > 0 \quad \forall t \leq 0\,.$$

(11.6)

Examples of weighting functions $w_Z(t)$ are the rectangular window defined as $w_Z(t) = 1$ for $t \in [-Z, 0]$ and 0 elsewhere and $w_Z(t) = \exp[-|t|/Z]\,\mathcal{U}(-t)$.

The technical details related to the weighted FOT distribution $F_{T,Z}(t;\xi)$ are presented in Leśkow and Napolitano (2002). Here we would like to focus on econometric and financial data applications of $F_{T,Z}(t;\xi)$. This function, even for large T, depends on the local time $t + T$ and this makes such function an attractive object of analysis for nonstationary data. We can say that

$F_{T,Z}(t;\xi)$ provides a nonstationary characterization of the time series $x(t)$. Additionally, such weighted FOT distribution depends also on the weight Z. An interesting open research problem is the optimal choice of the weight Z for a given data set. Such problem can be solved similarly to optimal window choice problem known in nonparametric time series analysis (see e.g. Leśkow (2001)). In this study, we will simply take a rectangular window function w_Z with a fixed length Z and show how to predict Value-at-Risk with a weighted FOT distribution.

Before studying the practical applications, we will define a modified Value-at-Risk prediction algorithm using the concept of the weighted FOT distribution. The details of such algorithm are presented in Leśkow and Napolitano (2002), so here we focus on the main facts only.

VaR prediction algorithm (nonstationary case)

Step 1. Given the observation $x(u)$, $u \in [t, t+T]$, calculate $F_{T,Z}(t;\xi)$
Step 2. Calculate

$$\hat{q}_\alpha^{T+\delta}(1) = \inf\left\{\xi \in \mathbf{R} \ : \ F_{T,Z}(t;\xi) \geq \alpha - \frac{\delta}{Z}\right\} \qquad (11.7)$$

and

$$\hat{q}_\alpha^{T+\delta}(2) = \inf\left\{\xi \in \mathbf{R} \ : \ F_{T,Z}(t;\xi) \geq \alpha + \frac{\delta}{Z}\right\} . \qquad (11.8)$$

The interval $[\hat{q}_\alpha^{T+\delta}(1), \hat{q}_\alpha^{T+\delta}(2)]$ is the prediction interval for the VaR.

In the next Section we illustrate our ideas with analysis of 4 typical financial data sets.

11.4 Applications

This Section contains Figs. 11.1-11.3, illustrating the application of our VaR predictor to real data sets.

In all calculations illustrated below, we took the two-day prediction horizon, i.e. $\delta = 2$, the rectangular window w_Z and the window width $Z = 20$. This correspond to calculating the two-day forecast of Value-at-Risk using the last 20 observations. The lower 5% and upper 95% VaR predictors have been considered. In all the figures, the shadowed strips represent the *predicted* VaR intervals whose lower and upper extremes are $\hat{q}_\alpha^{T+\delta}(1)$ and $\hat{q}_\alpha^{T+\delta}(2)$ evaluated by (11.7) and (11.8), respectively. The predicted VaR intervals are reported as functions of the sample size. The bright and the dark strips refer to the 5% and 95% VaRs, respectively. Moreover, in order to perform a comparison, for each strip the corresponding *(a posteriori) estimated* VaR is represented by a

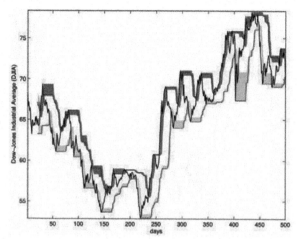

Fig. 11.1 The VaR predictors for Dow Jones Industrial Average

thin line. Finally, the true value of the observed time series is represented by a thick line.

Figure 11.1 contains calculation of VaR using our algorithm for the time series of returns on 6 months US Treasury Bills. The data set consists of 30 years monthly observations of the phenomenon in the period January 1972 to March 2001. Figure 11.2 presents the predictor of VaR for the Dow Jones Industrial Average Index for the period of 500 days following the January 2, 2000. The third picture presents the values of stocks of the Polish copper conglomerate, KGHM, in the period of 500 days following January 2, 2000. Finally, the fourth picture contains VaR prediction for the Techwig index of the Polish stock market, 250 observations following January 2, 2000.

The pictures in Figs. 11.1-11.4 represent different time series from the point of view of trend and volatility. However, due to the nonparametric and general nature of the FOT approach, the accuracy of VaR prediction is not altered by different data behavior. Some technical difficulty in obtaining the pictures above was related with appropriately choosing the smoothing parameter for calculating the prediction algorithm. For the details on this we refer the reader to the paper of Leśkow and Napolitano (2002).

Concluding remarks. The presented method provides an essentially new tool in important financial problem that is the Value-at-Risk prediction. The standard techniques involve the stochastic approach and some distributional assumptions on the history of the assets. The new method that we propose relies on analytical approach with very mild and general assumptions on the

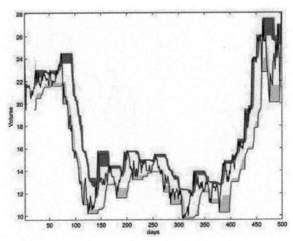

Fig. 11.2 The VaR predictors for prices of the Polish
copper conglomerate KGHM

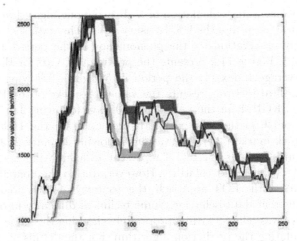

Fig. 11.3 The VaR predictors for prices of the Polish
blue chips index Elektrwig

class of functions that may be used as a model for the observed time series.

Acknowledgement. The authors acknowledge the support of the NATO
grant PST.CLG. 978068 and the matching grant of WSB-NLU of Nowy Sącz,
Poland that made this research possible.

References

1. Besicovitch, P. (1932), *Almost-periodic functions*, Cambridge University Press.
2. Engle, R. and Manganelli, S. (1999), *CaViar : conditional autoregressive Value-at-Risk by regression quantiles*, discussion paper 99-20, Department of Economics, University of California, San Diego.
3. Gardner, W.A. (1994), *An introduction to cyclostationary signals*, Chapter 1 in *Cyclostationarity in Communications and Signal Processing*, W.A. Gardner, Ed., pp. 1-90, IEEE Press, New York.
4. Izzo, L. and Napolitano, A. (1998), *The higher-order theory of generalized almost-cyclostationary time-series, IEEE Trans. Signal Processing*, vol. 46, pp. 2975-2989, November 1998.
5. Leśkow, J. (2001), *The impact of stationarity assessment on studies of volatility and Value-at-Risk*, Mathematic and Computer Modelling,vol. 34, pp. 1213- 1222.
6. Leśkow, J. (2000), *Value-at-Risk calculation and prediction in the fraction-of-time probability framework*, preprint, Institute of Mathematics, Technical University of Wrocław, Poland.
7. Leśkow, J. and Napolitano, A. (2002), *Quantile prediction for time series in the fraction-of-time probability framework,* Signal Processing, vol. 82, pp. 1727-1741.
8. Reiss, R.D. (1989), *Approximate Distributions of Order Statistics, with applications to nonparametric statistics.* Springer Series in Statistics. Urbanik, K. (1958), *Effective processes in the sense of H. Steinhaus, Studia Mathematica*, XVII.
9. Wiener, N. (1930), *Generalized harmonic analysis, Acta Mathematica*, vol. 55, pp.117-258.

Spectral Analysis for Economic Time Series[*]

Alessandra Iacobucci

OFCE, 69 quai d'Orsay, 75340 Paris Cedex 07 (France)
CNRS – IDEFI, 250 rue Albert Einstein, 06560 VALBONNE (France)
alessandra.iacobucci@sciences-po.fr

Summary. The last ten years have witnessed an increasing interest of the econometrics community in spectral theory. In fact, decomposing the series evolution in periodic contributions allows a more insightful view of its structure and of its cyclical behavior at different time scales. In this paper, the issues of cross-spectral analysis and filtering are concisely broached, dwelling in particular upon the windowed filter [15]. In order to show the usefulness of these tools, an application to real data — namely to US unemployment and inflation — is presented. By means of cross spectral analysis and filtering, a correlation can be found between these two quantities (i.e. the Phillips curve) in some specific frequency bands, even if it does not appear in raw data.

12.1 Introduction

The first appearance of spectral analysis in the study of macroeconomic time series dates from the middle 1960s, motivated by the requirement of a more insightful knowledge of the series structure and supported by the contemporaneous progress in spectral estimation and computation. The first works focused on the problem of seasonal adjustment procedures (see e.g. [20]) and on the general spectral structure of economic data [12]. Cross spectral methods were pointed out from the outset as being important in discovering and interpreting the relationships between economic variables [11, 13]. After the early years, the range of application of such analysis was extended to the study of other econometric issues, among which the controversial trend-cycle separation, the related problem of business cycles extraction and the analysis of co-movements among series, useful in the study of international business cycles. It has been clear from the beginning that spectral analysis is purely

---

[*] I thank Francesco Saraceno and an anonymous referee for their precious comments and suggestions.

descriptive and cannot be straightforwardly used for forecasting; it is nevertheless a powerful tool for inspecting cyclical phenomena and highlighting lead-lag relations among series. It also provides a rigorous and versatile way to define formally and quantitatively each series components and, by means of filtering, it provides a reliable extraction method. In particular, cross spectral analysis allows a detailed study of the correlation among series.

In this synthetic overview I will focus on both filtering and cross spectral analysis, which are often two stages of the same procedure. As a matter of fact, besides the definition and extraction of the different components of a series – typically trend, business cycle and seasonality – frequency filters can also be applied to perform a more targeted and efficient cross spectral analysis.

Time-frequency approaches — which represent the frequency content of a series, while keeping the time description parameter to give a three-dimensional time-dependent spectrum — will not be tackled in this paper. This is for essentially two reasons: first, they would require more than a simple section; second, and more importantly, because evolutionary spectral methods and wavelets are suited when dealing with very long time series, like those found in geophysics, astrophysics, neuroscience or finance. But their application to short series — the norm in macroeconomics — is difficult and may give unstable parameter-dependent results. For such series, traditional spectral analysis is probably more suitable.

The paper is organized as follows: the first section contains a concise description of spectral estimation and filtering issues[1] together with a recall of discrete Fourier analysis; in the second section I present the cross spectral analysis procedure, with a very short account of the genuinely technical yet central issue of estimation; in the third section I show an application of the techniques to the US Phillips curve. Some remarks and the conclusion can be found in the fourth and last section.

12.2 Spectral Estimation and Filtering: a Brief Review

At a first glance, the overall behavior of time series may be decomposed in three main parts: long, medium and short run behavior. These three parts are respectively associated with slowly evolving secular movements (the trend), a faster oscillating part (the business cycles) and a rapidly varying, often irregular, component (the seasonality). As it is often the case when no testable *a priori* hypothesis on the data generating process (i.e. on the model) is available, this separation is very phenomenological.

Modern empirical macroeconomics employs an assortment of *ad hoc* detrending and smoothing techniques to extract the business cycle, like moving averages to eliminate the fast components, first-differences to cut out the long

---

[1] For an more extensive and detailed treatment the interested reader may refer, among others, to the celebrated book by Jenkins and Watts [16].

term movements, or even the simple subtraction of the linear trend, to cancel the slow drift variables. Though conceptually not wrong, these methods lack a formal decomposition of the series and are incapable of giving a definition of the business cycle based on some required and adjustable characteristics. This is why the Fourier decomposition remains one of the most insightful ways of performing the separation of a signal into different purely periodic components.

Consider a finite series $u(j)$ of length $T = N\Delta t$, where N is the number of data and Δt the sampling periodicity; the frequency $\nu_k = k/(N\Delta t)$ and the time $t_j = j\Delta t$ are indexed by k and j respectively.

The discrete Fourier transform (DFT) $U(k)$ of $u(j)$ and its inverse (IDFT) for finite series are

$$U(k) = \frac{1}{N} \sum_{j=0}^{N-1} u(j) e^{-i2\pi jk/N} , \tag{12.1}$$

$$u(j) = \sum_{k=-\lfloor N/2 \rfloor}^{\lfloor (N-1)/2 \rfloor} U(k) e^{i2\pi jk/N} , \tag{12.2}$$

where $\lfloor \cdot \rfloor$ denotes the largest integer smaller or equal than the operand, $k \in [-\lfloor N/2 \rfloor, \lfloor (N-1)/2 \rfloor]$ and $j = 0, \ldots, N-1$. Of course, the discretization of the signal (i.e. its sampling with some finite period Δt) implies a limitation of its spectrum to the band $\nu \in [-(2\Delta t)^{-1}, (2\Delta t)^{-1}[$, where $(2\Delta t)^{-1}$ is the *Nyquist frequency*, as frequencies outside that range are folded inside by the sampling (an effect known as *aliasing* [7]). On the other hand, the finiteness of the signal in time implies a discretization of the spectrum, the interval between two successive values being $(N\Delta t)^{-1}$.

Equations (12.1) and (12.2) can only be an approximation of the corresponding real quantity, since it provides only for a finite set of discrete frequencies. The quantity $P_u(k) = |U(k)|^2$ is the signal (power) spectrum and its "natural" estimator would be Schuster's *periodogram*

$$P_u(k) = \Delta t \sum_{J=-(N-1)}^{N-1} \gamma_{uu}(J) e^{-i2\pi Jk/N}$$

$$= \Delta t \sum_{J=-(N-1)}^{N-1} \gamma_{uu}(J) \cos \frac{2\pi Jk}{N} , \tag{12.3}$$

where $\gamma_{uu}(J) = \gamma_{uu}(-J) = N^{-1} \sum_{j=-(N-J)}^{N-J} (u(j) - \bar{u})(u(j+J) - \bar{u})$ is the standard sample estimation at lag J of the autocovariance function.

The periodogram is a real quantity – since the series is real and the autocovariance is an even function – and is an asymptotically unbiased estimator of the theoretical spectrum. Yet, in the case of finite series, it is *non-consistent* since the power estimate at the individual frequency fluctuates with N, making difficult its interpretation. To build a spectral estimator which is more stable – i.e. has a smaller variance – than $P_u(k)$, we turn to the technique of *windowing* (see [8, 16, 21] among others). This technique is employed both in time and in frequency domain to smoothen all abrupt variations and to minimize the spurious fluctuations generated every time a series is truncated. The result of windowing is the *smoothed spectrum*

$$\hat{S}_u(k) = \Delta t \sum_{J=-(N-1)}^{N-1} w_M(J)\,\gamma_{uu}(J)\,\cos\frac{2\pi Jk}{N}\ , \qquad (12.4)$$

where the autocorrelation function is weighted by the *lag window* $w(j)$ of width M [1]. It can be shown that this is equivalent to splitting the series in N/M sub-series of length M, computing their spectra and taking their mean.

Since $P_u(k)$ and $\gamma_u(J)$ are related by DFT (equation 12.3), equation (12.4) can also be written as

$$\hat{S}_u(k) = \Delta t \sum_{k'=-\lfloor N/2 \rfloor}^{\lfloor (N-1)/2 \rfloor} P_u(k')\,W_{M'}(k-k')\ , \qquad (12.5)$$

the Fourier transform of $w_M(j)$, the *spectral window* $W_{M'}(k)$ of width $M' = M^{-1}$. Thus the smoothed spectrum at k is nothing but the periodogram seen through a window opened on a convenient interval around k. Equations (12.4) and (12.5) represent two perfectly equivalent ways to compute the smoothed spectrum. Usually, the multiplication, as in equation (12.4), is chosen because it is easier to compute. Nevertheless, sometimes the convolution may be more convenient as we shall see in the section devoted the windowed filter.

The choice of the lag window width M is performed by choosing a "reasonably" narrow window, i.e. a small initial value of M, and then widening it until a good spectral stability is obtained, i.e. until the spectral density remains roughly unchanged as M increases. Widening the lag window $w_M(j)$ corresponds to narrowing the band covered by its Fourier transform, the spectral window $W_{M'}(k)$. This is why the procedure is called *window-closing* [16]. This method allows to learn progressively about the shape of the spectrum. The initial choice of a wide bandwidth usually masks some details of the spectrum. By decreasing the bandwidth, more significant details can be explored. The choice of M is rather tricky since it has to be large enough to let all the fundamental details of the spectrum appear, but not too large, to prevent the generation of spurious peaks.

Windows can be chosen among those already existing in the literature (rectangular, triangular, Bartlett, Parzen, Tuckey, Blackman, Hamming,...) or can be built *ad hoc* for the specific problem treated. The research of the optimal window involves a compromise between *accuracy* and *stability* of the estimator (see [8, 16, 21] among others). Moreover, windows are used both in time and in frequency domain, according to the researcher needs. Both lag windows and spectral-windows can be used either as multiplying window, like the lag window in equation (12.4), or as convolving windows, like the spectral window in equation (12.5). Since a convolution in the time domain becomes a multiplication in the frequency domain and *vice versa*, a multiplying (convolving) lag window becomes by Fourier transform a convolving (multiplying) spectral window.

Filtering

The filtering operation can be performed either in time or in frequency domain since both approaches are equivalent by

$$v(j) = h(j) \circledast u(j) \equiv \sum_{n=0}^{N-1} h(n) \, u(j-n) \, _{\text{mod } N} \, ,$$

$$= \sum_{k=-N/2}^{(N-1)/2} H(k)U(k)e^{-i2\pi jk/N} \, . \tag{12.6}$$

The previous relation is nothing but the finite discrete version of the *convolution theorem* [16], where the linear convolution has been substituted by the *circular convolution* ($\circledast$) of length equal to the number of data N. Thus filtering simply consists in multiplying $U(k)$ by the filter frequency response $H(k)$ or, equivalently, in convolving the signal $u(j)$ with the filter time response $h(j)$, obtained from $H(k)$ by IDFT. In particular, the *band-pass filter* selects a frequency range, so that $H(k) = 1$ for $k_l \leq |k| \leq k_h$ (*pass-band*) and zero elsewhere (*stop-band*). Of course, the *low-pass filter* has $k_l = 0$ and selects all frequencies lower than k_h, while the *high-pass* filter has $k_h = N/2$, correspondent to the Nyquist frequency $\nu_N = 1/(2\Delta t)$, and selects all frequencies higher than k_l. Notice that the filter $H(k)$ is not causal in the time domain because it requires future values as well as past ones (see equation (12.6)). Asymmetrical (one-sided) filters using only past values may seem interesting because they allow forecasting [5, 18]; but, unless special care is taken in designing them — e.g allowing for a complex time response function — they are dangerous to use because they induce frequency-dependent phase shifts and may thus change the causality relations among different frequency components [15, 21]. This would make cross-correlation analysis useless.

A filter which is real in time domain $(h(j) = h^*(j))$, is symmetric in frequency domain $(H(k) = H(-k))$ and *vice versa*. Therefore, if we want real signals to remain real after filtering, both time *and* frequency response functions have to be real and symmetric, to avoid time and phase shifts. Indeed, it is easy to see that if the filter $H(k)$ is a complex function different frequencies undergo different phase shift and timing relations among components are destroyed (*dispersive filter*).

In the circular convolution the finite signal is replaced by its periodic version $u(N + j) = u(j \bmod N)$, the maximum period length being implicitly assumed by the Fourier transform to be $T = N\Delta t$. This amounts to assuming that the only frequencies present in the signal are integer multiples of T^{-1}, which is in general false and affects the analysis. Indeed the "forced" periodicity introduces an artificial discontinuity at the edges of the time series, that is reflected by spurious oscillations in the series DFT, the so-called *Gibbs phenomenon*. These oscillations are due to the form of the DFT of a rectangular window of width $T = N\Delta t$, which goes like $\sin(\pi\nu T)/(\pi\nu T)$ (see Figure 12.1).

The only way to prevent this effect, would be to choose T (or equivalently N) as a multiple of the largest period that is likely to occur. Unfortunately, this is feasible only if we have some idea of the frequencies involved in the process and would in any case entail some loss of data at one or both sample ends. As for the cutoff frequencies $\nu_l = k_l/(N\Delta t)$ and $\nu_h = k_h/(N\Delta t)$, given the value of N, they must be chosen to be multiples of T^{-1}, otherwise the filter does not completely remove the zero frequency component (i.e. the signal mean) and cannot help in eliminating unit roots (see below).

The time coefficients of an *ideal band-pass filter* are

$$h_{\text{ideal}}(j) = \frac{\sin(2\pi\nu_h j\Delta t) - \sin(2\pi\nu_l j\Delta t)}{\pi j} , \qquad j = -\infty, \ldots, \infty ,$$
$$h_{\text{ideal}}(0) = \lim_{j \to 0} h_{\text{ideal}}(j) = 2\Delta t(\nu_h - \nu_l) , \qquad\qquad (12.7)$$

and are obtainable only in the case of infinite series, since this could be the only way to precisely select the frequency band $[\nu_l, \nu_h]$. Indeed the filter must distinguish between frequencies ν_h (or ν_l) and $\nu_h + d\nu$ (or $\nu_l - d\nu$) when $d\nu \to 0$, that is, $N \to \infty$. This is why this filter is called "ideal". In the case of finite series of length N only the first N coefficients can be calculated

$$h(j) = \frac{\sin\left[\frac{2\pi j}{N}\left(k_h + \frac{1}{2}\right)\right] - \sin\left[\frac{2\pi j}{N}\left(k_l - \frac{1}{2}\right)\right]}{\sin\left(\frac{\pi j}{N}\right)} \qquad j = 1, \ldots, N-1 ,$$
$$h(0) = \lim_{j \to 0} h(j) = 2\left(k_h - k_l + 1\right) \qquad\qquad (12.8)$$

The same result is obtained by multiplying the coefficients (12.8) by the coefficients of a rectangular lag window of width N. As we saw above, the effect of this truncation is the Gibbs phenomenon, i.e. the appearance of spurious oscillations in the frequency response (see Figure 12.1). This causes the so-called *leakage*: the component at one frequency "contaminates" the neighboring components by modifying their amplitude. Thus, frequency components which are contiguous to the band limits, are allowed to leak into the band. Again the application of an appropriate window is the most straightforward way to bypass this problem and obtain a smoother response, as shown below, in the section devoted to the windowed filter.

Since the Fourier theory and the definition of the spectrum only apply to stationary time series, it is necessary to detect non-periodic components prior to the analysis of a series spectrum. First and foremost, it should be established whether the series has a trend, and, if so, whether the trend is

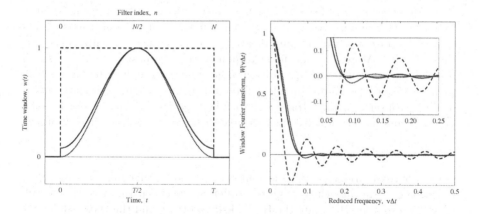

Fig. 12.1. Window Functions and Their Frequency Response. The rectangular (dashed line), Hanning (dotted line) and Hamming (full line) time windows (left panel) and their respective Fourier transforms (right panel). Note in the zoom (right panel, inset) the reduced side lobe amplitude and leakage of the Hanning and Hamming windows with respect to the rectangular one, the Hamming window performing better in the first side lobe.

stochastic or *deterministic*. Unfortunately there is no direct method to distinguish between the two categories in the case of raw data with no underlying model, so that the choice may often depend on the researcher's insight (see, e.g. [6]). If the trend is deterministic, e.g. a polynomial function of time, the Fourier basis decomposition is not unique, since the polynomial term and the periodic one are not orthogonal (a polynomial term contains all possible frequency components). Therefore, the operations of detrending and filtering do not commute and the trend must be preliminarily removed. It is also *necessary* to remove the artificial discontinuity introduced at the edges of the interval by the combination of the trend and the periodicity induced by the Fourier representation. In the case of a linear deterministic trend — that should be established beforehand by looking at the correlation coefficient of the signal with time —, the subtraction of the ordinary least-squares linear fit from the original series is performed, more or less explicitly, by some filtering procedures [2, 5, 15].

If the trend is stochastic, and the observed signal is an $I(p)$ process, i.e. the result of p integrations of a stationary process, it has a spectrum that goes as ν^{-p} for small ν. Thus, a filter whose frequency response function goes like ν^p makes the filtered series stationary. In particular, a ν^2-like response is sufficient for the elimination of two unit roots. The typical way of treating $I(p)$ signals would be to apply p times the first-difference operator to remove the p unit roots. The main drawback of this procedure is that the difference operator is an asymmetric filter, thus it has a complex response $H_{L^p}(k) = (-i)^p e^{ip\pi k/(N\Delta t)} \left(2 \sin \frac{\pi k}{N\Delta t}\right)^p$ which introduces a frequency-dependent phase shift. Moreover it amplifies all frequencies larger than one third of the Nyquist frequency (see among others [8, 15]). This means that p applications of this filter will cause a dramatic amplification of high-frequency components and thus of noise. Moreover, the filter response varies almost linearly for small frequencies, so that low-frequency components are strongly attenuated. It is then very hard to obtain an ideal filter after differencing, especially when dealing with series with a Granger-shaped [11] spectrum, in which much of the power occurs at very low frequencies, like those common in macroeconomics.

The Windowed Filter

Good approximations of the ideal filter — "good" referring to some optimization criteria, like the (weighted) difference between the desired and the effective response [21] — are the Hodrick-Prescott [9] and the Baxter-King [2] filters[2]. These procedures make stationary at least $I(2)$ processes. In particular, the HP-filter can eliminate up to four unit roots. As for these filters, which are widely known, the reader is referred to the original papers [9, 2]. Here I will focus on a filter recently proposed by Iacobucci and Noullez [15], which is obtained by the technique of *windowing* [21], a well-known technique in signal theory. This filter performs better than the others, since it has minimum leakage, a significantly flatter frequency response function in the pass-band and involves no loss of data.

As we have previously seen, the filter obtained by truncation has two main drawbacks: large amplitudes and a slow decay of the spurious lobes in the response function (see Figure 12.1). These can be ascribed, as previously said, to the discontinuous shape of the above-mentioned lag window, whose $\sin(\pi\nu T)/(\pi\nu T)$-profile Fourier transform disturbs the ideal frequency response. It seems then natural to try to adjust the shape of the rectangular window to obtain a gain that goes to zero faster. For this purpose, the "adjusted" window should be chosen to go to zero continuously with its highest possible order derivatives, at both ends of the observation interval [15].

Among a certain number of possible windows, Iacobucci and Noullez choose the Hamming window

$$w^{\mathrm{Ham}}(j) = 0.54 - 0.46\cos\left(\frac{2\pi j}{N}\right) , \qquad (12.9)$$

which is obtained by a combination of the Hanning window $w^{\mathrm{Han}}(j) = \frac{1}{2} - \frac{1}{2}\cos\left(\frac{2\pi j}{N}\right)$ and the rectangular window, to minimize the amplitude of the side lobes (see Figure 12.1). Its Fourier transform

$$W^{\mathrm{Ham}}(k) = \left[0.08 + \frac{0.92}{1 - k^2}\right]\frac{\sin(\pi k)}{\pi k} \qquad (12.10)$$

---

[2] Christiano and Fitzgerald [5] have also designed a band-pass filter, which is more complicated than the previous ones but, in my opinion, also more questionable (see [15]).

decreases like $(\nu T)^{-1}$ for large ν, but with a much smaller amplitude than the rectangular window. Moreover, it has non-zero components only at $k = 0$ and $k = \pm 1$.

The windowed filter algorithm is the following:

— subtract, if needed, the least-squares line to remove the artificial discontinuity introduced at the edge of the series by the Fourier representation;
— compute the discrete Fourier transform of $u(j)$

$$U(k) = \frac{1}{N} \sum_{j=0}^{N-1} u(j) e^{-i2\pi jk/N} , \qquad k = 0, \ldots, \lfloor N/2 \rfloor ,$$

where $U(k) = U^*(-k) = U(-k)$;

— compute the DFT of the Hamming-windowed filtered series

$$V(k) = [W(k) * H(k)] \, U(k) = \sum_{k'=-\lfloor N/2 \rfloor}^{\lfloor N/2 \rfloor} W(k')H(k - k')U(k)$$

$$= [0.23H(k - 1) + 0.54H(k) + 0.23H(k + 1)] \, U(k) \quad k = 0, ., \lfloor N/2 \rfloor ,$$

where $H(k)$ is defined by the frequency range as

$$H(k) = H(k)^{\text{ideal}} \equiv \begin{cases} 1 & \text{if} \quad \nu_l N \Delta t \leq |k| \leq \nu_h N \Delta t \\ 0 & \text{otherwise} \end{cases} ;$$

— compute the inverse transform

$$v(j) = \left[V(0) + \sum_{k=1}^{\lfloor N/2 \rfloor} \left(V(k) e^{i2\pi jk/N} + V(k)^* e^{-i2\pi jk/N} \right) \right] j = 0, ., N-1$$

Notice that windowing is performed in the frequency domain by convolution of the window Fourier transform with the ideal filter response. This is computationally more convenient than time domain multiplication, since the Hamming window Fourier transform has only three non-zero components, as I have already stressed. This procedure ensures both the best possible behavior in the upper part of the spectrum and the complete removal the signal mean. In the application I propose in Section 12.4, I make use of this filter because of its many advisable properties compared to the others, namely a flat, well-behaved response function and the fact that it involves no loss of data.

12.3 Cross Spectral Analysis: the Bivariate Extension

While univariate spectral analysis allows the detection of movements inside each series, by means of bivariate spectral analysis it is possible to describe pairs of time series in frequency domain, by decomposing their covariance in frequency components. In other words, cross spectral analysis can be considered as the frequency domain equivalent of correlation analysis. The definition of the (smoothed) *cross spectrum*, analogously to that of the (smoothed) spectrum (see equation (12.4)), is obtained by substituting the cross covariance function for the autocovariance function. Thus, if we have two time series $u_1(j)$ and $u_2(j)$ and their crosscovariance $\gamma_{12}(J) = \gamma_{21}(-J)$, the cross spectrum is

$$\hat{S}_{12}(k) = \Delta t \sum_{J=-(N-1)}^{N-1} w(J)\,\gamma_{12}(J)\,e^{-i2\pi Jk/N} = \hat{C}_{12}(k) - i\hat{Q}_{12}(k) \quad (12.11)$$

and is in general *complex*, since the crosscovariance is not an even function. The real part $\hat{C}_{12}(k)$ is the *cospectrum* and the imaginary part $\hat{Q}_{12}(k)$ the *quadrature spectrum*. Keeping the time-frequency analogy, I introduce the typical cross spectral quantities and indicate in parentheses the time domain equivalent:

— the *coherency spectrum* (correlation coefficient)

$$\hat{K}_{12}(k) = \frac{|\hat{S}_{12}(k)|}{\sqrt{\hat{S}_1(k)\hat{S}_2(k)}} = \frac{\sqrt{\hat{C}_{12}(k)^2 + \hat{Q}_{12}(k)^2}}{\sqrt{\hat{S}_1(k)\hat{S}_2(k)}}, \quad (12.12)$$

which measures the degree to which one series can be represented as a linear function of the other (sometimes its square is used, whose time domain equivalent is the R^2);
— the *phase spectrum* (time-lag)

$$\hat{\Phi}_{12}(k) = \arctan\left(-\frac{\hat{Q}_{12}(k)}{\hat{C}_{12}(k)}\right), \quad (12.13)$$

which measures the phase difference between the frequency components of the two series: the number of leads ($\Phi_{12}(k) > 0$) or lags ($\Phi_{12}(k) < 0$) of $u_1(k)$ on $u_2(k)$ in sampling intervals at frequency ν_k is given by the so-called *standardized* phase $(2\pi\nu_k)^{-1}\Phi_{12}(k)$;
— the *gain* (regression coefficient)

$$\hat{G}_{12}(k) = \frac{|\hat{S}_{12}(k)|}{\hat{S}_1(k)}, \quad (12.14)$$

which indicates the extent to which the spectrum of $u_1(k)$ has been modified to approximate the corresponding frequency component of $u_2(k)$.

The analysis of these three quantities together with the (auto) spectra of each series and with the amplitude of their cross spectrum gives us an overall view of the frequency interaction of the two series. As anticipated at the

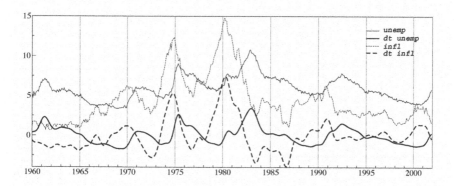

Fig. 12.2. US series. Raw series used in the application and the corresponding detrended ones: unemployment (raw: thin line, detrended: thick line) and inflation (raw: thin dashed line, detrended: thick dashed line). The data are monthly and cover the period Jan60-Dec01.

beginning, filtering procedures are often coupled to cross spectral analysis, either as preliminary or as a consequential step. In fact, it is sometimes evident from spectral peaks investigation that most of the power is contained in one or more bands. In particular, many macroeconomic time series (in level) have the typical Granger-shaped spectrum [12]. Such peaks may leak into nearby components and corrupt spectral and cross spectral investigation in low-power bands. That is why it may be advantageous to "pre-filter" the data. On the other hand, filtering can also be required afterwards when the spectral power concentration occurs in the coherency spectrum. This would involve that only some bands are important for the "interaction" between the series, all the remaining frequency components being useless.

12.4 An Application to the US Phillips Curve

In this section the methods just described are applied to the analysis of the US Phillips curve in the frequency domain[3].

We start by looking at the raw data (Figure 12.2): neither unemployment nor inflation show any obvious trend. Nevertheless, since the data cover a period of 42 years, we could expect the existence of a low-frequency trend, unobservable by simple visual inspection. Moreover, as the data are not seasonally adjusted, we risk to find an effect on correlation we are not interested in. I thus perform a filtering operation by means of the windowed filter described above which eliminates all periodicities smaller than one year and higher than 21 years, which, as we saw, may be fictitiously introduced by the

---

[3] This section builds on [10]. The issue of the Phillips curve historical behavior at different frequencies is also broached in [14, 17].

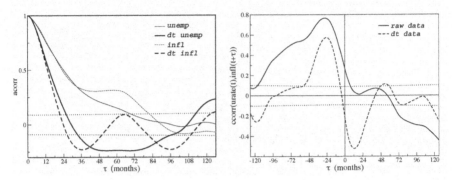

Fig. 12.3. Auto and Cross Correlograms. The dotted lines represent the approximated two standard error bound from the null value, computed as $\pm 2/\sqrt{N - \tau}$.

Fourier approach. The filtering operation on this particular band has the effect of detrending and smoothing our 42-years original series and hereafter I will refer to the resulting series as to the *detrended* series. Figure 12.3 reports raw and detrended data autocorrelation functions for both unemployment and inflation and their cross-correlation function. Only about $N/4$ lags (10.5 yr) are reported, as suggested by [4], in order to have enough lagged products at the highest lag, so that a reasonably accurate average is obtained. It may be seen that autocorrelation functions (left panel) drop to zero more quickly in the detrended than in the raw series. Furthermore a sort of oscillating behavior emerges in the case of detrended inflation, while it was absent in the raw case. We also observe significant negative autocorrelation values for both series which appear only after the detrending operation. These values correspond approximatively to lags between 3 and 7.5 years for unemployment and between 2.5 and 4 years and 7 and 9 years for inflation, the intermediate values being not significantly different from zero. Finally and more crucially, we notice that following the detrending operation a negative cross correlation (right panel) emerges in the short-to-medium run (0 to 36 months lag), which was absent in the raw case. This justifies our fears about the "hiding" effect of low-frequency high-power spectral components on short-to-medium term correlation visibility. The negative cross correlation between the detrended series means that: (a) in the (wide) band of frequency $\nu \in [21^{-1}, 1]$ yr^{-1} that we extracted it does exist a contemporaneous Phillips curve, as shown by the negative cross correlation at lag zero, which did not appear by the sole visual inspection of raw data; (b) there is a retarded negative effect of unemployment on inflation in this frequency range, which reaches its maximum at about 1 year, meaning that a rise in inflation will follow of an year a fall in unemployment, in other words a delayed Phillips curve.

Obviously, we are not able to establish a lead-lag relation by the simple study of the cross-correlation function. In fact, if we look at its negative

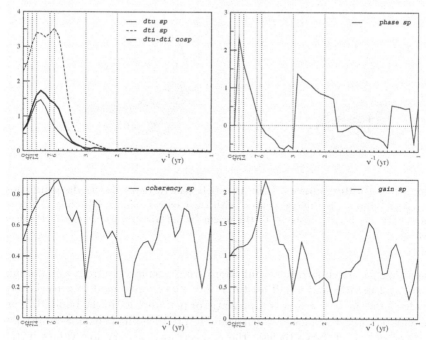

Fig. 12.4. Cross spectral analysis. Main quantities relative to US unemployment and inflation: auto and cospectrum (top left), standardized phase (top tight), coherency (bottom left) and gain (bottom right). For a greater legibility, on abscissae I report the period, which is the inverse of the frequency and is expressed in years.

lags part, we find a positive correlation between inflation and retarded unemployment, which reaches its maximum at a 2-year lag. This implies a positively sloped Phillips curve. We would need additional information to establish which of the two is leading the other. Cross spectral analysis (Figure 12.4) can answer this purpose and more. Notice that the residual spectral power at frequencies lower than $(21 \text{ yr})^{-1}$ is an effect of the smoothing. The same happens to the other quantities, thus their value outside the band should be disregarded. The parameter M was set to 140 after the preliminary window closing procedure. We notice (top left panel) that the spectrum of inflation is higher than that of unemployment, confirming the former's higher variance. Moreover, inflation shows two non-harmonic peaks at the periodicities ν^{-1} of 14 and 6 years, while the only unemployment spectral peak is at 10.5 years. The cospectrum (top left panel) shows a concentration of the two series covariance approximatively in the periodicity band $[21, 6]$ yr. In point of the standardized phase (top right panel), the plot shows a leading behavior of unemployment on inflation in the *same* band ($[21, 6]$ yr). This is consistent with the findings of the cross correlation inspection, which showed a negative correlation between unemployment and lagged inflation (Figure 12.3). The

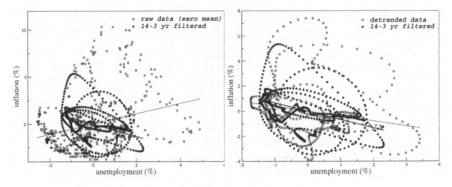

Fig. 12.5. Phillips curves. Raw data (left panel), detrended (right panel) and 14 − 3 yr filtered US Phillips curves, with their corresponding OLS line. The raw data curve has been translated to the origin by subtracting from both series their means.

leads (lags) of unemployment (inflation) components vary from a maximum of about 2 months and a half for the $\nu^{-1} = 21$ yr component to zero (i.e. coincident) for the $\nu^{-1} = 6$ yr component. For periodicities in the band $[6, 3]$ yr, we remark a negative phase, which would imply a leading behavior of inflation on unemployment. Nevertheless, the cospectrum has very low values in the second half of this band, where the phase is more significantly different from zero. We may thus conclude that these components do not account for much of the series covariance and that the prevailing trend is the former, i.e. unemployment leading inflation. The coherency plot (bottom left panel) shows a maximum frequency domain correlation at 5.25 yr, suggesting that filtering around this frequency would yield a more pronounced Phillips relation. The same information is given by the gain function (bottom right panel), whose maximum is found at the same frequency as the coherency.

Regarding the overall behavior of coherency, gain and phase, there is a transition at $\nu^{-1} = 3$ yr. In fact, while their behavior for $\nu^{-1} < 3$ yr can be considered reliable for cross spectral analysis, their non negligible value for $\nu^{-1} > 3$ yr might be the effect of some divergence (little denominators), since the cospectrum and the individual spectra above this value are nearly zero. To see more clearly at high frequencies, that part of the spectrum should be studied separately, that is we should extract the frequency components corresponding to the band $[3, 1]$ yr or $[3, 0.5]$ yr and perform the cross spectral analysis again. The interested reader is referred to [10].

Turning to the Phillips curve analysis, Figure 12.5 shows the raw data, the detrended and the 14 − 3 yr filtered curve. The first thing we notice is that the raw curve OLS line has a positive slope of 0.54 ($t_{\text{Student}} = 6.30$), with a low correlation coefficient of 0.27. If we look at the detrended curve, the slope becomes negative (-0.39 with $t_{\text{Student}} = -4.9$), with a lower correlation coefficient of -0.21. This is in agreement with the information given by the cross correlation (Figure 12.3). Finally, I filtered both unemployment and inflation on the band [14, 3] yr, containing the coherency maximum and indeed we find a negative slope of -0.66 with the highest correlation coefficient $\rho = -0.38$ ($t_{\text{Student}} = -9.32$). Thus, cross spectral analysis guided us in finding the band where we can detect a stronger Phillips relation.

12.5 Conclusion

This paper highlighted the main features of spectral analysis and their practical application. After a general theoretical introduction, it approached the issue of filtering for the extraction of particular components, mostly those related to the business cycle. In fact one of the advantages of the method is that it allows a quantitative definition of the cycle, and the extraction of long, medium or short term components, according to the researcher's wish. Then, the theory and practice of cross spectral analysis were sketched, introducing some typical concepts, like coherency and phase spectrum, which may provide some essential information, complementary to that given by time domain methods. Finally, these tools are applied to the study of US Phillips curve. This combined analysis managed to show that a Phillips relation arises between inflation and unemployment, at the typical business cycle components ([14, 6] yr), even if there is no hint of it in raw data. Moreover, by means of phase spectrum analysis, it was shown that unemployment leads inflation, the latter being delayed of about one year.

To conclude I would like to comment on the *fact-without theory* technique of bandpass filtering, which, after a period of great glamour, has lately come under attack (see e.g. [3]).

A major limit of this approach is that it is impossible to say anything about the evolution in time of the frequency content of the series, since the spectrum depends on the frequency but not on time. Thus, if a particular frequency component remains "switched on" only during a subsample, it is impossible to detect this interval by means of the inspection of sole series spectrum. To keep the information about time-localization some other kind of analysis is required, like time-frequency or wavelets, which, as we said, are tricky to apply to short series and for this reason may give unstable results.

Another issue is the fact that spectral analysis does not allow to disentangle the data generating process of a series from its spectrum. For instance, it cannot help in separating the cycle from the trend in a model where the latter is nonstationary, e.g. an $I(1)$ process [3, 19]. But in fact, because of the spectral form of an $I(1)$ process (non-null at all frequencies), this would require a separation *within* the individual frequency components, something that this purely descriptive method is not meant to do. Thus, dismissing the whole approach because of that would be like dismissing correlation analysis because it fails to detect causality between two variables.

In other words, the tool is far from being perfect, but, as all other methods in time series analysis, it has limits which have to be known and thoroughly explored to ensure a proper utilization; and, as any other method or model, it can not be expected to be universal.

References

1. Bartlett, M. S., *An Introduction to Stochastic Processes with Special Reference to Methods and Applications*, Cambridge University Press, Cambridge (1953).
2. Baxter, M. and R. G. King, "Measuring Business Cycles: Approximate Band-Pass Filters for Economic Time Series", The Review of Economics and Statistics vol. 8, no. 4 (November 1999), pp. 575–93.
3. Benati, L., "Band-Pass Filtering, Cointegration and Business Cycle Analysis", Bank of England, working paper (2001).
4. Box, G. E. P., G. M. Jenkins and G. C. Reinsel, *Time Series Analysis. Forecasting and Control*, third edition, Prentice Hall, Upper Saddle River, New Jersey (1994).
5. Christiano, L. and T. J. Fitzgerald, "The Band-Pass Filter", International Economic Review, vol. 44, no. 2, May 2003.
6. Hamilton, J. D., "Time Series Analysis", Princeton University Press, Princeton, New Jersey (1994).
7. Hamming, R. W., *Numerical Methods for Scientists and Engineers*, second edition, Dover Publications, Inc., New York, 1973.
8. Hamming, R. W., *Digital Filters*, third edition, Dover Publications, Inc., New York, 1998.
9. Hodrick, R. J. and E. C. Prescott, "Postwar US Business Cycles: An Empirical Investigation", Journal of Money, Credit, and Banking, vol. 29-1 (1997), pp. 1–16.
10. Gaffard, J. L. and A. Iacobucci, "The Phillips Curve: Old Theories and New Statistics", mimeo.
11. Granger, C. W. J.and M. Hatanaka, *Spectral Analysis of Economic Time Series*, Princeton University Press, Princeton, New Jersey (1964).
12. Granger, C. W. J, "The Typical Spectral Shape of an Economic Variable", Econometrica, vol. 34(1), (1966), pp. 150–161.
13. Granger, C. W. J, "Investigating Casual Relations by Econometric Models and Cross-Spectral Methods", Econometrica, vol. 37(3), (1969), pp. 424–438.
14. Haldane, A. and D. Quah, "UK Phillips Curves and Monetary Policy", Journal of Monetary Economics, Special Issue: *The Return of the Phillips Curve*, vol. 44, no. 2, (1999), pp. 259-278.
15. Iacobucci, A. and A. Noullez, "Frequency Filters for Short Length Time Series", Working Paper IDEFI–IDEE 2002, n. 1.
16. Jenkins, G. M. and D. G. Watts, *Spectral Analysis and Its Applications*, Holden-Day, San Francisco, (1969).
17. Lee, J., "The Phillips Curve Behavior Over Different Horizons", Journal of Economics and Finance, 19 (1995), pp. 51-69.
18. Mitchell, J. and K. Mouratidis, "Is There a Common Euro-Zone Business Cycle?", presented at the coloquium on *Modern Tools for Business Cycles Analysis*, EUROSTAT, Luxembourg, 28-29 November 2002.

19. Murray, C. J., "Cyclical Properties of Baxter-King Filtered Time Series", The Review of Economics and Statistics, May 2003, 85(2): 472-476.
20. Nerlove, M., "Spectral Analysis of Seasonal Adjustment Procedures", Econometrica, vol. 32(3), (1964), pp. 241–286.
21. Oppenheim, A. V. and R. W. Schafer, *Discrete-Time Signal Processing*, second edition, Prentice-Hall, New Jersey, 1999.

Policy Analysis Using a Microsimulation Model of the Italian Households [*]

Carlo Bianchi[1], Marzia Romanelli[2], and Pietro A. Vagliasindi[3]

[1] University of Pisa cbianchi@ec.unipi.it
[2] Sant'Anna School of Advanced Studies romanelli@sssup.it
[3] University of Parma pvagli@unipr.it

Summary. In this paper, we apply a dynamic microsimulation approach, which allows us to examine the evolution of the system as a whole and at the same time to focus our attention on the different typologies of workers and pensioners. The latter objective is achieved by simulating individual reactions to systemic changes, while taking into account the regional dimensions. This technique also enables us to perform a general micro-analysis of the effects of past reforms on family pension-income distribution and average individual pension-benefits. The analytical framework used is the dynamic microsimulation model MIND, jointly developed by the University of Parma and Pisa, that incorporates behavioural analysis of individual choices of retirement age, derived from the Stock-Wise [24] option value model. We also perform sensitivity analysis on the model. This represents a valuable technique for treating uncertainty in input variables and for testing the robustness of the simulation results to possible changes in the macroeconomic scenario. In particular, we measure the economic impact resulting from alternative values of the income growth rate and the real interest rate as well as the effect of different distributions of the education degree reached by the individuals.

Key words: pension reforms, income distribution, sensitivity analysis

13.1 Introduction

Over the past few years, pension treatments have been significantly modified by the introduction of new laws that are reforming the Italian pension system and modifying its structure and technical parameters. The dynamic microsimulation approach, by simulating individual reactions to systemic changes, allows us both to examine the evolution of the system as a whole and to focus

[*] MIUR financial supports are gratefully acknowledged. We also thank an anonymous referee and the participants at the workshop New Tools for Qualitative Economic Dynamics held at the CIMAT, Guanajuato - Mex (October 2002) and VII WEHIA Workshop held in Trieste (May 2002) for helpful comments and suggestions.

our attention on the different types of workers and pensioners, while taking into account the regional dimensions. This technique also enables us to perform a general micro-analysis of the effects of past reforms on family pension-income distribution and average individual pension-benefits. Furthermore we can compare the evolution of income concentration and poverty both in the short and long run.

The MIND (Micro Italy National Dynamics) model[4] is particularly useful for these purposes, since it considers both the national socio-demographic dynamics and the different Italian regional geographical areas (North, Centre, South and Islands).[5] Specifically, it allows us to examine the increase in the pensioners/workers ratio, determined by the present demographic evolution in Italy, and the pension/wage ratio, that may cause strains in the social security budget.

The main purpose of this paper is to deal with social security issues, presenting new evidence on the effect of alternative policy options relative to the past pension and social security reforms on inequality and poverty, abstracting from the recent measures of the Berlusconi government in favour of lowest pensioners, which recently created a sort of selective safety net. Specifically, the paper discusses some basic policy experiments and describes some simulation results; furthermore we perform sensitivity analyses on the model. This may, in fact, represent a valuable technique for treating uncertainty in input variables and for testing the robustness of the simulation results to possible changes in the macroeconomic scenario. In particular, we measure the economic impact resulting from alternative values of the income growth rate and the real interest rate as well as the effect of different distributions of the education degree reached by the individuals.

In the following, we microsimulate the future evolution of pension treatments, that in the '90s have been substantially modified by the introduction of two new laws.[6] In this framework we will assess the consequences of linking the lowest pensions to real wages (or to per capita income) through an explicit in-

---

[4] Its analytical framework is based on the dynamic aging model discussed in Bianchi et al. [6]

[5] MIND is inspired by the model developed by Cannari and Nicoletti Altimari [9]. The main differences between the two models refer to the inclusion in MIND of a geographically differentiated demographic evolution and a behavioural analysis of individual choices of retirement age, derived from the Stock-Wise [24] option value (OV) model.

[6] The Amato law (L.503/92) raised the old age pension retirement age, lowered future pensions for workers subject to the old defined benefit regime (called "retributivo" and based on final wages, i.e. wages received in the last years of the working period) and modified the indexation rule. The Dini law (L.335/95) introduced a new defined contribution regime (called "contributivo" and based on social contributions) for the new entrant workers and a long transition phase (the so called "mixed regime") for mature ones. It also suppressed minimum pensions, shifting vertical redistribution outside the pension system and implying a

dexation mechanism, distinguishing between the rules applied to (private and public) dependent workers and to self-employed ones.[7] Indexation to prices[8], introduced before the 1992 reform, has important drawbacks mainly because it transforms the lowest pensions from a social minimum into a biological one. Moreover it reintroduces differences between pensions depending on the year of retirement (the so called "vintage pensions") thus setting the premises for future discretional interventions aimed at eliminating such problems on an equity basis.[9] Considering behavioural reactions to changing incentives, we will appraise (i) the intra-generation and inter-generation consequences on inequality and poverty as well as on the cumulated average values of wages and pensions of the new defined contribution system, and (ii) the restoration of wage indexation for the lowest (welfare) pensions.

The paper is organised as follows. In section 13.2, using a Monte Carlo experiment, we study the consequences of social security reforms on inequality and poverty. In particular, considering a time period of 50 years, from 1996 to 2045, we follow the evolution of the phenomena of distributive inequalities and poverty in the Italian families with at least one pension income recipient (section 13.2.1) and of average cumulated values of wages and pensions (section 13.2.2), discussing the issue of indexing lowest pensions to wages. Section 13.3 concludes with some final remarks.

---

larger role for welfare benefits (i.e. the social pensions) in the future. Moreover, it envisaged some sort of wage indexation of the lowest pensions from 2009.

[7] In particular we consider the two lowest types of Italian pensions; the "assegni sociali"(i.e. the social pensions) and the "pensioni integrate al minimo"(i.e. minimum pensions that received a social integration to reach a minimum threshold, which is slightly above the level of social pensions). In this way we strongly restrict the application of indexation to wages and reintroduce an automatic vertical redistribution inside the pension system. This put us out of the wider debate on the issue of price versus wage indexation of pensions, that has taken place even before Italian reforms and dates back at least to Aaron [1].

[8] In 2001 the indexation rule has been modified. A larger number of pensions is now fully indexed to price dynamics. In our simulations we consider this new indexation regime and apply it to the new indexed lowest pensions, without considering the discretional benefit increase taking place from 2001 onwards.

[9] Non-welfare pensions have different purchasing power depending on the year of retirement (since they depend on the past wages) and ,differently from future wages, do not grow in real terms, during the retirement period. Starting from "high pensions"individual pension treatments, trade unions accepted a partial indexation to prices. In this way, since wages and incomes grow in real terms, there is no certainty about the future value of the pensions during the retirement period. This uncertainty stems from the inflation level, normally incorporated in nominal wage and income growth. In the transition phase, this mechanism is expected to consistently reduce the relative purchasing power of "privileged"pension treatments when real wages grow. However, by lowering the purchasing power of ongoing pension it creates "vintage pensions"and a periodical stimulus to rise the level of pensions a few years after the retirement.

13.2 Trends in Inequality, Poverty and Wealth Under Alternative Policy Experiments

In what follows, we will use a small well-aligned sample[10] to derive future scenarios, that can be described using alternative indexes, considering the mean values of 20 Monte Carlo replications.[11] We compare possible developments of the existing state of affairs (**B** basic scenarios) with the ones following the indexation of lowest pensions to wages (**A** alternative scenarios). Around 15% of pensioners are entitled to wage indexation but also other pensioners may indirectly benefit due to the pressure of higher indexation to prices (e.g. 100% instead of 75%). Furthermore, we consider different "retirement behavioural rules", depending on whether workers (case **1**) decide their retirement age maximising the expected pension benefits or (case **2**) look for the maximisa-

---

[10] The database we use in our analysis has been based on the 1995 Surveys of Households Income and Wealth (SHIW) of the Bank of Italy [3]. These surveys are the most updated and reliable source of information on demographic and economic characteristics of the Italian families. A brief description of the information included in the SHIW surveys can be found in Cannari and Nicoletti Altimari [9]. The 1995 survey includes 23,294 individuals belonging to 8,135 families representative of the universe of 19.6 million of Italian families at December 31, 1995. We made several modifications to the original sample. In particular, we first adjusted the sample weights to match some of the basic features of the Italian population in 1995. We then reconstructed the gross income for all labour income recipients and pensioners, starting from the net income given in the survey, according to law provisions. Finally, using a statistical procedure, similar to the one discussed in Cannari and Nicoletti Altimari [9], we generated a representative sample, consisting of 41,941 individuals belonging to 15,068 families. This sample keeps the maximum heterogeneity as it includes, at least once, all the interviewed families.

[11] As discussed in Bianchi, Romanelli and Vagliasindi [6], MIND is a stochastic model in the sense that it uses dynamic aging to generate, probabilistically, the various events in individuals' lives. The drawback of this procedure is the uncertainty it introduces into the model's results - if one re-seeds the model's random number generator, one gets a different result for the "same" simulation (see Morrison [11]). By using mean values of Monte Carlo replications, we reduce the uncertainty introduced by the dynamic aging approach. 20 replications seem to be enough to ensure reliable results. In fact, considering mean values of 20 replications we substantially reduce the variance by 95% even if, according to Wolf [17], it must be recalled that "in survey-data item-imputation applications of the multiple-imputation technique, a small number (say 3-6 replications) has been viewed as sufficient". The reduction of variance can be performed either considering the mean values of Monte Carlo replications, or using "ad hoc" techniques. Unfortunately, as discussed in Neufeld [12], the use of "ad hoc" techniques - such as *selective sampling* or *side walk methods* - to reduce the variance may have some drawbacks, e.g. altering the probabilities of the events. The use of Monte Carlo replications, even if costly in term of computational time, allows to (a) improve the precision of the sensitivity analysis of the results with respect to alternative policies and (b) quantify the uncertainty associated to the simulation results.

tion of their whole benefits flow coming from labour and pension incomes. In conclusion, we simulate four different scenarios: **B1**, **A1**, **B2** and **A2**. [12]

13.2.1 Gross Cumulated Values of Wealth from Wages and Pensions and Gender Inequalities

In what follows we perform a Monte Carlo experiment comparing the average gross wealth from wages and pensions accumulated in the working life and retirement period by generations born between 1950 and 1980.[13] We denote with W^A and W^B (P^A and P^B) wealth from wages (pension), respectively under **A** and **B**, according to whether there is (or not) indexation to wages. We define the cumulated values of salaries Y_j, capitalised in the contribution years a before retirement p, and of discounted pensions B_s, perceived up to the S age, as the wealth from wages and pensions of individual i as follows[14]:

---

[12] The additional cost per year of indexing lowest pensions to wages is quite negligible, ranging from 1% around 2020 to reach 4% in 2045 under **A1**. The cost is is substantially lower (about one half) under **A2**. Specifically, we obtain the two maximising behaviours (**1** and **2**) by assuming different values of α in the indirect utility function of labour income in the option value (OV) model, discussed in Bianchi et al. [6]; in case **1** we assume $\alpha = 0$ while in case **2** we set $\alpha = 0.75$. The two alternatives mean that workers either consider only the benefits from pensions or appraise the net benefit of 1 euro from their wage as equivalent to 0.75 euro from their pension. When $\alpha = 0.75$ workers tend to retire later to enjoy for a longer period the higher wage incomes, hence, to avoid corner solutions, we have set a social constraint to retire at 66 for dependent workers at 67 for public and 68 for self employed.

[13] In this way we can refer to the defined benefit system (considering the individuals born until 1959), to the mixture (considering those born between 1960 and 1976) and to the defined contribution (considering the individuals born from the 1977). Linking the age of citizens to the pension regime is somewhat arbitrary, e.g. some individual born in 1958 or 1959 will be under the mixed regime. However, this distinction is necessary for comparing wealth from wages and pensions of different generations and to attribute them to the influence of different regimes. It must be recalled, the model considers 18 typologies on the basis of three characteristics, geographical area (north, centre and south), sex (males, female) and occupational sector (public dependent, private dependent, self-employed worker).

[14] S has been set equal to 74 years for males and 81 for females; the real rate of interest are those on the Ordinary Treasure Bill BOT until 1994 (until 1979 see Homer and Sylla [10]); then it equals 2.5%. W and P are calculated setting 1999 as the starting year (considering individuals born from 1950 till 1979). They are aggregated in periods of 5 years and made comparable by discounting them with the real rate of growth of per capita GDP. This procedure is related to intergenerational accounting. However, our comparison between present and future generations doesn't imply the constancy of relative profiles, since we consider behavioural changes (in accordance with the OV hypothesis) as fiscal and contributive rules are modified.

$$W_{ip} = \sum_{j=p-a}^{p-1} Y_{ij}(1+r)^{p-1-j} \qquad P_{ip} = \sum_{s=p}^{S} B_{is}(1+r)^{p-s-1}$$

In this way, we can evaluate the regional impact on wealth from wages and pensions considering the regional dimension (North, Centre and South). To appraise in more details the results we also consider the employment status (Dependent worker, Self Employed, Public).

The average wealth (in thousands of euros) per worker (pensioner) of five generations of Italian citizens (males and females) are reported under the scenario *1* in table 13.1 and under the scenario *2* in table 13.2.

In general, the estimates of wealth from wages show how the differences between sexes decrease: (i) under *B*, because, while female wealth increases till the beginning of '70, the male's one stops rising during the '60s, (ii) introducing indexation, since female wealth grows more than male's one. On the other hand, under *B*, wealth from pensions decreases for males more than for females. Indexation slightly reduces percent differences (new pension treatments increase around 1 and 1.5%, respectively for males and females).[15]

Wealth from wages is higher in case *2* (when α equals 0.75), depending on the desire to retire later in order to enjoy wage incomes for a longer period. This also accounts for the lower pension wealth (given the shorter period in which the benefit is enjoyed, and the hypothesis of pension wealth maximisation in case *1*, for $\alpha = 0$). On the other hand, wealth from pensions slowly increases both for males and females. Apart from the last generations of dependent workers in the North and of self-employed in the South, indexation slightly reduces percent differences (new pension benefits tend to increase more for females than males). The profiles emerging from the different hypotheses of retirement behaviour are also very similar: distinctions emerge only in the absolute values.

13.2.2 Inequality, Poverty and Cumulated Wealth in Alternative Macro-scenarios

Considering five-year intervals in the period 2000-2045 and employing the indexes suggested in the recent literature[16], the MIND model allows us also to examine the evolution of poverty and the concentration of gross incomes (net of capital incomes) among families with at least one pensioner (generally the

---

[15] Indexation drives females to postpone retirement, increasing wealth from wages and affecting in this way the level of the average female pension benefit. However, absolute values could greatly reflect critical approximations and hypotheses based on the working of the model.

[16] Following Atkinson [2] and Champernowne and Cowell [8], for income concentration we use the Gini index while for poverty we compute the diffusion index, the intensity index and the Sen index modified by Shorrocks [13].

Table 13.1. Wealth from wages and pensions under *scenario 1*

AGE CLASS	D W^A	W^B	SE W^A	W^B	PUBL W^A	W^B	D P^A	P^B	SE P^A	P^B	PUBL P^A	P^B
NORTH MALES												
50-54	993	993	864	865	1047	1047	225	222	187	186	106	105
55-59	1075	1075	859	858	1076	1076	224	223	169	167	131	129
60-64	1067	1067	1051	1048	1040	1042	207	206	160	158	101	99
65-69	1008	1009	968	965	1011	1012	184	183	116	114	95	91
70-74	980	979	969	967	841	841	176	174	110	109	83	81
75-79	881	883	928	923	769	777	154	153	95	94	83	83
NORTH FEMALES												
50-54	635	635	461	461	889	889	233	228	186	185	250	249
55-59	659	658	668	669	868	869	210	209	176	174	231	229
60-64	672	671	558	561	927	937	205	203	163	161	211	209
65-69	799	796	634	630	906	873	190	188	182	180	193	189
70-74	841	831	604	609	889	870	146	146	180	178	186	184
75-79	815	797	650	650	856	850	131	131	170	167	180	178
CENTRE MALES												
50-54	940	940	759	757	947	948	229	228	135	133	200	199
55-59	936	937	863	861	1086	1085	215	214	166	163	208	206
60-64	849	850	1168	1170	980	980	177	176	155	154	178	178
65-69	958	958	906	921	966	963	174	173	115	114	174	172
70-74	918	912	873	874	871	863	159	158	90	89	150	148
75-79	974	974	901	873	784	790	162	161	89	86	134	134
CENTRE FEMALES												
50-54	482	481	502	502	764	764	140	139	107	106	234	232
55-59	561	564	399	400	753	756	148	147	94	92	204	203
60-64	668	673	500	495	851	844	159	157	88	85	197	195
65-69	664	658	550	550	1011	1002	148	146	86	83	221	218
70-74	779	764	659	659	838	824	164	162	86	86	179	176
75-79	738	721	678	702	809	801	154	152	87	90	172	170
SOUTH MALES												
50-54	675	675	576	576	849	849	153	152	103	101	168	167
55-59	655	657	605	606	880	880	133	132	87	85	166	165
60-64	809	810	691	693	867	866	147	146	84	82	149	147
65-69	934	936	664	655	676	678	162	162	75	72	117	117
70-74	909	909	627	621	674	663	154	153	67	65	113	112
75-79	917	906	617	618	668	672	152	149	64	63	112	112
SOUTH FEMALES												
50-54	324	325	234	234	687	689	102	99	60	57	176	175
55-59	532	530	306	305	825	828	134	132	57	55	194	193
60-64	653	649	433	442	721	722	149	146	65	64	162	161
65-69	653	653	422	419	824	820	144	143	58	57	177	177
70-74	787	777	497	494	788	777	166	164	65	64	167	165
75-79	834	821	512	502	812	787	174	172	64	63	171	165

Table 13.2. Wealth from wages and pensions under *scenario 2*

AGE CLASS	D W^A	W^B	SE W^A	W^B	PUBL W^A	W^B	D P^A	P^B	SE P^A	P^B	PUBL P^A	P^B
NORTH MALES												
50-54	1043	1045	930	932	1125	1125	198	198	149	146	193	192
55-59	1129	1130	953	952	1161	1161	200	198	122	120	179	177
60-64	1163	1164	1230	1236	1173	1176	174	172	102	99	163	162
65-69	1157	1158	1164	1171	1210	1219	144	143	74	72	137	137
70-74	1137	1140	1163	1178	999	1021	141	139	72	71	108	109
75-79	1024	1033	1132	1104	929	935	125	125	66	64	100	100
NORTH FEMALES												
50-54	669	669	551	550	1066	1065	176	175	82	80	195	193
55-59	698	698	780	785	1035	1036	168	165	97	96	185	183
60-64	737	740	653	656	1053	1053	151	150	79	77	191	189
65-69	892	892	711	716	974	968	168	166	83	78	180	176
70-74	896	900	663	667	940	940	170	168	73	72	174	173
75-79	835	836	698	699	895	898	162	160	74	74	168	167
CENTRE MALES												
50-54	1088	1089	893	906	1009	1010	159	157	75	72	169	169
55-59	1075	1078	1060	1081	1187	1188	156	153	83	82	167	165
60-64	1002	1000	1415	1449	1105	1103	127	125	93	93	137	136
65-69	1143	1151	1140	1098	1134	1136	139	140	76	71	128	126
70-74	1096	1095	1086	1093	1034	1022	132	132	65	64	112	110
75-79	1137	1153	1105	1098	946	954	137	139	65	64	101	102
CENTRE FEMALES												
50-54	529	529	545	545	883	887	123	122	93	90	192	188
55-59	618	620	422	418	864	862	134	132	88	86	174	171
60-64	734	733	541	543	932	934	148	146	75	72	182	181
65-69	708	714	600	604	1081	1082	141	139	73	70	206	205
70-74	808	807	710	720	884	871	158	156	74	75	168	164
75-79	754	753	717	751	849	843	148	147	74	77	161	158
SOUTH MALES												
50-54	758	759	625	628	938	940	111	110	86	84	124	121
55-59	733	735	649	660	996	998	104	102	69	67	122	117
60-64	935	934	795	802	991	993	117	116	54	53	109	107
65-69	1099	1101	799	791	811	810	133	133	48	46	88	87
70-74	1075	1077	749	777	809	793	131	130	44	45	87	85
75-79	1066	1068	758	745	809	813	129	129	43	42	87	87
SOUTH FEMALES												
50-54	341	341	252	252	776	778	95	91	60	54	149	149
55-59	561	560	340	338	920	922	126	123	61	48	176	175
60-64	694	692	482	484	785	788	142	139	62	56	151	151
65-69	685	691	458	450	875	878	139	138	63	49	167	167
70-74	809	810	534	525	833	825	159	159	64	55	158	155
75-79	845	844	541	537	847	825	166	165	65	56	159	153

head of the family or the spouse).[17] Our experiments[18] both follow the distributive consequences along the two alternative paths (**A** and **B**) and asses the robustness of the simulation results to possible changes in the macroeconomic variables. The use of deterministic relationships, ignoring the uncertainty about the future and the imperfect knowledge of the present, may induce to accept inappropriate policies.[19] In this respect, emerges also the problem of choosing a policy which minimizes the uncertainty associated with macroeconomic variables.[20] The analysis we propose, assuming different values of some crucial variables is useful to verify whether the selected policy is also sufficiently robust to the uncertainty associated with their possible alternative realizations. Dealing with concentration and poverty indexes - varying between 0 (equality) and 1 (maximum inequality) - we assume the family as the unit of comparison, and an absolute poverty threshold anchored to per capita income, set in 1999 equal to 7.64 thousand euros for a family with 2 components. The income equivalence scales[21] used for poverty (*absolute equivalence scale*) and concentration (*relative equivalence scale*) are both derived from ISTAT.

Table 13.3. ISTAT (Italian statistical agency) equivalence scales

Number of components	1	2	3	4	5	6	7
ISTAT absolute eq.scale	0.67	1.00	1.42	1.80	2.27	2.61	2.95
ISTAT relative eq.scale	0.60	1.00	1.33	1.63	1.90	2.15	2.40

[17] It must be recalled that: a) between 2000 and 2020 the system of defined benefit prevails, b) subsequently mixed calculation takes over (coexisting with pensions already liquidated with the defined benefit), c) finally, after 2035 only the defined contribution survives (persisting however individuals retired with the two former regimes).

[18] We assume a probability to find occupation of 8% in the Centre-North and 6% in the South and the same career profiles in the Centre-South. It is not possible to estimate the possibility to loose or change occupation, as this information is not included in the SHIW database.

[19] For instance, policies may be unsustainable. "The awareness of the stochastic nature of the environmental variable may induce to reject policies which, deterministically sustainable, are associated with a dangerously low γ [ex-post sustainable development probability]. Differently put, the set of 'stochastically sustainable development' is included in the set of 'deterministically sustainable development'." Cf. Tucci [15], pag. 155.

[20] "The problem of choosing an economic policy which minimizes the uncertainty associated with the actual output growth and its effect on sustainability becomes fundamental. The explicit introduction of stochastic policy variables needs to be done to select the 'optimal' (or 'acceptable') policy mix minimizing the uncertainty associated with output growth." Cf. Tucci [15], pag. 162.

[21] These coefficients, increasing with the number of individuals, are used for eliminating the heterogeneity of incomes due to the different numerousness of family units.

In figures 13.1 up to 13.6 we report some experiments under the hypothesis of the maximisation of expected pension benefit flow (case *1*), using different macro scenarios. In particular, under the hypotheses *A* and *B* we analyse (i) the distribution of gross incomes and poverty in families with at least one pensioner and (ii) gross wage and pension wealth.

We consider three alternative macroeconomic scenarios, comparing the base simulation (case *1*) with real income per-capita growth rate g equal to 1% and a future real discount rate r of 2.5%[22] with other two simulations, *G* and *R*:

a) under *G* , we set the growth rate g at 2%;
b) under *R*, r is equal to 4% while the other parameters are left unchanged.

Additional simulations are also realised assuming a different probability distribution of the educational levels reached by the individuals (*E*) assessing a stronger tendency to gain a higher level of education (increasing by 20% the probability to obtain the higher degrees).

As shown in figure 13.1, under all the different hypotheses, indexation of the lower pensions to wages seems very effective in reducing income concentration and poverty.

The differences emerging among the various scenarios, either a higher growth rate or a greater discount rate, determine an increase of income concentration and poverty intensity (with respect to the baseline solution). In particular, considering the Gini index, case *R* presents a profile very similar to the baseline simulation (case *1*), both with or without indexation. On the other hand, *G* lays systematically above (it might be interesting to notice that, till 2030, the indexation of lowest pensions (*AG*) is able to keep the concentration stable and at the same level reached under *B1* and *BR*).

Summing up, indexation of lowest pension to wages reduces income concentration also with higher real per-capita growth and discount rates.

The effect of indexation on poverty is also analogous to the basic case both under *R* and *G* and the profiles have quite similar behaviours compared to case *1*, although they present some dissimilarities. The scenario with different growth rate diverges till 2030 and after converges toward case *1*. On the other hand, poverty diffusion (intensity) under case *R* follows the same trend of case *1* until 2020 (2010) but ends up with a higher level. Thais happens also because a higher discount rate seems to induce the younger cohorts to anticipate exit from the labour market (about 2 years earlier than in the basic case).

---

[22] Specifically, the annual average wage growth rate we use in the baseline case is equal to that one realized in the period 1995 - 2000 plus g (obviously, for each individual, the wage varies also according to her/his career, etc). The discount rate, represents the average after tax interest rate consumers can obtain in the market. Accordingly, the used value (2.5%) may perhaps represents an optimistic hypothesis on the average interest rate consumers can obtain from their assets in the long run.

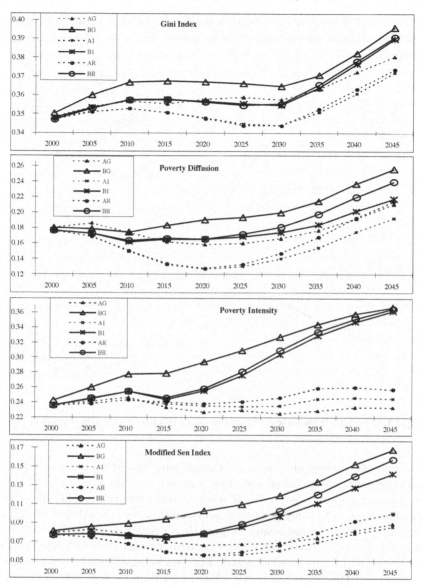

Fig. 13.1. Concentration and poverty indexes

Figure 13.2 reports the average wealth per worker of five generations of Italian citizens (males and females) under the three alternative macroscenarios.

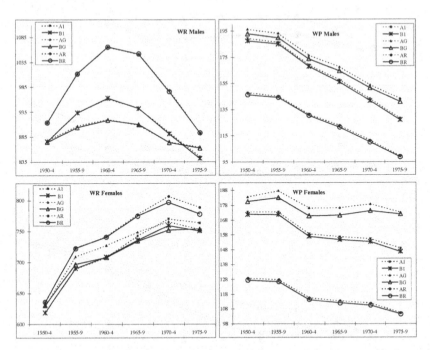

Fig. 13.2. Wealth from wages (W) and pension (P) per worker (thousands of euros)

Under **G** workers anticipate retirement with analogous pension benefits, since they accumulate faster, but enjoy a greater pension wealth being retired for a longer period. Also under **R** workers anticipate retirement, since they are more anxious to enjoy pension benefits. In this case, even if they work less, they enjoy a greater wage wealth, wages being cumulated with a higher interest rate. Obviously, workers have longer retirement periods and lower pension treatments and actual pension wealth decreases being discounted with a higher interest rate. This behaviour is even clearer for female workers (who on average anticipate retirement of two years, instead of one year). Accordingly, they have lower pension treatments and pension wealth.

Considering total expenditure and the average individual pension treatment till 2045, figure 13.3 shows a strong increase in aggregate and average pension expenditure under **G** (while the profiles are quite the same under **R**), so that with a higher growth rate, the indexation policy has greater costs.

Finally, let us consider a scenario characterised by a sharper trend toward a higher education level (**E**). Concerning the analysis of income concentration and poverty, from figure 13.4, it emerges that giving the possibility to reach

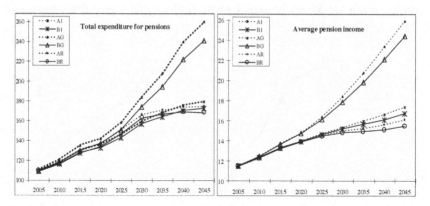

Fig. 13.3. Aggregate expenditure for pensions and average treatment (**A** and **B**)

higher education levels to a wider group of people has a slight incidence on the Gini index, which shows a higher level of concentration starting from 2025. On the contrary changes in poverty are really minor.

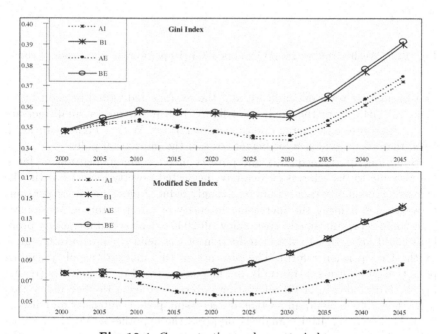

Fig. 13.4. Concentration and poverty indexes

Figure 13.5 shows that on the males' side there are no major effects on wealth from wages and pensions, while females, especially the younger generations, reach significantly higher levels and in this way differences among genders are reduced.

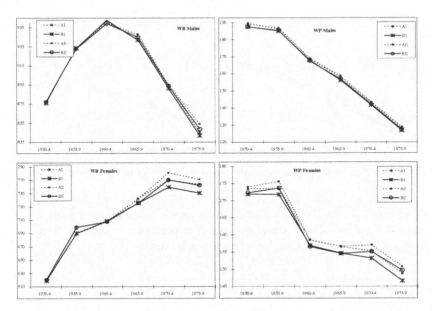

Fig. 13.5. Wealth from wages (W) and pension (P) per worker (thousands of euros)

Considering total expenditure and the average individual pension treatment till 2045, from the analysis of figure 13.6 it emerges a small increase in the aggregate and average pension expenditure. With a higher education level, there is no evidence for bigger costs of the indexation policy.

Summing up, (i) inequality and poverty increase among pensioners' families also due to the reforms (which have neglected poverty issues) but (ii) the increasing inequality trend could be strongly reduced indexing lowest pensions to wages, i.e. fighting the increasing intensity of poverty among pensioners. This could represent an effective policy till 2030 when several inequality problems could arise, due to the introduction of the defined contribution regime by the Dini pension reform. The presence of this indexation policy against poverty slows down substantially the increasing trend of inequalities (reducing the Gini index in 2045) alleviating poverty (reducing the Sen index even more than 40% in 2045), in particular decreasing poverty intensity and disparities. Thus, our sensitivity analysis supports the hypothesis that the increasing trends of inequality and poverty among pensioners are mainly owed to the price-indexation system and shows the stability of the results with respect to alternative values of the macroeconomic variables.

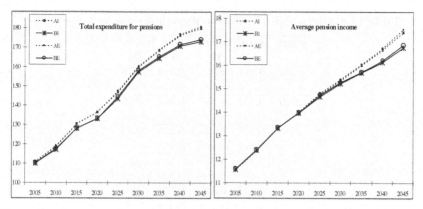

Fig. 13.6. Aggregate expenditure for pensions and average treatment (*A* and *B*)

The analysis uses Monte Carlo methodology to ensure: (i) more reliable results, since they are not based on one experiment, but are derived by the mean values of 20 replications, and (ii) the possibility to clearly distinguish, on the basis of the adopted indexes, the outcomes obtained under the alternative scenarios. In fact, the variability of the outcomes may be so high that our conclusions may loose significance, not allowing a clear discrimination between the effects of the two options. This discrimination is possible using Monte Carlo method, that allows the estimate of the variances of the simulation results.[23]

As an example of the reliability of the simulation results, in table 13.4 we report the standard errors (values x 10^3) of the mean values of th Gini index and the modified Sen index for the baseline case (*A1* and *B1*). What emerges are the very small values of the standard errors which are never higher than 0.7% of the mean value of the respective indexes.

---

[23] In order to be able to distinguish the two policies on the basis of the adopted indexes we must estimate the probability that the value of an index for a given year with a given option (*A* or *B*) under a given macro-scenario could be attributed to another alternative (and vice versa). With normal distribution of the Monte Carlo replications it would be easy to determine the confidence intervals at the 99% or at 90% levels- simply by adding and subtracting the standard error n (= 2.32 or 1.28) times the mean values - and verify if they overlap. Since we don't know the form of the probability density functions of the indices, we cannot test the significance of the difference between the two means. However, we can use the following simple n-sigma criterion. In particular we assume the two options as distinct when the intervals, calculated by adding and subtracting twice the standard error to the mean values, are separated. According to this criterion, in our analysis, for each macro-scenario, the indexes obtained under the two policy options (*A* and *B*) are always distinguishable. For a numerical illustatrion of the criterion, see Bianchi, Romanelli and Vagliasindi [4].

Table 13.4. Gini and Sen indexes; in parentheses standard errors x 10^3

years	Gini Index		Sen Index	
	A1	B1	A1	B1
2005	0.351	0.353	0.074	0.078
	(0.537)	(0.537)	(0.425)	(0.447)
2010	0.353	0.357	0.067	0.076
	(0.626)	(0.648)	(0.380)	(0.402)
2015	0.351	0.358	0.058	0.074
	(0.827)	(0.805)	(0.402)	(0.425)
2020	0.348	0.357	0.055	0.078
	(0.984)	(1.006)	(0.380)	(0.469)
2025	0.345	0.356	0.056	0.086
	(1.163)	(1.163)	(0.402)	(0.425)
2030	0.344	0.355	0.061	0.097
	(1.140)	(0.961)	(0.402)	(0.447)
2035	0.351	0.364	0.070	0.111
	(0.760)	(0.783)	(0.447)	(0.425)
2040	0.361	0.377	0.079	0.128
	(0.693)	(0.872)	(0.469)	(0.537)
2045	0.372	0.390	0.086	0.143
	(0.805)	(0.693)	(0.425)	(0.581)

13.3 Concluding Remarks

The microsimulation experiments allowed us to follow the evolution of the average cumulated wages and pensions as well as inequality and poverty trends for families of pensioners. Furthermore, we compared the redistributive consequences of price indexation (current system) with those relative to the introduction of indexation to wages of the lowest pensions.

In the Italian case, the analysis showed how demographic trends and the adopted reforms produce the increase of inequality and poverty among pensioners' families over time, notwithstanding the greater disincentives to anticipated retirement, the decrease of unemployment, the presence of more regular careers and the emerging hidden economy. Moreover, the absence of indexation (of social and lowest pensions to per capita income) speeds up the rising trend of inequalities by increasing poverty, suggesting that the increase of inequality and poverty is largely due to the current indexation system.

The Monte Carlo analysis confirmed the reliability of our results.

Of course, there are still problems to be solved by further research. For example, if we want to consider incomes incomes and pensions on after-tax basis, we must estimate future taxes and contributions (in relation to different types of employment) and wealth arising from the accumulation of private pension funds inclusive of reformed TFR (the Italian pension fund cumulated at the end of the working period as dependent worker, see Vagliasindi [16]). Furthermore, it could be interesting to: (a) examine economic individual behaviours in more detail, setting family choices in connection to the labour market[24] and (b) introduce immigration, creating a database of families and extending the relative behavioural differences to the various socio-economic and cultural choices.

References

1. Aaron H (1966) The Social Insurance Paradox, Canadian Journal of Economics 32: 371-379
2. Atkinson AB (1983) The Economics of Inequality. Clarendon Press Oxford
3. Aaron H (1966) The Social Insurance Paradox, Canadian Journal of Economics 32: 371-379
4. Banca d'Italia (1997) I bilanci delle famiglie italiane nell'anno 1995. Supplemento al bollettino statistico 14, Banca d'Italia Roma
5. Bianchi C, Romanelli M, Vagliasindi P (2001) Inequality and poverty among pensioners: microsimulating the role of indexing lowest pensions to wages, ECOFIN Discussion Papers, Università di Parma
6. Bianchi C, Romanelli M, Vagliasindi P (2003) Validating a dynamic microsimulation model of the Italian households. In this volume
7. Cannari L, Nicoletti Altimari S (1998) A microsimulation model of the Italian household's sector. In: Le previsioni della spesa per pensioni, ISTAT, Annuali di statistica, Serie 10, 16:103-134
8. Champernowne DG, Cowell FA (1998) Economic inequality and income distribution, University Press Cambridge
9. De Sandre P, Pinnelli A, Santini A (eds) (1999) Nuzialità e fecondità in trasformazione: percorsi e fattori del cambiamento, Il Mulino Bologna
10. Homer S, Sylla R (1991) A History of Interest Rate. Rutgers University Press London
11. Morrison R J (2000) DYNACAN, the Canada Pension Plan Policy Model: Demographics and Earnings Components. In: Gupta A, Kapur S (eds) Microsimulation in Government Policy and Forecasting, North-Holland Amsterdam
12. Neufeld C (2000) Alignment and Variance Reduction in DYNACAN. In: Gupta A, Kapur S (eds) Microsimulation in Government Policy and Forecasting, North-Holland Amsterdam
13. Shorrocks AF (1995) Revisiting the Sen poverty index. Econometrica 63:1225-1230
14. Stock JH, Wise DA (1990) Pensions, the option value of work, and retirement. Econometrica 58:1151-1180
15. Tucci MP (1998) Stochastic Sustainability. In: Chichilninsky G, Heal J, Vercelli A (eds) Sustainability: Dynamics and Uncertainty, Dordrecth, Kluwer ed., pp. 151-169

---

[24] In this way, the decision to get married and to have children could be related to other characteristics (e.g. social status, life style) or choices (e.g. the decision to work or not, or to ask for special working condition, such as part-time). On these issues, related to fertility decline, interesting consideration and evidences can be found in De Sandre, Pinnelli, Santini [9].

16. Vagliasindi P (1999) Riforma del TFR e futuro del sistema pensionistico. ECOFIN, Discussion Paper Series, n.2, Università degli Studi di Parma
17. Wolf DA (2001) The role of microsimulation in longitudinal data analysis. Papers in Microsimulation Series, N6, Syracuse University, Syracuse

Validating a Dynamic Microsimulation Model of the Italian Households *

Carlo Bianchi[1], Marzia Romanelli[2], and Pietro A. Vagliasindi[3]

[1] University of Pisa cbianchi@ec.unipi.it
[2] Sant'Anna School of Advanced Studies romanelli@sssup.it
[3] University of Parma pvagli@unipr.it

Summary. The recent literature – including among others Redmond et al. [22], Gupta and Kapur [13], Mitton et al. [19]– has highlighted model alignment and validation as crucial issues to be tackled when microsimulating the consequences of public policies. This paper discusses some preliminary validation experiments performed on the model inputs, procedures and simulation results. The validation process that we use involves external checks, such as the ex-post comparison (from 1996 since 1999) of aggregated key macro variables (e.g. dependent workers' incomes, and demographic variables) with official data (e.g. supplied by the Italian National Institute for Statistics ISTAT) at national and at regional level (North, Centre and South). We test the appropriateness of the main assumptions and specification of the model and policies' effectiveness also using Monte Carlo simulations. Specifically, our analysis allows us to test MIND's ability in forecasting demographic and economic trends and in capturing socio-economic dynamics for regional areas.

Key words: microsimulation, retirement choice, validation

14.1 Introduction

The recent literature – including among others Redmond et al. [22], Gupta and Kapur [13], Mitton et al. [19]– has highlighted model alignment and validation as crucial issues to be tackled when microsimulating the consequences of public policies. In particular, in order to analyze the effects of social security policies, a great attention should be paid to the study of the evolution of: (a) demographic and family structures, (b) incomes and pension treatments considering the main sectors: private and public dependent workers and self

* MIUR financial support is gratefully acknowledged.We are also grateful to the participants at the workshop New Tools for Qualitative Economic Dynamics held at the CIMAT, Guanajuato - Mex (October 2002) and VII WEHIA Workshop held in Trieste (May 2002). All errors and omissions are our own.

employed workers. Microsimulation models, once validated, enable us to carefully consider the effects of social policies on the distribution of individual and family incomes and to tackle issues related to inequality and poverty.

The main aim of this paper is to describe our modelling approach and in particular the validation and calibration experiments. Its analytical framework is based on MIND (Micro Italy National Dynamics), a dynamic aging model similar to the one developed in Cannari and Nicoletti Altimari [9], and it incorporates behavioral analysis of individual choices of retirement age, derived from the Stock-Wise [24] option value model.

The $MIND$ model considers both the national socio-demographic dynamics and the different Italian regional geographical areas (North, Centre, South and Islands). Similarly to models currently developed for other countries,[4] the $MIND$ structure is organized in modules, which simulate the evolution of: (a) demographic and family structures, (b) socio-economic phenomena related to work and incomes and (c) optimal retirement choice and pension treatments.

The paper discusses the basic validation experiments performed on the model inputs, procedures and simulation results. The $MIND$'s input consists mainly of microdata derived from the 1995 Bank of Italy's households' survey $(SHIW)$ [3].[5] It is widely recognised that self-selection and underreporting can seriously affect the quality and reliability of the data included in the survey. In order to tackle these issues we calibrate the sample with respect to the actual population, attaching appropriate household weights, according to several dimensions, such as the demographic structure, the education level of workers, sex and geographic areas. The validation process that we use involves external checks such as the ex-post comparison (from 1996 since 1999) of grossed up aggregation of the key macro variables (e.g. dependent workers' incomes, and demographic variables) with official data (e.g. supplied by the $ISTAT$ - Italian National Institute for Statistics) at national and at regional level (North, Centre and South). Specifically, our analysis allows us to test $MIND$'s ability in forecasting demographic and economic trends, in capturing socio-economic dynamics for regional areas.

The paper is organised as follows. In section 14.2 we introduce the principle of microsimulation and we present the basic structure of the model. In particular, section 14.2.1 describes the demographic module and validates it, comparing our results with official $ISTAT$ data and with IRP (Institute for Research on Population) forecasts [16] from 1996 to 2045. Section 14.2.2 discusses socio-economic phenomena related to work incomes, presenting some validation experiments on economic trends (i.e. incomes and income tax revenues from 1996 to 2000), in which we compare our results with national

---

[4] See, for example, $NEDYMAS$ [20], $DYNAMOD$ [2] and $DESTINIE$ [15].

[5] The model (jointly developed in the Universities of Parma and Pisa) is based on annual periods, mainly because the official data, involved in the building and use of the model, are supplied on an annual basis. Starting from the $SHIW$ data, we build alternative samples and use them to microsimulate the $MIND$ model. The software is written in FORTRAN95 and includes about 12000 code lines.

account data. Finally section 14.2.3 illustrates the structure of the optimal retirement choice and pension treatments, moreover performing calibration exercises relative to the functional form of the individual withdrawal decision. Some conclusive remarks are included in section 14.3.

14.2 Structure of the Model and Experiments

The dynamic microsimulation technique, used in the $MIND$ model, consists in replicating for each individual and family in the initial database the decisional processes that reproduce the functioning of the Italian economic system, on the base of defined behavioral models, as shown in figure 14.1.

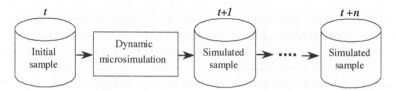

Fig. 14.1. A scheme of dynamic microsimulation

We consider families that differ with respect to the number of components and related characteristics, allowing for a better evaluation of the reform paths and income/wealth redistribution between families, generations, etc.. In particular, dynamic microsimulation, by means of dynamic aging,[6] modifies demographic and socio-economic structures through the evolution of individuals and families. This allows us not only to study a socio-economic system with heterogeneous individuals, but also to model demographic and socio-economic evolution and to introduce simple individual and family reaction-functions. MIND model has a temporal unity equal to the solar year and it is based on three main modules: 1) "demographic"; 2) "work and income"; 3) "social security".

14.2.1 Validation of "Demographic"Module

The "demographic"module includes the subsections on mortality, marriage, divorce, singles, and birth. Our model is quite complex and exhaustive. Not only we keep track of the status of individuals in relation to families, and describe the evolution of existing families as well as the formation of new ones, but we explicitly consider regional differences. Therefore we differentiate

---

[6] It must be recalled that in order to reduce the variance induced by the dynamic aging approach briefly discussed in section 14.2.1, the validation results reported in the paper are mean values of 20 Monte Carlo replications.

for the three main Italian areas a number of events, such as divorce and remarriage, new marriages, the transformation of young dependent adults into singles. In practice, in each period we update for all existing individuals, (and we generate for the new ones) all the demographic and socio-economic characteristics needed to determine their future income and pension benefits claims, considering all possible changes that can occur in the whole families after their exposure to the main demographic events (death of the head of the family, divorce, marriage, etc.) included the transfers of real capital and financial wealth between households.

Based on the survey of the population in 1995 ($SHIW$), $MIND$'s demographic modules determines the number of deaths; subsequently, on the basis of different transition rates, survivors can get married, divorce or become single. Finally, it simulates births.[7]

In what follows, we report the ability of the model to accurately forecast the evolution of the demographic pattern. To get microdata as close as possible to the universe of the Italian population, we re-calibrated family weights considering $ISTAT$ data for age structure, level of education and occupational sector (because of their impact on the income generation process). The expanded sample ($MIND_E$ proportional to the new weights) is used to validate the demographic modules.[8]

---

[7] Marriage-rates, birth-rates and divorce-rates depend on the geographical area and individuals' age. Every year the model attributes to each individual a probability of death, depending on his sex, age and living area. Following the dynamic aging approach, we generate a random number r with uniform distribution in the interval $[0 - 1]$. If r is greater than the relevant probability the individual lives and his age is increased, otherwise he is taken out of the database. In this case, if he was married, the civil state of the partner is modified; if was head of the family, the partner (or the eldest relative) is the new family head and inherits his wealth; the number of family components is decreased. The marriage is simulated in two phases. In the first one $MIND$ selects candidates (aged between 17 and 48), based on a given probability table. In the second phase, each female candidate gets married to the best candidate (for whom the difference between the extracted random number and the theoretical probability is minimized). Inside each marriage, births depend on the mother's age, the geographical area and the number of existing children. The new individuals enter the database modifying the characteristics of their families. Each partner can divorce on the basis of a probability table. Children and 2/3 of the family wealth (half in the absence of children) go to the wife with a 92% probability, as in [9]. Individuals aged between 28 and 35 become singles with a probability equal to 20% of their probabilities of marriage. Transition probabilities are derived from $ISTAT$ data and from the $SHIW$ sample. All demographic probability tables are reported in [25].

[8] In order to tackle these issues we have calibrated the initial sample to be used in simulation with respect to the actual population, attaching appropriate household weights, according to several dimensions, such as the demographic structure, the education level of workers, sex and geographic areas, see [25]. The $SHIW$ survey includes 8,135 families and 23,294 individuals; setting the minimum weight equal

The module's alignment consists in adjusting the probabilities, derived in the demographic module, to bring them closer to the corresponding historical value ($ISTAT$ data) and forecasts (IRP projections) considering a period of 50 years. In validating the model, we verify the reliability of the databases and the model specifications and their suitability for policy purposes.[9]

Specifically, our re-sampled data and simulation results are compared with the $ISTAT$ and $SHIW$ data, as well as the IRP forecasts and the output obtained using the same transition probabilities (NA) along the whole country (without geographical differentiation) as in Cannari and Nicoletti Altimari [9].[10]

Table 14.1. Average age by geographical area

	AGE								
	NORTH			CENTRE			SOUTH		
year	GA	IRP	NA	NA	IRP	NA	GA	IRP	NA
1999	44.24	43.11	43.87	43.35	42.68	43.41	39.05	37.77	39.60
2004	45.60	44.62	44.31	44.55	44.09	43.86	40.11	38.87	40.50
2009	46.91	46.23	45.02	45.66	45.57	44.37	41.13	40.05	41.50
2014	48.32	47.96	45.97	46.77	47.11	45.10	42.13	41.30	42.49
2019	49.68	49.58	46.92	47.93	48.56	46.01	43.16	42.51	43.58
2024	50.94	50.98	47.63	49.01	49.83	46.84	44.15	43.59	44.71
2029	52.05	52.17	48.02	49.96	50.93	47.38	45.13	44.54	45.74
2034	53.12	53.21	48.32	50.81	51.90	47.73	45.99	45.40	46.61
2039	54.15	54.13	48.60	51.65	52.77	48.07	46.73	46.17	47.32
2044	55.08	54.93	48.87	52.46	53.51	48.42	47.29	46.82	47.93

In table 14.1, GA are our own estimates with rates specific by geographic area, NA are the estimates taken from Cannari and Nicoletti Altimari [9] with national data while IRP are the forecasts supplied by IRP [16]; average age is calculated on classes of five years, attributing the frequency to the central value of the class.

to one we get an initial expanded sample $MIND_E$ of 104,017 families and 271,208 individuals.

[9] First, we consider the single modules correcting possible discrepancies, controlling for outputs to replicate historical values, calibrating probability tables and using statistical indicators to verify internal and external microdata compatibility. Then, we examine the different modules jointly in order to validate the whole model, comparing aggregated variables (or simulated frequency distributions) with official ones or with forecasts derived using alternative methods. On these issues, see, among others, [22], [13] and [19].

[10] The validation of the demographic output is carried out, first, up to 1999, comparing it with the $ISTAT$ resident population data and then, up to 2045, using IRP projections; these data, like ours, do not include migrations but assume a gradual increase in average life up to 2018. In order to reach our aims, we apply the usual validation procedures discussed in [6] and in [22].

Table 14.1 shows a strong aging process, more marked in the Centre-North than in the South. While the GA values are very closed to the IRP projections, the NA values underestimate the IRP projections for the Centre-North while they overestimate the IRP values in the South.

Considering different geographical areas, to validate our model in the short (1995-1999) and long run (1999-2044), with respect to $ISTAT$ and IRP data, we compute $\Delta_1 = \frac{1}{k}\sum_{i=1}^{k}|f_{Ai} - f_{Bi}|$, the average difference of frequency distributions in the various age classes, and z_1 and z_1^c, the relative dissimilarity indexes of order one.[11] The age classes on which the indexes have been calculated are the groups considered in table 14.4.

Table 14.2. Dissimilarity measures by geographical area (1995-1999)

	1995	1996		1997		1998		1999	
NORTH	**BI**	**NA**	**GA**	**NA**	**GA**	**NA**	**GA**	**NA**	**GA**
Δ_1	.007	.007	.007	.029	.030	.008	.009	.008	.008
z_1	.023	.023	.024	.101	.104	.029	.033	.027	.028
z_1^c	.013	.011	.012	.049	.052	.006	.010	.009	.012
	1995	1996		1997		1998		1999	
CENTER	**BI**	**NA**	**GA**	**NA**	**GA**	**NA**	**GA**	**NA**	**GA**
Δ_1	.007	.008	.008	.010	.010	.008	.009	.009	.009
z_1	.023	.028	.028	.035	.035	.029	.031	.032	.033
z_1^c	.009	.008	.008	.009	.010	.008	.009	.008	.009
	1995	1996		1997		1998		1999	
SOUTH	**BI**	**NA**	**GA**	**NA**	**GA**	**NA**	**GA**	**NA**	**GA**
Δ_1	.007	.006	.006	.008	.007	.006	.005	.006	.005
z_1	.022	.020	.020	.027	.025	.021	.017	.021	.018
z_1^c	.007	.014	.012	.015	.011	.014	.009	.013	.007

In table 14.2, the degree of dissimilarity of the frequency distributions (Δ_1) presents values are around 0.78% (with a maximum in the North in 1997 equal to 3%). In general the results are very satisfactory using either our separate tables for each geographic area GA or national tables NA reported in Cannari and Nicoletti Altimari [9]. This is confirmed by the low values of dissimilarity indexes (maximum in the North in 1997 with the relative simple dissimilarity index $z_1 = 0.104$ and the cumulated $z_1^c = 0.052$). Similar results have been obtained using second order dissimilarity indexes but, for brevity sake, they are not reported here.

The dominance of the results obtained using GA over the NA ones already emerges in table 14.2 but is still clearer in the long run, as shown in table 14.3.

---

[11] The formulae of the relative dissimilarity measures of order r are:
$$z_r = \sqrt[r]{\frac{1}{2}\sum_{i=1}^{k}|f_{Ai} - f_{Bi}|^r}; \qquad z_r^c = \sqrt[r]{\frac{1}{k-1}\sum_{i=1}^{k-1}|F_{Ai} - F_{Bi}|^r}$$
where f_{Ai} and f_{Bi} are the frequency distributions of A and B and F_{Ai} and F_{Bi} are the cumulated frequencies of A and B. For more details see Leti [17])

Table 14.3. Dissimilarity measures by geographical area (1999-2044)

	1999	2004	2009	2014	2019	2024	2029	2034	2039	2044
NORTH										
$\Delta_1 GA$	.010	.012	.006	.005	.008	.012	.016	.011	.008	.009
$\Delta_1 NA$	.009	.013	.011	.016	.018	.024	.034	.037	.042	.051
$z_1 GA$	.021	.023	.012	.010	.016	.024	.032	.023	.015	.019
$z_1 NA$	.018	.025	.022	.031	.035	.048	.069	.075	.084	.101
$z_1^c GA$	.004	.005	.003	.002	.002	.003	.005	.003	.002	.003
$z_1^c NA$	.003	.004	.005	.008	.010	.011	.015	.017	.019	.022
CENTER										
$\Delta_1 GA$	.007	.006	.002	.003	.006	.013	.021	.017	.012	.011
$\Delta_1 NA$	.006	.007	.009	.014	.017	.024	.034	.037	.037	.041
$z_1 GA$	.014	.012	.003	.005	.012	.026	.041	.035	.025	.023
$z_1 NA$	.012	.013	.018	.028	.034	.048	.069	.073	.075	.082
$z_1^c GA$	.003	.002	.001	.001	.002	.004	.007	.006	.005	.004
$z_1^c NA$	.002	.003	.004	.007	.009	.011	.014	.016	.017	.018
SOUTH										
$\Delta_1 GA$	.009	.008	.012	.013	.011	.011	.010	.011	.008	.007
$\Delta_1 NA$	.010	.013	.017	.018	.014	.013	.013	.013	.012	.010
$z_1 GA$	.019	.017	.023	.026	.022	.022	.020	.023	.017	.014
$z_1 NA$	.020	.026	.033	.037	.027	.025	.026	.026	.023	.020
$z_1^c GA$	.005	.004	.005	.005	.005	.004	.004	.004	.003	.003
$z_1^c NA$	.006	.006	.007	.007	.006	.005	.005	.005	.005	.005

Using the $MIND_E$ database both with the GA and NA tables, we can also compare the trends of each age class with the IRP forecasts. For each geographical area, the R^2 coefficients reported in table 14.4 show the high linear correlation between the relative frequencies of our simulation output (for every age class) with IRP ones and it is evident the better fit of the estimates obtained using GA.

Table 14.4. R^2 for different age classes

AGE CLASS	NORTH			CENTRE			SOUTH		
	$MIND_E$		$MIND_3$	$MIND_E$		$MIND_3$	$MIND_E$		$MIND_3$
	GA	NA	GA	GA	NA	GA	GA	NA	GA
0-4	.99	.61	.99	.97	.52	.94	.99	.79	.93
5-19	.99	.27	.99	.99	.46	.98	.94	.87	.96
20-59	.95	.96	.96	.93	.93	.94	.92	.93	.93
+60	.99	.98	.99	.98	.98	.98	.99	.99	.98
+80	.98	.97	.99	.90	.81	.97	.92	.92	.97

Results very closed to the IRP ones are obtained also using GA tables and samples of smaller size ($MIND_1$, $MIND_2$ and $MIND_3$), alternative to $MIND_E$,[12] and, as an example, table 14.4 reports the R^2 associated with the use of the sample $MIND_3$. Hence, we can conclude that the model has a robust and relatively stable structure, so that hen performing Monte Carlo replications we may also employ samples with smaller size.

14.2.2 Validation of "Work and Income"Module

The "work and income"module is organised as follows: the work sub-module simulates the entry in the labour market while the income one estimates workers' incomes.

In consequence of the demographic events (death, divorce, marriage, etc.) we need not only to transfer wealth between households but also to generate or update for each agent all the socio-economic characteristics (such as education levels, labour force participation, occupational status, unemployment, retirement etc.) necessary to determine individuals' incomes. As already said we have preliminarily carefully calibrated our initial database in order to correctly reproduce the characteristics of the Italian population by geographical areas with respect to the available official projections. In fact, it is widely recognised that self-selection and underreporting can seriously affect the quality and reliability of the data included in the survey.[13] However, it is worth to mention that, notwithstanding our reweighing procedure, the sample still presents some serious differences in occupational data, that underestimate self employed and pensioners.

In our model, one of the key characteristics for determining individuals income is education. We consider three basic education degrees (elementary,

---

[12] In policy analysis $MIND_E$ sample is too large to perform Monte Carlo experiments for a period of 50 years. Accordingly, using a procedure similar to the one discussed in [9] we have reduced the size of the expanded sample. Using the original Bank of Italy weights, we get the sample $MIND_1$ or the sample $MIND_2$ using our modified weights. Finally, we also consider the sample $MIND_3$ (of 41,941 individuals and 15,068 families), which maintains the maximum observed heterogeneity as it includes, at least once, each family of the $SHIW$ sample.

[13] As emphasized in Brandolini and Cannari [5], given the reluctance of interviewed families to provide actual income data, estimations of income (from financial activities, owned residences and secondary works) are lower than the aggregated values of the national accounts and Census data. The average number of hours worked during the week is instead higher than the corresponding $ISTAT$ values. Cannari and D'Alessio [7] calculate the bias due to self-selection, estimating the real probability of inclusion in the sample. The 1987 aggregate income increases by 5% but there is still a strong discrepancy with $ISTAT$ surveys, because of fiscal evasion, estimated by Cannari and Violi [10] at a level greater than 20%. Cannari and D'Alessio [8] show that the self employed, the elderly, and the less educated heads of the family are more likely to under-report financial activities.

high school, university) and the associated number of years (EDU) are attributed through a stochastic procedure which simulates the probability of achieving a certain degree, for all the individuals entering the labour market.

Students do not participate to the labour force. For individuals that are no longer studying, the participation to the labour force is randomly determined and is conditioned on their characteristics such as position inside the family (e.g. head of household), age and sex. In this way unemployed can find a job with probabilities depending on being a new entrant in the labour force, the head of the household, and other characteristics such as age, sex and region. Individuals withdraw from the labour force when they retire from work.

The choice between the three different categories of work is stochastic, depending on the probabilities to be dependent workers (private or public) or self-employed.

Following Andreassen et al. [1], the estimation of income levels (y), based on cross sectional data, has the standard log-linear specification suggested by Mincer [18].[14] For the individual i, the forecast value ($\hat{y}_{it}$) at time t is obtained by replacing the values assumed by the characteristics of the individual in the vector $\mathbf{x}_{it}$ of the equation $\hat{y}_{it} = e^{(\mathbf{x}_{it}b + \hat{u}_i)}$ where, to avoid biases induced by the non-linear transformations, the estimated residuals $\hat{u}_i$ are also included.[15]

We verify the forecasting precision of the socio-economic variables mainly focusing on the education degrees (elementary, high school, university) reached in different geographical areas and aggregate trends of incomes and personal income tax revenues.

Among socio-economic variables the most important ones are represented by the aggregated incomes. Restricting ourselves to dependent workers' aggregated income Y (net of social contributions), we multiply the sum of incomes (Y_{DEP}) by the ratio of number of dependent workers estimated by ISTAT ($TotDep_{ISTAT}$) to that of our database ($TotDep_{MIND}$). Hence:

$$Y = \left[\sum_{DEP} Y_{DEP} \right] \frac{TotDep_{ISTAT}}{TotDep_{MIND}}$$

---

[14] In particular we consider a structure diversified for AGE, AGE^2, SEX, EDU and $HOURS$ for three occupational sectors (dependent, self-employed, public) and geographical areas (North, Centre, South). Almost all the coefficient are of the right sign and statistically significant at the 99% confidence level. For further details, see Bianchi, Romanelli and Vagliasindi [4]

[15] Real income at period t is derived from: $\tilde{y}_{it} = \hat{y}_{it}(1 + g)^{(t-t_0)}$, where g, the real growth rate of incomes, is assumed to be to 1% after 2000 and t_0 is the initial simulation period. The number of hours worked by entrants ($HOURS$) are estimated on the basis of the average number of hours worked ($HOURSA$) and of the respective standard error ($STDERR$), depending on the sector and the geographical area: $HOURS = HOURSA(AGE, SECT, AREAGEO) + (STDERR(AGE, SECT, AREAGEO) \times rd)$ where rd is a random number with normal standard distribution which introduces a stochastic element.

Table 14.5. Dependent worker's aggregated income (billions of current euros)

	1996	1997	1998	1999	2000
ISTAT	290.108	302.340	313.197	325.087	341.562
MIND	291.175	302.151	314.694	328.064	341.602
$\Delta\%$	0.37%	-0.08%	0.07%	0.54	0.01%

Table 14.5 shows that the differences between the two aggregates are really minor, so that the alignment of 1995 incomes has overcome self-selection and under-reporting problems.

14.2.3 Validation of "Social Security" Module

The "social security" module determines the retirement age and pension treatments using a reaction function based on the option value (OV) model from Stock and Wise [24]. In this way workers are allowed to choose to postpone retirement when the expected value of their utility to retire is growing over time. As in Stock and Wise, individuals, given the available information, calculate the expected values of the utility of retiring today and in future as follows:

$$V_t(R) = \sum_{s=t}^{R-1} (1+r)^{t-s} U_y(Y_s) + \sum_{s=R}^{S} (1+r)^{t-s} U_b(B_s(R)) =$$

$$= \alpha \sum_{s=t}^{R-1} (1+r)^{t-s} Y_s + \sum_{s=R}^{S} (1+r)^{t-s} B_s(R)$$

where $V_t(R)$ is the value at time t to retire at time R and depends on the actualized flows of incomes Y_s [16] and pensions $B_s(R)$ [17].

---

[16] In order to reconstruct the history of wages and social contributions of each worker in our database, necessary to calculate his pensions, we used the following simplifying hypothesis: 1) a future real rate of discount, equal for all to 2.5%, 2) a real rate of per capita growth of 1%, 3) absence of individuals that temporarily suspend the payment of their contributions, 4) payment of pension and incomes at each end of year, 5) indexation of pensions to the rate of inflation 1.8%: total (i.e. 100 %) if inferior to the double of the value of the social pension (equal to 3357 euro in 1999), partial at 90% if inclusive between the double and the triple of this value and at 75% otherwise, 6) workers face a social constraint to retire set at 66 for dependent workers, 67 for public and 68 for self-employed (in this way we generate a corner solution).

[17] The indirect utility specification of labour and pension incomes are respectively $U_y(Y_s) = \alpha Y_s + \varpi_s$ $(0 \le \alpha \le 1)$ and $U_b(B_s(t)) = B_s(t) + \xi_s$, with ϖ_s and ξ_s zero mean random variables whose variance can be estimated as in [24]. For recent work in this area see [12], [21] and [23]. In our paper the comparison involves flows of gross incomes. Obviously, being individual interested to disposable incomes, the tax system and the income tax progressivity may affect the results, but it is quite impossible to envisage how these will evolve in the next 45 years.

The differential benefit $OV_t(t)$ of postponing retirement to the next year is then considered. If OV is positive the retirement will be postponed, otherwise (negative OV) the worker retires immediately. OV is estimated each year until the maximum age of retirement is reached. In the calculation of OV we consider the different computing procedures (defined benefit, "mixed"or defined contribution)[18] depending on the characteristics and contribution of each individual and on the benefits and requisites related to workers' type and period.

In our simplified specification of the model, in each period t workers calculate the differential benefit $OV_t(t+1)$ of postponing retirement to time $(t+1)$ and decide to retire if $OV_t(t+1) \leq 0$.

$$OV_t(t+1) = E_t V_t(t+1) - E_t V_t(t) =$$
$$= \alpha Y_t - B_t(t) + \sum_{s=t+1}^{S} (1+r)^{t-s}(B_s(t+1) - B_s(t))$$

Given our simplifying assumptions, each worker derives the age that maximizes the value of retiring (V). In fact, the differential benefit monotonically decreases, reaching the maximum when it becomes null, since the variation, Δ_t, of the option value from time t to time $t+1$ is: [19]

$$\Delta_t = \alpha(Y_{t+1} - Y_t) - (\frac{2+r}{1+r})(B_{t+1}(t+1) - B_t(t)) +$$
$$+ \sum_{s=t+2}^{S} (1+r)^{t-s}[(1+r)(B_s(t+2) - B_s(t+1) - B_s(t))]$$

Assuming that the absolute increment of the pension benefit $(B_j(s+1) - B_j(s))$ decreases at a rate equal or grater than $(r/(1+r))$,[20] Δ_t is negative when: $\alpha \leq \frac{2+r}{1+r}\frac{B_t(t)}{Y_t}$.

Hence, being r relatively small, this condition becomes $\alpha \leq 2\frac{B_t(t)}{Y_t}$ and it is satisfied by a ratio $B_t(t)/Y_t = 50\%$ (independently from the value of α, whose range is between 0 and 1).

---

[18] In the Italian $PAYG$ system future treatments are based: (i) on final wages in the old defined benefit regime ("regime retributivo"), (ii) on social contributions in the future new defined contribution regime ("regime contributivo"introduced by the Dini law L.335/95) for the new entrant workers and (iii) on a mixture of final wages and social contributions in the so called mixed regime ("regime misto") for mature workers, during a long transition phase leading toward the new contributive regime.

[19] Considering real quantities, we have $B_{t+1}(t) \leq B_t(t)$. In what follows, we assume that equality holds. In fact, a fortiori the results still hold if strict inequality is supposed.

[20] In this way we have $(1+r)(B_s(t+2) - B_s(t+1)) - (B_s(t+1) - B_s(t)) \leq 0$.

For validating the social security module, we want to calibrate the parameter values of the module as to reproduce, at least in the short term, the official dynamics of the socio-economic phenomena.

Given the difficulties in comparing our database with the entire universe of pensioners, our goal is simply to see whether our database evolves along the same lines of reality. Accordingly, we try to mimic the trend of the percentage change in the quota of workers retired between 1996 and 1999. In order to gain this result, we change the parameter's value of the retirement choice function till the simulated output fits as good as possible with the official data.

As reported before, the reaction function included in our model is based on the comparison between the present value of the expected utility of the flow of the pension benefits obtained by an agent opting for an immediate retirement with the one obtained postponing it to the next period. However, when retirement is postponed, the problem arises whether (and to which extent) we should take into account the wage income earned in the following period. It's important to consider the functional form of the indirect utility functions and in particular the chosen value of α.[21] In particular, we assume for the same amount of money, the utility of income from labour is lower than the utility gained from pension benefits. This justifies a value of α less than one.

The validation of the social security module is carried out trying different values for the parameter α until reaching the lowest distortion between simulated output and official data. In particular we consider the time period between 1996 and 1999 and compare the simulated and official percent change in the quota per age class of the individuals retired and still alive in 1999. In calibrating α we have disaggregate the simulation results, as far as possible, by considering sex and occupational sector (dependent, self-employed, public) of the individuals and we have also used the alternative samples $MIND_1$, $MIND_2$ and $MIND_3$. From the experiments it emerges that $\alpha = 0$ is the value for which we observe (for all the samples) the best fit of the simulation results to the official ISTAT data and a brief illustration of the observed evidence is given in figures 14.2, 14.3 and in table 14.6.

In the figure 14.2, setting $\alpha = 0$ and using the alternative samples, we report the simulated results and official ISTAT data(dark diamond). In general, for all the samples, the simulation produces profiles very similar to the ones supplied in the $ISTAT$ statistics, except for the case of the public sectors (especially for females). This may be due in part to the tendency, in a period perceived as uncertain, to retire early in order to avoid future restrictions as well as to take advantage of economies of scope of the family and of possibilities offered by the informal sector.

Our model seems to concentrate the values around the modal classes. In particular the model does not replicate completely the phenomenon of

---

[21] We remember that the indirect utility specification of labour and pension incomes are respectively $U_y(Y_s) = \alpha Y_s$ ($0 \le \alpha \le 1$) and $U_b(B_s(t)) = B_s(t)$.

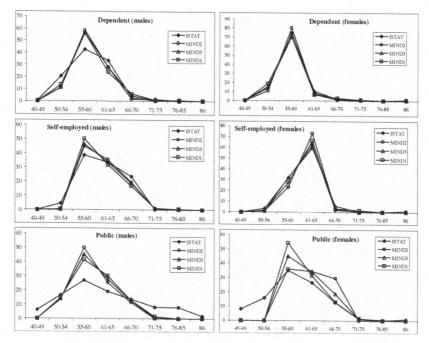

Fig. 14.2. Change in quota (%) for different sectors $\alpha = 0$

young pensioners (workers who decide to retire before the age of 50). We ascribe this discrepancy to the common sense of uncertainty coming from the approval in 1995 of the Social Security System reform, which characterises the temporal interval used in our validation process and probably represented an accelerating factor toward the choice of retirement. However, we can assume that this feature disappears after the early phase of the reform.

Using the sample $MIND_3$ and different values of the parameter α, figure 14.3 reports the simulation profiles for the occupational sectors, while table 14.6 includes the R^2 coefficients with respect to the official ISTAT data. From the analysis of the results the following seems to emerge: the lower the value of α the better the fit.

Table 14.6. R^2 for different values of α

α	MALES			FEMALES		
	D	SE	P	D	SE	P
0.000	0.900	0.965	0.916	0.997	0.993	0.816
0.125	0.899	0.964	0.915	0.997	0.991	0.815
0.250	0.896	0.960	0.910	0.993	0.980	0.810
0.375	0.891	0.981	0.900	0.996	0.915	0.473
0.500	0.890	0.987	0.711	0.950	0.890	0.282
1.000	0.196	0.238	0.010	0.720	0.847	0.003

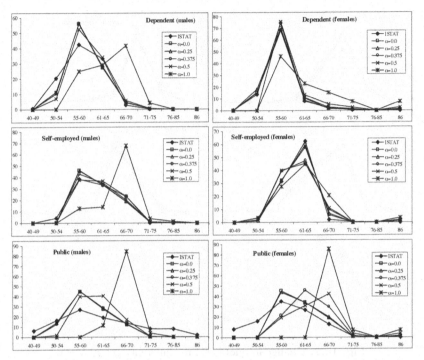

Fig. 14.3. Change in quota (%) for different sectors - $MIND_3$

14.3 Concluding Remarks

At the beginning of the '90s, the US National Academy of Science highlighted the crucial importance of the research on validation methods for microsimulation models and on specific criteria to evaluate the uncertainty of simulation results (see Cohen et al. [11]). Nevertheless, as observed by Wolfson [26], in the last ten years such recommendations have been neglected, maybe also because of the high costs involved in validation experiments. Furthermore, given the intrinsic complexity of the model, its construction and validation required the involvement of different types of expertise.[22] Accordingly, we validate the demographic and the socio-economic input and output of our microsimulation dynamic model ($MIND$), at a national and regional level, using $ISTAT$, IRP and $SHIW$ data. It must be recalled that in carrying out external checks, we have used the 1995 $SHIW$ sample in order to have a reasonable interval (1996 - 1999) for performing ex-post comparison with official $ISTAT$ data whereas it was crucial the availability of the IRP forecasts from 1999 up to 2045 for the ex-ante long run validation analyses.

---

[22] On this aspect, we acknowledge, the precious support of researchers from *Bank of Italy* (Giuseppe Bruno, Luigi Cannari, Ivan Faiella), $ISTAT$ (Emanuele Baldacci, Luca Inglese, Gaetano Proto) and $SECIT$ (Fernando Di Nicola).

By validating microdata and calibrating the model specification, we obtained a reliable tool for simulating, at a disaggregated level, the evolution of the distribution of incomes and pensions and for highlighting the redistributive consequences of economic policies or reforms.[23]

The development and validation of $MIND$ allowed us to use alternative ways to represent individual abilities in problem-solving and decisional processes, which are at the basis of individual "reaction functions" to the policies. In this context, we plan to undertake future research, incorporating and corroborating behavioral functions – such as the ones which endogenise retirement choice (the base of the option value) – to improve the model's forecasting abilities with respect to policy changes.[24] A more accurate specification of the Italian reality would also require the introduction of a "migration" module, but, as far as we know, the data that would allow us to accurately model the phenomenon are not yet available.

References

1. Andreassen L, Fredriksen D, Ljones O (1996) The future burden of public pension benefits: a microsimulation study. In: Harding A (ed) Microsimulation and Public Policy, North-Holland Amsterdam
2. Antcliff S, Grusskin A, Harding A, Kapuscinski C (1996) Development of DYNAMOD: 1993 and 1994, Dynamic Modelling Working Paper, No.1, National Center for Social and Economic Modelling, University of Canberra, Australia
3. Banca d'Italia (1997) I bilanci delle famiglie italiane nell'anno 1995. Supplemento al bollettino statistico 14, Banca d'Italia Roma
4. Bianchi C, Romanelli M, Vagliasindi P (2001) Inequality and poverty among pensioners: microsimulating the role of indexing lowest pensions to wages, ECOFIN Discussion Papers, Università di Parma
5. Brandolini A, Cannari L (1994) Methodological Appendix: The Bank of Italy's Survey of Household Income and Wealth. In: Ando H, Guiso L, Visco I (eds) Saving and Accumulation of Wealth. Cambridge University Press Cambridge
6. Caldwell SB (1996) Content, validation and uses of CORSIM 2.0, a dynamic microanalytic model of the United States. In: Harding A (ed) Microsimulation and Public Policy. North-Holland Amsterdam

---

[23] Our dynamic microsimulation results encourage us to face the future challenges to incorporate and to verify the heuristic value of individual "reaction functions" (e.g. in the family, in the processes of creation and distribution of the wealth and in the labour market). This modeling strategy seems to be supported by recent research in the field. As observed in [14] transition probabilities are being replaced by hazard models.

[24] In particular, it could be worthwhile to examine in more detail the individual choices in family setting, not merely as a stochastic process, but dependent also on socio-economic variables. In this way, the decision to get married and to have children could be related to other characteristics (e.g. social status, life styles) or choices (e.g. the decision to work or not, or to ask for special working conditions, such as part time).

7. Cannari L, D'Alessio G (1992) Mancate interviste e distorsione degli stimatori. Temi di discussione 172. Banca d'Italia Roma
8. Cannari L, D'Alessio G (1993) Non-reporting and Under-reporting Behavior in the Bank of Italy's Survey of Household Income and Wealth. In: Proceedings of the ISI 49th session. ISI Firenze
9. Cannari L, Nicoletti Altimari S (1998) A microsimulation model of the Italian household's sector. In: Le previsioni della spesa per pensioni, ISTAT, Annuali di statistica, Serie 10, 16:103-134
10. Cannari L, Violi R (1991) Under-reporting del reddito nell'indagine sui bilanci delle famiglie condotta dalla Banca d'Italia, In: Personal Income Distribution, Inequality and Poverty. Siena
11. Cohen M, Billard L, Betson DM, Erickson EP (1991) A Validation Experiment with TRIM2. In: Citro CF, Hanushek EA (eds)Improving Information for Social Policy Decisions - The Uses of Microsimulation Models National Academy Press Washington DC
12. Gruber J, Wise DA (eds) (1999) Social Security and Retirement around the World. University Chicago Press Chicago
13. Gupta A, Kapur S (eds) (2000) Microsimulation in Government Policy and Forecasting. North-Holland Amsterdam
14. Harding A (2000) Dynamic Microsimulation: Recent Trends and Future Prospects. In: Gupta A, Kapur S (eds)
15. INSEE (1999) Le modèle de mocrosimulation dynamique, S érie des documents de travail de la Direction des Etudes et Synthè ses Économiques, No. G9913. INSEE Paris
16. IRP, Golini A, De Simoni A and Citoni F (1995) (eds), Tre scenari del possibile sviluppo della popolazione delle regioni italiane al 2044 (base 1994), IRP Roma
17. Leti G (1997) Statistica Descrittiva. Il Mulino Bologna
18. Mincer J (1974) Schooling, Experience and Earnings. Columbia University Press New York
19. Mitton L, Sutherland H, Weeks M (eds) (2000) Microsimulation Modelling for Policy Analysis: Challenges and Innovations. Cambridge University Press Cambridge
20. Nelissen JHM (1994) Income redistribution and social security, an application of microsimulation. Chapman & Hall London
21. Peracchi F (1999) Specification and Estiamtion of Microeconomic Retiremenet Models, presented at the Workshop on "Economic Demographic Factors and the Future of Welfare State in Italy". Messina
22. Redmond G, Sutherland H, Wilson M (1999) The arithmetic of tax and social security reform. A user's guide to microsimulation methods and analysis. Cambridge University Press Cambridge
23. Spataro L (2000) Le scelte di pensionamento in Italia: un'applicazione (ed estensione) del modello Option Value. Studi Economici, N° 72. Dipartimento di Scienze Economiche Pisa
24. Stock JH, Wise DA (1990) Pensions, the option value of work, and retirement. Econometrica 58:1151-1180
25. Vagliasindi P, Romanelli M, Bianchi C (2002) MIND a Dynamic Microsimulation Model for Italy: Probability and Validation Tables. ECOFIN Discussion Papers, Università di Parma
26. Wolfson M (2000) Socio-economic microsimulation and public policy. In: Gupta A, Kapur S (eds)

15

Recent Advances in Micromodeling: The Choice of Retiring[*]

Luca Spataro

Dipartimento di Scienze Economiche, Università di Pisa, Italy.
l.spataro@ec.unipi.it

Summary. Recently, much attention has been devoted to econometric models as a new tool for dynamic microsimulation. In particular, microeconometric approaches to retirement decisions have been increasingly adopted for "calibrating" dynamic microsimulation frameworks aiming at endogenizing retirement choices. By doing this, both the understandment and the prediction of the effects of policy reforms (for istance, of Social Security systems) can be significantly improved. In this work an overview of the most recent developments in micromodeling retirement decisions is carried out. In particular, as for the choice of the estimation strategy, special emphasis is posed on the trade-off between the degree of realism of hypotheses, on the one hand, and on data tractability and/or estimation performance, on the other hand. Finally, some issues which represent a challenging avenue for future research are discussed.

Key words: retirement choices, econometric methods, survey.

15.1 Introduction

In this paper I present an overview of the most recent developments in the field of micromodeling retirement decisions. The literature on this topic has been flourishing steadily over the last two decades, as both the demand for information and the set of theoretical and computing tools at hand have been growing significantly.

[*] A more extended version of this work can be found at the address: http://cerp.unito.it/Pubblicazioni/archivio/WP_CeRP/WP_28.pdf. I am grateful to Carlo Bianchi, Elsa Fornero, Wilbert Van der Klaauw, Ugo Colombino, the participants at the 2002 ESPE Congress, at the Eighth CEF Conference, at the Third Latin American Workshop in Guanajuato, and two anonymous referees for useful comments and suggestions. Funding from Murst: "Progetti di interesse nazionale" is gratefully acknowledged. The usual disclaimers apply.

Originally, one of the crucial issues boosting this stream of literature was the decline of the older cohorts activity rates since the middle of the last century, which only partially has been compensated by the increase of the women participation rates over the same period. In fact, from 1960 till the middle 1990s, the fraction of the working lifetime in OECD countries has fallen from about two thirds to almost one half ([40]).

Currently, a central issue driving the research is the analysis of the worrying scenarios that most developed countries will be facing in the next future, characterized by high "dependency ratios" (that is the fraction of the old relative to the young generations): this tendency is likely to have strong implications on relevant policy issues such as the sustainability of Social Security (SS from now on) systems and on policies tackling poverty. Thus, beside being relevant per se, understanding how individuals decide the timing of retirement is crucial for conceiving reforms of Social Insurance systems which can effectively correct the consequences of the demographic transitions described above.

As regards this point, it is worth mentioning that recently retirement has been a matter of interest also for dynamic microsimulation[2]. In particular, the 1990s have experienced important developments in the way transition probabilities of micro-units (i.e. from work into retirement, from unemployment into employment status) can be calculated. For example, in the most recent dynamic models annual transition probabilities are being calculated through hazard models: the DYNAMOD-2 model uses survival functions to predict the time at which selected possible monthly changes of status will occur[3]. Similarly, [3] endogenize individual retirement choices by applying the Option Value decision rule to their dynamic microsimulation model calibrated for Italy. Despite the promising improvements in this field, still few are the works adopting these new approaches.

The paper is organized as follows: in the first section I select some of the major characteristics of retirement choices so as to classify the existing theoretical and empirical models according to the degree to which they can account for such features. Next, I present a review of the literature, starting from the Linear Model, the Logit and Probit specifications and the Duration Model. As for the more recent structural models, I lay out the Option Value and the Dynamic Programming model. A discussion of the main prospects of the research and final remarks end the work.

15.2 The Main Features of Retirement Choices

In principle retirement choices could be studied both from a "static" and a "dynamic" standpoint. according to the former, which has been adopted es-

---

[2] See [39] and [21].

[3] See [24] for an exposition of such point and [2], [32], [16] and [18] for some applications.

pecially for modeling labor supply decisions, individuals need not take into account the future, since the solution to their choice problem (i.e. allocation of resources between consumption and leisure) relies on or concerns current period variables only. Such approach can also deal with uncertainty affecting any aspect of the allocation problem. The dynamic standpoint, on the contrary, by reckoning that retirement choices have an intertemporal, possibly life-cycle nature, considers individuals that in each period have to solve a maximization problem and/or to compare accumulation opportunities which will occur in the future and are, in general, uncertain. While both approaches can account for the evidence that retirement is typically a discrete and dichotomous choice, the dynamic one can better match the fact that very often retirement is an absorbing state.

These preliminary arguments may help understanding why researchers have been treating retirement differently from the standard choice of labor supply. For the sake of completeness, I provide a list of some of the main elements characterizing (although not exclusively) the choice of ceasing to work. In short, it can be said that retirement:

1) is a discrete choice, i.e. implies the assessment of two or more mutually exclusive alternatives (or states);

2) typically, is an absorbing state (although with a few exceptions): workers ordinarily retire only once in a lifetime;

3) is a decision which can be made (or can be made only) in an age interval defined by the law;

4) implies forward looking behavior (i.e. assessment of future economic opportunities);

5) depends on both individual and institutional characteristics- i.e. family composition and wealth, pension formula and eligibility rules, but also on labor market rules, health care provision and so on;

6) is a sequential (or dynamic) choice;

7) it is taken in an uncertain context.

Although in many real situations the distinction between labor-supply and retirement decision can be very cumbersome as workers get older, the points mentioned above can help discerning which model is more suitable and satisfactory for representing agents' decisions. For example, when points 2) and 3) apply, an ad-hoc retirement model should be used: in fact, the irreversibility of the decision of leaving the workforce is a characteristic shared by many SS systems; this, in turn, raises dramatically the need for a forward looking behavior (point 4) and makes uncertainty more relevant as well (point 7); similarly, eligibility rules do matter, since they exactly define the moment (or period of life) in which the problem of whether going on working comes into play. Also other institutional settings (point 5) play a crucial role in making retirement choice "unique": for example, the possibility of switching from a full-time into a part-time job may suggest treating retirement as a "process" rather than as a mere discrete choice; finally, the severance payment often associated to the choice of retirement, the presence of other SS facilities (health care provision conditioned on being a pensioner), may contribute to amplify the difference between the two approaches.

Concluding, the different nature and same complexity of the institutional frameworks makes it necessary to handle retirement through specific models; only to the extent a researcher is akin of the peculiarities of retirement, will he/she be able to decide properly which model better fits the circumstances under investigation.

In the light of this the classification criterion adopted in this work is the following: I ordinate the existing models according to their ability to address the points mentioned above and I also discuss those cases in which the boundary between these models appears blurred. In the section that follows I present the main features of the more traditional models[4].

15.3 The Linear Model

As mentioned, in the first applied works retirement was modeled by adopting static models of labor supply, whereby the event of retirement was simply a special case (i.e. hours supplied or worked equal to zero). For example, [15], [5] and [8] all make use of the one-period labor supply model without uncertainty in order to examine the role of SS in retirement choices.

More sophisticated works tried to estimate structural equations handling the lifetime dimension of the optimization problem. In particular, these authors estimated equations representing the solution of the life cycle labor supply problem. Such approach selects either yearly or the whole life time labor supply (or, equivalently, the age of retirement) as the dependent variables[5]. Following [7], who is among the first authors explicitly addressing the lifetime dimension of the consumption-leisure allocation problem, the linear model can be represented by the following equation:

$$R = \alpha X + \beta Z + \varepsilon \qquad (15.1)$$

where R is the retirement age, α and β vectors of parameters to be estimated, X a set of observed variables including personal characteristics affecting retirement, such as health status or marital status, and Z a collection of age specific variables qualifying the lifetime budget constraint; finally, ε is a suitably distributed error. A very similar approach is used by [27] in his work on retirement decisions in a family contest: he estimates a structural model in which the utility function arguments are consumption of goods and number of retirement years (of both husbands and wives).

Despite being a significant contribution in modeling retirement choices, the linear model presents several drawbacks: firstly, as for the original one period version, it does not take into account the fact that retirement is related

---

[4] Other useful overviews on retirement literature are [34] and, more recently, [36].
[5] See, for example, [22] and [7].

to a lot of important institutional features (such as mandatory retirement, eligibility rules) which do not pertain to the ordinary labor supply decisions. Secondly, it fails to capture the discrete nature of retirement choice (which, moreover, means quite often entering an "absorbing state"). Third, as for the dynamic specification, it is worth noting that dealing with complete life-cycle solutions for labor supply choices means imputing the compensation path of individuals over the entire life-cycle, a characteristic which is common to the more sophisticated and more recent structural models; however, accounting for such aspect in the linear framework has the serious shortcoming of producing highly nonlinear and even nonconvex individual budget constraints, which does not guarantee the uniqueness of the optimal retirement age.

15.4 Multinomial Probit and Logit Models

An alternative method adopted for modeling retirement decisions is the probit or logit analysis. These models share the characteristic of treating the participation to the labor force as a discrete choice among a set of alternatives (such as full time, part-time job). I present the case of a multivariate (three alternatives) version, but the example can be readily generalized to more or less than three states.

Suppose an individual faces a three-way choice: i) retiring, ii) working part-time or iii) working full-time. For each individual, define three random utilities corresponding to each choice: $U_i = u_i(X, \beta) + \varepsilon_i$, $i = 1, 2, 3$, where $u()$ is the deterministic part of the function depending on X, a vector of exogenous variables, β a vector of corresponding coefficients to be estimated, while ε_i represents a random shock to utility meant to catch imperfect optimization by the individual and/or the inability of the econometrician to measure exactly all the relevant variables. Finally, suppose Ω to be the covariance matrix of the errors. In fact U_i is unobservable and what is observed is y_i, which is a dummy variable such that: $y_i = 1$ if $U_i > U_j$, $\forall j \neq i$; $y_i = 0$ otherwise. In other words, we do not observe the utility levels, but only the individual's choice, that is the case in which the utility of an opportunity exceeds the others. Therefore, the probability that the first alternative is chosen is given by: $P(y_1 = 1) = P(\varepsilon_i < \varepsilon_1 + u_1 - u_i, \forall i \neq 1)$. Now, if the errors ε_i's are independently and identically distributed according to the Type I extreme value distribution in standard form (also known as a Weibull or Gumbel distribution, with cumulative density function $F(\varepsilon) = exp(-exp(-\varepsilon)))$, in the multinomial logit model the probability given above will have the analytic form: $P(y_1 = 1) = \frac{e^{u_1}}{\sum e^{u_i}}$. Finally, by assuming that $U_i = \beta'X + \varepsilon_i$, one gets:

$$P(y_1 = 1) = \frac{e^{\beta'X_1}}{\sum e^{\beta'X_i}}. \tag{15.2}$$

While the multinomial logit is popular in the literature, it suffers from the well-known property of independence from irrelevant alternatives (IIA), since the odds ratio concerning alternatives ith and jth (i.e. $exp(u_i)/exp(u_j)$), does not depend on the total number of the choices considered. Since this assumption does not fit many real situations, an alternative model is the multinomial probit.

This model assumes that the ε_i are jointly normally distributed with mean vector zero and covariance matrix Ω. Since only differences in utilities can be considered, define $V_{i1} = U_i - U_1 = u_i - u_1 + \varepsilon_i - \varepsilon_1 = u_i - u_1 + v_{i1}$, $\forall i \neq 1$, so that the outcome conditions may be rewritten as: $y_i = 1$ if $V_i < 0$, $\forall i \neq 1$; $y_i = 0$ otherwise. Therefore, one may write the probability that the first alternative is chosen: $P(y_1 = 1) = P(v_{i1} < u_1 - u_i, \forall i \neq 1)$. Since v_{21} and v_{31} have a bivariate normal distribution with mean vector zero and covariance matrix Ω_1 where $\omega_{kl1} = E(\varepsilon_k - \varepsilon_1)(\varepsilon_l - \varepsilon_1) = \omega_{kl} - \omega_{k1} - \omega_{l1} + \omega_{11}$ is the generic element of Ω_1, then the probability that alternative one will be chosen is given by:

$$P(y_1 = 1) = \int_{-\infty}^{u_1 - u_2} f(v_{21}, v_{31}) \, dv_{21} dv_{31} \qquad (15.3)$$

where $f()$ has a bivariate normal distribution with mean vector zero and covariance matrix Ω_1.

A typical problem arising within such approach is that, as the choice set grows, inference requiring exact evaluation of such integrals rapidly becomes infeasible. A few applications, in particular for the probit model, have tried to overcome this issue by performing (quasi) Monte-Carlo simulations and numerical integration of the choice probabilities and by substituting these simulated probabilities into likelihood functions or moment conditions[6]. Other limits of such models are that, as reduced form estimation methods they are independent of any particular behavioral theory[7] and unable to take into account properly the uncertainty affecting future events (like earnings, health status and so forth).

On the other hand, one advantage is the possibility to evaluate the role played by forward looking behavior, that is by the assessment both of the "reward" comprised in the choice of retirement at different future ages and of the information updating process as individuals get older. This has been worked out, respectively, by allowing for specific forward looking measures of Social Security accumulation opportunities entering the X vector[8] and by

---

[6] On this point see [17].

[7] However, [11] provides a dynamic interpretation of such a static model, by obtaining it as the solution of an intertemporal optimization problem

[8] I.e. the Option Value, the SS Accrual, the Peak Value, etc. See [20] for several estimations and applications of these measures. See also [44] for a complete presentation of the measures of SS incentives. As for the role of SS systems, while there appears a general consensus on the generosity of the systems to be the leading explanation of the sizable exit rates from labor force at early ages occurred in many developed countries, there is still a debate on the degree of workers' sensitivity to the State-provided incentives to early retirement.

using longitudinal data, by which agents' behavior can be observed over a time interval. In this respect, it is worth noting that the use of pooled cross sections of different years raises no problem if observations in different cross-sections are mutually independent. However, if this condition does not hold, the choice of pooling them for estimation is unsatisfactory, even accounting for some "correlation" between observations over time; in fact, in this case estimates are more properly carried out by modeling retirement as a sequential choice. The reason is that the probability of observing a state i at a certain date t for an individual depends on the sequence of choices made in past periods $(t-1,\, t-2$ and so on). As a consequence, it turns out clearly that the retirement decision has a sequential nature, which is better handled by the Survival Analysis or, more generally, by the more recent Structural models.

In the analysis that follows I will distinguish between structural and reduced form models, according to whether the specification does or does not stem directly from the maximization of a value function in the presence of uncertainty affecting the future. To the reduced form (or non-structural) models belong the Survival Analysis (or Duration Models), while the Option Value and the Dynamic Programming model belong to the other category.

15.5 The Survival Analysis

The seminal applications of Survival Analysis (or Duration Models, DM) to retirement were carried out by [13] and [25]: both works relied on a version of hazard models, first applied in economics to the problem of measuring the duration of an unemployment spell (see [33]). In short, the major characteristics of the duration analysis can be summarized as follows: 1) the dependent variable is the "waiting-time" until the occurrence of a well-defined event: in our specific case, "time-to-retirement" is taken as a (positive) continuous random variable; 2) observations are censored, that is for some individuals the event of interest has not occurred at the time the data are analyzed and 3) there are predictors or explanatory variables which affect the waiting time. Similarly to the Logit and Probit models, these covariates can be time varying and accounting for forward looking behavior. Moreover, the distribution of the random variable can be fully parametric, semi-parametric or non-parametric and also time-gaps in observations can be easily tackled[9].

To lay out the model, assume that T is a continuous random variable with probability density function $f(t)$ and cumulative distribution function $F(t) = P(T \leq t)$, giving the probability that the event of retirement has occurred by duration t. Similarly, the complement of F, the survival function, can be defined as: $S(t) = P(T > t) = 1 - F(t) = \int_t^\infty f(x)\,dx$, which gives

---

[9] Notice that this approach is particularly suitable when retirement is an absorbing state, that is, an irreversible decision which can be analyzed through epidemiology methods of studying the risk of diseases occurrence (such as death).

the probability of being "alive" at duration t. Then, by defining the hazard function as: $\lambda(t) = \lim_{\Delta \to 0} \frac{P(t < T \leq t + \Delta | T > t)}{\Delta}$, which expresses the instantaneous rate of occurrence of the event[10], one can easily verify that $\lambda(t) = \frac{f(t)}{S(t)}$. In other words, the hazard at duration t equals the ratio between the density of the event at t and the probability of being at t without having experienced the event. Finally, from the definitions given above, the following relationships hold: $\lambda(t) = -\frac{d}{dt} \log S(t)$, and $S(t) = \exp\left(-\int_0^t \lambda(x)\, dx\right)$. These results show that the distribution of T can be characterized equivalently both in terms of the survival and the hazard function. Now, suppose there are individuals $i = 1, ..., N$, entering a state (e.g. employment) at duration $t = 0$ and some of them retiring at $t_i > 0$ (that is, some of them exit the spell while others are censored): if censoring does not provide any information regarding the prospects of survival, one can write the likelihood function for this sample as: $L = \prod_{i=1}^N L_i = \prod_{i=1}^N \lambda(t_i)^{d_i} S(t_i)$, where d_i is an indicator variable taking the value 1 if individual i retires at duration t_i and zero otherwise. Taking logs and recalling the definition of $S(t)$ one obtains:

$$\log L = \sum_{i=1}^N \left[d_i \log \lambda_i - \int_0^t \lambda(x_i)\, dx_i \right] = \sum_{i=1}^N [d_i \log \lambda_i - \Lambda(t_i)] \qquad (15.4)$$

where $\Lambda(t_i)$ is defined as the cumulative hazard (or cumulative risk).

Usually, econometricians are interested in the effect that different variables may have on the probability of retiring. For this reason a typical estimation strategy is to express the (logs of) duration as a function of (observed) covariates X_{it} and (unobserved) parameters $\beta(t)$ which may vary or not over time. For example, an usual specification is: $\log T_i = X_i'\beta + \varepsilon_i$, with ε_i a suitably chosen error term, so that $T_i = \exp\left(X_i'\beta + \varepsilon_i\right)$. Moreover, different kinds of parametric models can be obtained by assuming certain distributions for the error term (for example, the Tobit model or the exponential regression model) the parameters of which are all estimable by maximizing the log-likelihood for censored data described above.

However, since economic theory does not necessarily produce these functional forms, researchers prefer not to impose too much structure to the (unknown) duration function. In the light of this, reduced forms are adopted by specifying directly the hazard function. The most popular is the Box-Cox proportional hazard model of the form: $\lambda(t|X_{it}) = \lambda_0(t) \exp\left(X_{it}'\beta\right)$, where $\lambda_0(t)$ is the baseline-hazard which describes the risk for individuals with $X_{it} = 0$ and serves as a reference cell, while the second term is the relative risk, that

---

[10] Precisely, the numerator is the conditional probability of the event occurrence by duration interval $(t, t + \Delta)$ given that it has not occurred before. The ratio to the interval width gives the rate of occurrence per unit of time, and, finally, by taking the limit we get the instantaneous rate of occurrence.

is a proportionate increase or reduction in risk associated to the set of characteristics X_{it}. Again, a number of specifications are obtainable according to
the hypothesis on the shape of the baseline-hazard: parametric (i.e. Weibull[11],
Gamma and generalized F distributions), semi-parametric, with mild assumptions on $\lambda_0(t)$, and non parametric, which leaves the baseline-hazard completely unspecified[12].

DMs are quite useful in that, by reproducing the sequential nature of the
retirement choice can unveil the agent's decision rules and allow to test the
significance of a number of sociological, demographic and economic variables
all potentially entering the decision process, which, instead, in structural models are more difficult to handle[13]. Moreover, they are particularly suitable to
deal with time gaps or censoring in the data. On the other hand, they share
with the other reduced-form estimation models the limit of being independent of any particular behavioral theory. In particular, in these models the
role played by forward looking variables can only be captured indirectly (for
example, via SS incentive measures for anticipated retirement or by the estimation of foregone wages) since none of such models can reproduce the way in
which these variables enter the underlying maximization process of workers.
Similarly, uncertainty affecting the future (like income, health status) is not
suitably modeled. In fact, current decisions would depend on the complete
joint distribution of shocks affecting all future events and outcomes[14]. For
this reason researchers have been exploring alternative approaches, known as
Structural Models, which descend more directly from economic theory.

These approaches share the characteristic of modeling individual choices
by assuming that forward looking agents maximize a "utility" or "value"
function in presence of uncertainty affecting future events: in other words, the
choice of retirement is the solution to a lifetime maximization problem with
incomplete information concerning the future. In these frameworks estimated
parameters have a precise and intuitive economic meaning so that they can
be easily interpreted. On the other hand, the cost of such approaches is the
complication of computations and, often, the strong specification dependence
of the estimates.

---

[11] For example, in the Weibull specification the hazard function has the form $\lambda(t) = \lambda p(\lambda p)^{p-1}$, with λ and $p > 0$.

[12] This approach relies on the partial likelihood function proposed by [12] in his
original paper. See also [30], chap. 2, for an extensive presentation of other possible
specifications.

[13] The discrete time version of the DM was originally presented in [28]. [44] provides
a comparison between the performance of a discrete time version of the DM and
of a structural (Option Value) model for the Italian case.

[14] It is worth recalling that the Proportional hazard model can be obtained as a
special form of the Option Value model presented in the next section, which is
a structural framework. However this property holds under very special circumstances.

15.6 The Option Value Model (OVM)

In order to introduce the OVM, originally set up by [46], suppose an individual (i) is assessing the decision of retiring in the current year (s): according to the OVM, such individual will compare the value of retiring in that year with the (maximum) value of retiring in any of the subsequent years. The difference between the latter and the former value is called the Option Value (OV) of postponing retirement. The decision rule is straightforward: if the OV is positive, the individual will go on working, if negative, he/she will retire. More precisely, agents are supposed to have identical preferences and are assigned an intertemporal indirect utility function (or Value function), which for each individual i, at time s has the form: $V_{s,i}(\tau) = \sum_{t=s}^{\tau-1} \gamma^{t-s} U(W_{t,i}) + \sum_{t=\tau}^{D} \gamma^{t-s} U(B_{t,i}(\tau))$, where τ is retirement year, D is the year of death, W and $B(\tau)$ are, respectively, the real wage and the real pension benefit and γ is the intertemporal discount rate. Then, assume the instantaneous indirect utility function be as follows: $U_{t,i} = (W_{t,i})^{\alpha} + \vartheta_{t,i}$ and $U_{t,i} = (\theta B_{t,i}(\tau))^{\alpha} + \zeta_{t,i}$, where α is a risk aversion parameter, θ is the value of consumption while retired in terms of consumption while working (expected to be greater than 1), $\vartheta_{t,i}$ and $\zeta_{t,i}$ are individual-specific independent disturbances, representing, for example, maximization mistakes, incomplete information, health or job-related shocks. Now, calling R the set of years in which, according to the current institutional rules, it is possible to retire, agent i will compare the "value of retiring" in year s with the value of all the years $r > s$ belonging to R. Consequently, the complete decision rule is: retire in year s iff $V_{s,i}(s) > V_{s,i}(r)$, $\forall r \in R, \Leftrightarrow O_{s,i}(r_s^*) \equiv V_{s,i}(r_s^*) - V_{s,i}(s) < 0$, with $O_{s,i}(r_s^*)$ the OV, r_s^* $argmax$ $V_{s,i}(r > s)$ from year s standpoint. Expanding the expression of the OV according to equations of the utility given above and, as usual in the literature, supposing that the survival probabilities are independent of the earning streams and of the disturbances, one can write

$$O_{s,i}(r_s^*) = \sum_{t=s}^{r_s^*-1} \pi(t|s) \gamma^{t-s} E_{s,i}[(W_{t,i})^{\alpha} - (\theta B_{t,i}(s))^{\alpha} + (\vartheta_{t,i} - \zeta_{t,i})] +$$

$$+ \sum_{t=r_s^*}^{D} \pi(t|s) \gamma^{t-s} E_{s,i}[(\theta B_{t,i}(r_s^*))^{\alpha} - (\theta B_{t,i}(s))^{\alpha}] \tag{15.5}$$

with $\pi(t|s)$ the probability of surviving in year t conditionally on being alive in year s. Finally, as for the random process originating the disturbances, it is assumed that the errors follow a Markov chain such that: $\zeta_{t,i} = \phi\zeta_{t-1,i} + \xi_{i,\zeta}$ and $\vartheta_{t,i} = \phi\vartheta_{t-1,i} + \xi_{i,\vartheta}$, with $E_t(\xi_{i,\zeta})$, $E_t(\xi_{i,\vartheta}) = 0$, for $t = s+1, ..., D$. Now, defining $\omega_{t,i} = \vartheta_{t,i} - \zeta_{t,i}$ and exploiting the fact that $E_{s,i}(\omega_{t,i}) = \phi^{t-s}\omega_{s,i}$, eq. (15.5) can be written as: $O_{s,i}(r_s^*) = m_{s,i}(r_s^*) + e_{s,i}(r_s^*)\omega_{s,i}$, where $e_{s,i}(r_s^*) = \sum_{t=s}^{r_s^*-1} \pi(t|s)(\gamma\phi)^{t-s}$ and $m_{s,i}(r_s^*)$ is equal to the rest of the RHS of (15.5).

It is possible to write the probability that an individual retires in year s as the probability that OV is negative. Thus, manipulating (15.5) it follows

$$P\left(\tau = s\right) = P\left(O_{s,i}\left(r_s^*\right) < 0\right) = P\left(-\frac{m_{s,i}\left(r_s^*\right)}{e_{s,i}\left(r_s^*\right)} > \omega_{s,i}\right) = F\left(-\frac{m_{s,i}\left(r_s^o\right)}{e_{s,i}\left(r_s^o\right)}\right)$$
(15.6)

with F the normal cumulative probability and r_s^o is argmax $O_{s,i}(r_s^*)$. The superscript of r^o means that r is the optimum from year s point of view. Furthermore, notice that r_s^o is an estimator of r_s^*. The event of retiring in year s is thus a random variable which, under the hypotheses above, has mean zero and variance σ_w[15].

It is worth noting that in this framework the decision to cease working is taken according to the *maximum of the expected values, (MEV)* of future utility levels, while, typically, a stopping-rule framework would select for retirement the age associated to the *expected value of the maximum utility (EVM)*. Since the EVM is greater than MEV, this means that the OV model will yield a lower value of postponing retirement. However, the relevance of this approximation error is not univocal[16]. Finally, recent works dealing with reduced forms equations have used the OV (previously estimated or calibrated on the population) as a regressor of the probability of retiring, meant to represent a forward looking measure of SS provided incentives for early retirement. So far, however, the role played by this more sophisticated measure in reduced form estimations is not univocal[17].

15.7 Dynamic Programming

The Dynamic programming approach assumes that agents' behavior is the output of an "optimal decision rule" or the solution of a controlled discrete stochastic process[18]. Precisely, the sequence of decisions, taken under uncertainty, can be represented through a stochastic decision process; individuals are supposed to be rational and maximizing a life-time utility function of the form: $E\left\{\sum_{t=0}^{T} \beta^t u\left(d_t, s_t\right) | s_0 = s\right\}$ where E is the expectation operator, $u()$

---

[15] Obviously, the probability of continuing to work is the complement to one of $P(r = s)$. Notice that if one allows for the possibility to follow an individual's behavior for two subsequent years, the model can explicitly account for the sequential nature of the choice. See [46].

[16] For example, [35] show that their strategy has the same predictive power as the exact solution and, in any case, is much better that a model ignoring dynamics completely. On the contrary, [45] points out that the approximation above performs well only in limited situations, while the correct expected maximum value rule is satisfactory even when error distributions are misspecified.

[17] See [20].

[18] In this session I draw from [41] line of exposition.

is the instantaneous utility function, d and s sets of control and state variables respectively, $\beta \in (0,1)$ is the intertemporal discount rate. The problem can be solved by finding an optimal decision rule $d_t = \mu(s_t)$ that is solution to: $V_0^T(s) \equiv \max E_\mu \left\{ \sum_{t=0}^{T} \beta^t u(d_t, s_t) \,|s_0 = s \right\}$ where E_μ is expectation with respect to the controlled stochastic process (d_t, s_t). The sources of uncertainty are, on one hand, uncertain future states (such as health status, mortality, employment) on whose transitions agents have subjective (Markovian) transition probabilities $p(s_{t+1}|s_t, d_t)$; on the other hand, there are unexpected shocks (ε_t) occurring in each decision period and, typically, affecting the utility function.

Since in retirement decisions d is a discrete variable, the problem has a discrete decision process nature. Consequently, the optimal decision rule is determined by a system of inequalities rather than as a solution to a (Euler) first order condition. Besides that, since closed forms of μ are rarely available, most structural estimation methods for these kind of models require estimation of μ via numerical methods (although other techniques rely on Monte Carlo simulations of the controlled stochastic process (d_t, s_t)). A simplified version of the problem assumes that ε_t enters u in an additive separable fashion, is an IID extreme value variable and p satisfies a conditional independence condition; under such assumptions a dynamic generalization of the multinomial logit model can be obtained, whereby the conditional choice probability $P(d_t|x_t, \vartheta_t)$ (with ϑ the vector of parameters of p and u to be estimated) is given by:

$$P(d_t|x_t, \vartheta) = \frac{e^{[v_\vartheta(x_t, d_t)]}}{\sum_{\overline{d} \in D(x_t)} e^{[v_\vartheta(x_t, \overline{d})]}} \tag{15.7}$$

where x is the non-stochastic partition of variables in s observed by the econometrician. The relevant difference with the static logit model is that v is a one period utility function, linear in parameters ϑ, while in this contest v is the sum of all the expected discounted utilities in the (present and) future periods[19].

Some works have relaxed the hypotheses above, by allowing for the error terms to enter u in a nonlinear, non-additive fashion and be serially correlated[20]. Moreover, researchers would like to extend the level of realism of the models by introducing new variables (i.e. savings or the family dimension) and relaxing assumptions (i.e. rational expectations). However, it is worth recalling that typically in these models the "curse of dimensionality" problem (i.e. the exponential increase of the burden in terms of time/space needed to solve the problem) dramatically affects the level of detail which can be achieved.

---

[19] Also the functional form of v is not known a priori, so its values have to be computed numerically for any particular value of ϑ.

[20] See [26].

Furthermore, the more complex the model is, the heavier its dependence on the specification adopted. As pointed out by [48], to deal with this issues researchers may intervene over a number of dimensions: 1) the size of the choice set, 2) the size of the state space, 3) the functional form of the utility function and the (joint) distribution of unobservables[21].

However, despite a number of sufficiently detailed utility-based optimizing models can be calibrated to replicate observed economic variables, one can end up with very different predictions about the effects of policy changes on macroeconomic variables. For example, [14] evaluate the effects of fundamental tax reform by calibrating several different models and sets of model parameters -such as deterministic OLG, stochastic OLG, infinite horizons, each with several parameter set choices- to the same initial economy. In fact they obtain very different predictions according to the specification adopted. All this considered, it can be useful to conclude with the argument stressed by [48]: especially for policy analysis, *"...informed judgments* [concerning which model to use] *should not be based on methodological predispositions, but on evidence of performance"*. After all, a "fair pragmatism" is to be considered still a valid criterion of choice whenever the trade-off between computational complexity (or theoretical coherence) and predictive capability arises.

15.8 Open Issues

As mentioned, recent works have tried to improve the performance of the existing models by developing new SS incentive measures which account for the forward looking nature of retirement choices (see [44] for a complete presentation of these variables). This has been done also in order to control for the source of identification of retirement incentives in reduced form models. More precisely, [10] argue that the OV, for example, depends critically on an individual's wage, which in turn is likely to be correlated with underlying tastes for retirement: consequently, using cross sectional variation in the OV can lead to biased estimates of the effects of changes in SS provided incentives for early retirement. For this reason [10] propose a new measure (the "Peak value") by which they are able to better disentangle to role played by wages and the nonlinearities in the SS (and private pension) program rule. [9], on the other hand, choose a different strategy, in that they adopt a fixed effect specification with panel data and show that the coefficients are almost half the magnitude of similar OLS regressions. Thus, they bring evidence that the bias due to unobserved heterogeneity can be potentially large. The latter approach seems more effective in that, by explicitly controlling for fixed individual characteristics such as tastes for leisure, is in principle able to rule out the mentioned problem and, furthermore, represents a more general methodology, which can be extended to the structural model specifications.

---

[21] See also the results in [42] which break, via random Monte Carlo integration methods, the curse of dimensionality for discrete decision processes.

As for the new econometric models, a major goal currently on the researchers' agenda is to set-up an encompassing framework explaining the most relevant lifetime economic decisions of agents. For example, it is unlikely that individuals make retirement decisions independently of their lifetime savings plan. This is particularly true in the late years of working careers, as workers become more aware of their future income needs and of the adequacy of existing saving. So far almost no model has dealt simultaneously with saving and retirement, although there are many instances in which a feedback from saving to retirement (and vice versa) is present. For example, [43] include uncertainty in their model and imperfect markets, but no saving decision. Similarly, [46] assume that individuals consume all their income in every period.

Another issue is concerned with the effect of health insurance and private pensions; in fact, some empirical evidence suggests that retiree health insurance affects the timing of retirement by modifying both the budget constraint and preferences[22]. Also private pensions are documented to be significant in the retirement decision[23]: as a consequence, health insurance and/or private pensions should be incorporated in a model of retirement. Another development is the extension of these models to analyze the joint decision making within families or couples about retirement.

Intuitively, the primary motive for joint retirement decisions is the complementarity of leisure: for married individuals who wish to spend their leisure time together with her/his spouse, leisure time gives more satisfaction to one spouse as the leisure time of the other spouse increases. This may also be due to couples' desire to relocate to different regions of the country as they become older, which would be easier to accomplish if neither spouse were working. From another point of view, assortative mating may also lead a couple to have similar retirement patterns, in that spouses with similar preferences over combinations of leisure and consumption are likely to retire at similar ages. Furthermore, correlations in wages, SS benefits and pensions between husbands and wives can also lead to similar retirement decisions.

The existing applications of family decisions have adopted versions both of the probit/model approach and of the more sophisticated ones. Among the latter, both DM (see for example the work by [1]) and structural models have been explored (see, for example, [23], [4]). However, other applications are needed in order to shed light on the significance of couple coordination behavior in retirement decisions. Finally, to my knowledge the work by [47] is the very first attempt to estimate a structural model of joint savings and retirement decisions both by single and couples, and including public and private pensions and uncertainty of a host of future outcomes (like health status, survival and the generosity of the Social Security system).

Concluding, it can be said that the possibility of developing new applications and setting-up richer structural models is a fascinating challenge which

---

[22] See, for example, [43], [29], [19], [31].
[23] See the works by [37] and [38].

has been already undertaken by several authors; however, it is worth recalling that such a strategy is not costless: as I have pointed out, the adoption of more realistic and detailed models, at the moment, poses major problems, such as high computer-time consuming and specification dependent estimates and the need of high quality datasets as well. As for the latter, the availability of suitable longitudinal or panel data is claimed to be still a problem for some countries (and, to some extent, also for Italy)[24].

15.9 Conclusions

In this work I carried out an overview of the most recent developments in micromodeling retirement choices. I classify the models proposed by the retirement literature according to their ability to match the main characteristics of retirement choices. In particular, the major contribution to this branch of literature has been the recognition of the difference between the choice of ceasing to work and the traditional labor supply static model, due to the more explicit life cycle dimension and forward looking nature of the former. In fact, all the most recent works, both in the reduced forms (i.e. Logit/Probit or Duration models) and in the structural specifications (i.e. the Option Value model and the Dynamic Programming models) aim at taking into account the forward looking behavior comprised in the choice of retiring.

The degree of sophistication of economic theory and of econometric specifications of retirement models have increased dramatically. The promising results obtained by the works carried out so far are boosting further developments: on the one hand, explaining joint retirement and saving decisions, private pensions and health care interaction in an unified framework are some of the most challenging objectives on the researchers' agenda. On the other hand, very interesting steps into modeling joint labor supply decisions in the family context have been already taken: the issue is both interesting per se and can improve the level of realism of future applications, such as economic microsimulation. In any case, the increase in the models complication is not priceless: in general, the cost is lower "flexibility", high computer-time demanding procedures and specification dependence. Also good longitudinal or panel data availability is a necessary condition for implementing more complex and robust models[25].

---

[24] On this point see [6] and [44].

[25] Econometric estimates, reported in [44], show that the Duration Model performs better than the Option Value Model, despite the latter is more appealing in terms of economic significance and intuitive interpretation of parameters. Thus, in this work the trade-off between data tractability and model sophistication turns out to be even worse than expected, in that the more complex model does not produce better estimates.

References

1. An MY, Christensen BJ, Gupta ND (1999) A Bivariate Model of the Joint Retirement Decisions of Married Couples. Centre for Labour Market and Social Research, WP 99-10, December
2. Antcliff S (1993) An Introduction to DYNAMOD. DYNAMOD Technical Paper No. 1, National Centre for Social and Economics Modelling, University of Camberra
3. Bianchi C, Romanelli M, Vagliasindi PA (2003) Microsimulating The Evolution Of The Italian Pension Benefits: The Role Of The Retirement Choices And Lowest Pensions Indexing. Labour, Forthcoming
4. Blau DM (1998) Labor Force Dynamics of Older Married Couples. Journal of Labor Economics 16: 595-629
5. Boskin MJ, Hurd MD (1978) The effect social security on early retirement. Journal of Public Economics 10: 361-377
6. Brugiavini A, Peracchi F (2003) Micro Modeling of Retirement Behavior in Italy. In: Gruber J, Wise DA (eds), Social Security Programs and Retirement around the World: Micro-Estimation, The University of Chicago Press, forthcoming
7. Burtless G (1986) Social Security, Unanticipated Benefit Increases, and the Timing of Retirement. Review of Economic Studies LIII: 781-805
8. Burtless G, Hausman J (1980) Individual retirement decisions under employer provided pension plan and social security. Mimeo, MIT, Cambridge, MA
9. Chan S, Stevens A (2001) Retirement Incentives and Expectations. NBER Working Paper, n. 8082
10. Coile C, Gruber J (2000) Social Security and Retirement. NBER Working Paper, n.7830
11. Colombino U (2001) Intertemporal Interpretation of a 'Static' Model of Retirement. An Exercise with Italian Data. WP CHILD n. 18/2001
12. Cox DR (1972) Regression models and life-tables (with discussion). Journal of Royal Statistical Society, Series B 34: 187-220
13. Diamond P, Hausman J (1984) Individual Retirement and Savings Behavior. Journal of Public Economics 23: 81-114
14. Engen EM, Gravelle J, Smetters K (1997) Dynamic Tax Models: Why They Do the Things They Do. National Tax Journal 50: 657-82
15. Feldstein M (1974) Social security, induced retirement and aggregate capital accumulation. Journal of Political Economy 82: 905-926
16. Galler HP (1997) Discrete-Time and Continuous-Time Approaches to Dynamic Microsimulation. Paper presented at the Microsimulation in New Millennium: Challenges and Innovations Workshop. Cambridge, August
17. Geweke J (1995) Monte Carlo Simulation and Numerical Integration. Federal Reserve of Minneapolis Research Department Staff Report, n. 192
18. Gribble S (2000) LifePaths: A Longitudinal microsimulation model using a synthetic approach. In: Gupta A, Kapur V (eds) Microsimulation in Government Policy and Forecasting (Chapt. 19). Elsevier, Amsterdam & New York
19. Gruber J, Madrian BC (1995) Health Insurance Availability and the Retirement Decision. American Economic Review 84: 938-948
20. Gruber J, Wise DA (eds) (2003) Social Security Programs and Retirement around the World: Micro-Estimation. The University of Chicago Press, forthcoming

21. Gupta A, Kapur V (eds) (2000) Microsimulation in government policy and forecasting. Elsevier, Amsterdam & New York
22. Gustman AL, Steinmeier TL (1986) A disaggregated structural analysis of retirement by race, difficulty of work and health. Review of Economics and Statistics 67(3): 509-513
23. Gustman AL, Steinmeier LS (1994) Retirement in a Family Context: A Structural Model for Husbands and Wives. NBER Working Paper No. 4629. Cambridge, MA
24. Harding A (2000) Dynamic Microsimulation: Recent Trends and Future Prospects. In: Gupta A, Kapur V (eds) Microsimulation in Government Policy and Forecasting (Chapt. 15). Elsevier, Amsterdam & New York
25. Hausman JA, Wise DA (1985) Social Security, Health Status, and Retirement. In: Wise DA (ed) Pensions, Labor, and Individual Choice. University of Chicago Press, Chicago
26. Hotz VJ, Miller RA, Sanders S, Smith S (1993) A simulation Estimator for Dynamic Models of Discrete Choice. Review of Economic Studies 60: 397-429
27. Hurd MD (1990) The Joint Retirement Decision of Husbands and Wives. In: Wise DA (ed) Issues in the Economics of Aging, University of Chicago Press, Chicago
28. Jenkins S, (1995) Easy Estimation Methods for Discrete Time Duration Models. Oxford Bulletin of Economics and Statistics 57: 129-137
29. Johnson RW, Davidoff AJ, Perese K (2000) Health Insurance Costs and Early Retirement. Paper presented at the Allied Social Sciences Meeting, Boston, January 8
30. Kalbfleisch JD, Prentice RL (1980) The statistical analysis of failure time data. John Wiley and Sons, New York
31. Karoly LA, Rogowski JA (1994) The Effect of Access to Post-Retirement Health Insurance on the Decision to Retire Early. Industrial and Labor Relations Review 48: 103-123
32. Laditka S (1996) Individuals' Lifetime Use of Nursing Home Services: A Dynamic Microsimulation Approach. Papers in Microsimulation Series n.3, Maxwell Centre for Demography and Economics of Ageing
33. Lancaster T (1979) Econometric methods for the duration of unemployment. Econometrica 47: 939-956
34. Lazear EP (1986) Retirement from the Labor Force. In: Ashenfelter O Layard R (eds) Handbook of Labor Economics Vol. I. Elsevier, Amsterdam & New York
35. Lumsdaine RL, Stock JH, Wise DA (1992) Three models of retirement: computational complexity versus predictive validity. In: Wise DA (ed) Topics in the Economics of Ageing.University of Chicago Press, Chicago
36. Lumsdaine R, Mitchell O (1999) New developments in the Economic analysis of Retirement. In: Adhenfelter O Card D (eds) Handbook of Labor Economics Vol. 3. Elsevier, Amsterdam & New York
37. Madrian B (1993) Post-Retirement Health Insurance and the Decision to Retire. Unpublished manuscript. Graduate School of Business, University of Chicago
38. Madrian B, Beaulieu ND (1998) Does Medicare Eligibility Affect Retirement? In: Wise DA (ed) Inquiries in the Economics of Aging. University of Chicago Press, Chicago
39. Nelissen JHM (1994) Income redistribution and social security: an application of microsimulation. Chapman & Hall London

40. OECD (1995) The Labour Market and the Older Workers. Social Policies Studies n. 17. OECD, Paris
41. Rust J (1994) Structural Estimation of Markov Decision Processes. In: Engle RF McFadden DL (eds) Handbook of Econometrics, vol. 4. North-Holland, Amsterdam
42. Rust J (1997) Using Randomization to break the Curse of Dimensionality. Econometrica 63: 487-516
43. Rust J, Phelan C (1997) How Social Security and Medicare Affect Retirement Behavior in a World of Incomplete Markets. Econometrica 65: 781-831
44. Spataro L (2002) New tools in micromodeling retirement decisions: overview and applications to the Italian case. CeRP WP, n. 28/02
45. Stern S (1996) Approximate Solutions to Stochastic Dynamic Programs. Econometric Theory 13: 392-405
46. Stock SH, Wise DA (1990) Pensions, the option value of work, and retirement. Econometrica 58: 1151-80
47. Van der Klaauw W, Wolpin KI (2003) Social Security, Pensions and the Savings and Retirement Behavior of Households. Mimeo
48. Wolpin KI (1996) Public-Policy Uses of Discrete-Choice Dynamic Programming Models. American Economic Review 86: 427-32

16

Applied Econometrics Methods and Monetary Policy: Empirical Evidence from the Mexican Case[*]

Luis Miguel Galindo[1] and Horacio Catalán[2]

[1] Faculty of Economics, UNAM gapaliza@servidor.unam.mx
[2] Faculty of Economics, UNAM catalanh@correo.unam.mx

Summary. The main objective of this paper is to illustrate, using Mexican data, how the results yield by modern econometric methods are dependent upon each specific technique as well as upon the statistical properties of the series analyzed. The problems are even stronger and more evident in the case of economic series with structural changes and high variability as is the case of Mexico. Applied econometrics should be explicitly based upon a probability viewpoint, and different methods should be taken to produce only approximations to the actual data generation process. Thus, alternative techniques can only show distinctive features of the actual data that still need to be validated with the rest of empirical evidence. This indicates that applied econometricians have to look for maximum information by correctly applying different techniques without forgetting the relevance of economic reasoning. Using Mexican data, alternative econometric estimations are evaluated indicating that the formulation of a monetary policy only on the basis of some specific technique, without considering its potential pitfalls, should not be recommended.

16.1 Introduction

Over the last two decades, monetary policies have acquired a place of preeminence in under developing economies. In effect, monetary policy has become a major instrument in order to obtain price and exchange rate stability and also to achieve a continuous economic growth. The increasing importance of monetary policy in the case of developing countries has additionally been associated with important institutional and regime changes. The Mexican Central Bank, in particular, has acquired its formal independence during the decade of the nineties while concentrating its activity on the control of the inflation

---

[*] We appreciate the comments to an early version of this document by Lionello Punzo and Martín Puchet. The econometric estimations were performed by Eduardo Alatorre. The usual disclaimer applies here. This project was elaborated under the support of PAPIIT, UNAM: Monetary and financial policy and the effects of the opening of the external sector: an econometric approach, IN 304702.

rate, on a move from a regime based on control and on the supervision of the monetary aggregates, to another one that is based on an instrument called "el corto" that influences the nominal interest rates[3]. The main economic results of this new monetary policy include the control of the inflation rate but also a sluggish economic growth.

The arguments to support this change of monetary policy regimen in Mexico basically consist of three points: the endogeneity of the monetary aggregates (Gil- Díaz, 1997), the existence of an unstable demand for money that makes the transmission mechanism unstable (Garcés, 2002), the lack of a stable relationship between the monetary aggregates and output or inflation rate and the neutrality hypothesis (Carstens and Reynoso, 1997). Most of these arguments are related with the use of alternative econometric techniques. However, the empirical evidence on these topics is far from unique and indicates that the empirical results are a function of the selection of the econometric technique or method. In this sense, the existence of contradictory evidence that lead to completely different economic policy recommendations are, partially, a function of the selection of the econometric technique. This is partially the consequence of the use of a wide range of econometric techniques without carefully considering their possibilities and limitations (Smith, 2000). This situation is certainly worse in the Mexican economy where the main economic series show fundamental evidence of instability, variability and structural change.

In this context, the main objective of this essay is to show, using the Mexican data as an illustration, that the econometric results are highly dependant on each particular technique, especially in the case of structural change and the strong variability of the series. Under these circumstances it is not reasonable to constraint a monetary policy to follow only one set of results that are derived from one particular econometric method. Therefore, it is worth value to not only be aware of the limitations of each econometric technique, but also to try to use a broader set of methods. In this context, it seems appropriate to consider the robustness of the results across the whole range of econometric techniques. Therefore, this essay pretends to show that the appropriate selection of modern econometric tools is an essential part to obtain adequate empirical evidence.

This essay is organized as follows. After the introduction the following section includes the econometric discussion and the empirical evidence. The third section presents the conclusions and some general comments.

---

[3] Banco de México (1996) includes an explanation of the mechanics of this instrument.

16.2 Econometric Methods and Monetary Policy in the Mexican Economy

Applied monetary policy researchers use different methods to analyze or simulate economic data or even to contrast alternative hypthotesis. The range of options is rather large and includes as varied methods as the calibration of real business cycle models to very sophisticated econometric procedures and methodologies. The use of this variety of methods, together with the existence of different statistical properties in each economic series produces results that are highly variable and even contradictory. This situation is even worse in a situation where economic series show evidence of structural breaks and modifications in their variances.

In effect, econometric theory (Hendry and Clements, 1999) argues that estimations of the coefficients are unbiased and efficient only in cases of correct specification with constant parameters and stationary series. Unfortunately, in most cases the economic series are non-stationary, are subject to structural changes and the correct model is not known. Furthermore, the options and methods normally selected in order to obtain a well-specified model with stationary series are not robust enough and can generate rather misleading results. Under these circumstances the applied econometrician is not clear of which tests are superior to others (Phillips and Xiao, 1998).

For the Mexican case this problem can be illustrated with the aid of the money neutrality hypothesis. The neutrality of money is arguably the most important point of debate in monetary economics partly due to its crucial implications for monetary policy. This hypothesis sustains that money does not affect the trajectory of the real variables. There are two versions of this hypothesis: the strong one argues that there is no effect in either the short or the long run, and the weak case only sustains its validity in the long run (Walsh, 2000 or Obstfeld and Rogoff, 1996). The analysis of this hypothesis presents at least three potential problems that can be summarized in the following aspects:

16.2.1 The Unit Root Issue

One of the key features of any economic time series is the potential presence of a unit root in its trajectory. The relevance of this point is partly a consequence of the spurious regression problem (Granger and Newbold, 1974) and the possibility to identify the shocks as either permanent or transitory (Nelson and Plosster, 1982). Therefore, it is common practice in applied modern econometrics to initially test for the presence of unit roots using either the Dickey-Fuller (ADF) (1981) or the Phillips-Perron (PP) (1988) tests. However, these tests have very low power in those cases where the series follow a random walk, have structural breaks or have a deterministic trend (De Jong and Whiteman, 1991) and furthermore, it is important to remember that these tests require Gaussian errors (Phillips and Xiao, 1998). Under these circumstances some applied researchers might even consider that the unit root analysis is not necessary due to the low power of the tests and in particular in the case of a trend break (Phillips and Xiao, 1998). For example, from the model perspective, it does not matter if the β of the unit root hypothesis is .98 or 1.01 unless the analysis involves the very long run.

As a consequence of this situation there is more and more literature written on the search for potential solutions to the critic of the low power of the unit root tests. For example, there is an increasing number of tests that include the use of large time spans, different null hypothesis (Maddala and Kim, 1998) or even tools that allow for the identification of the presence of structural breaks in the framework of unit root tests (Perron, 1997). However, it is still difficult to distinguish between a genuine unit root and a structural change in the series (Maddala and Kim, 1998).

Therefore, in particular for the Mexican case, the use of a battery of tests is certainly relevant because, due the long tradition of high inflation rates and structural instability, it is difficult to distinguish between a stationary series with a structural break from a series with genuine unit roots. Moreover, there is strong evidence that the economic series in the Mexican economy are subject to the Perron phenomenon (Badillo, Belaire-France and Contreras, 2002). This phenomenon indicates that a real I(1) series with structural break could appear to be a I(2) process. It is highly likely that Mexican data with series such as the price index or the monetary aggregates should be a long run I(1) process despite the fact that the evidence suggest that they are I(2). Under these circumstances it is a common option to already use the linear combination of the I(2) and I(1) variable in the analysis which normally tends to be I(1) only, in order to reduce this Perron phenomenon (Pesaran and Smith, 1998).

The specification of the unit roots equations (with constant and trend or without constant and trend) and the number of lags included can be selected using different criteria. This selection has important consequences in the results derived from these tests. In principle, there is the general to specific procedure that estimates, initially, a regression with constant and trend and tests their significance while the lag length (k) is chosen using the t-sig or F-sig procedure or through the use of Lagrange multiplier statistics for autocorrelation (Ng and Perron, 1995 and Maddala and Kim, 1998) or some information criteria[4]. Additionally, it is becoming common practice to use the KPSS (Kwiatkowsky et al, 1992) test as a confirmatory tool or in some cases to identify the order of integration when there are structural breaks in the series (Charemza and Syczewska, 1998 and Chen, 2002). For example, Charemza and Syczewska (1998) propose a joint test using the ADF and

---

[4] It is worth value to mention that there is not one criterion that always performs better and it is therefore still a matter of the appropriate judgment of the applied researcher (Weber, 2001).

KPSS considering that the ADF has as the null hypothesis the presence of unit roots and the KPSS has as the null hypothesis that the series is stationary. Furthermore, Chen (2002) argues that the KPSS has better properties for detecting structural breaks at the beginning and end of the sample. Finally it is also possible to test for the presence of structural change using the Perron (1997) statistic.

Unfortunately, this battery of tests and selection criteria normally lead to different results. It is, therefore, indispensable and in particular for the case of Mexico to carefully consider that the application of the basic unit root tests includes several decisions that depend more on the judgment of the researcher (at least in most empirical work) than on the result of a particular behavior of the data. It is these decisions that have important consequences for the final outcome. In this sense, it seems advisable to consider a battery of correctly specified unit root tests that also includes the possibility of structural change. It is also important to consider the problem of test dependence and test conflict. That is, the true order of integration will not become more evident by means of adding more tests. It is, therefore, advisable to select for each case only a few relevant tests that can give inside information (Patterson and Heravi, 2003).

In the Mexican case the existence of structural instability increases the doubt about the true order of integration of the series. Ibarra (1998), for example, argues that the price index is I(1) while (Galindo and Cardero, 1997) sustained that it is I(2). It is also possible to find economic analysis that combines different orders of integration of the series in the same VAR (Martínez, Sánchez and Werner, 2001), a method that might generate misleading results.

As an illustration of this point we can use quarterly data from the Mexican economy over the 1980(I) to 2000(IV) period. The series are seasonally unadjusted[5] from real gross domestic product (Yt), consumer price index (P_t), the nominal interest rate represented by the three months interest rate of the CETES market (R_t) and the monetary aggregate M2 (M_t). Variables in small letters represent the natural logarithm of the series.

Table 16.1 includes the Augmented Dickey-Fuller (ADF), the Phillips and Perron (PP) and KPSS (Kwiatkowsky et al, 1992) tests for the monetary aggregate M2, real income, the price index and the nominal interest rate in Mexico. Kwiatkowsky et al (1992) test (KPSS) uses the null hypothesis that the series is stationary contrary to the traditional statistics of the ADF and PP (Maddala and Kim, 1998). Additionally, It might be also possible to consider the option of seasonally unit root and cointegration such as Franses and McAleer, (1998). Table 16.2 presents the joint test ADF and KPSS which tries to increase the power of the tests combining the two. Finally, Table 16.3 presents the Perron (1997) test. This statistic makes the break point endogenous by using the null hypothesis of a unit root process without any

---

[5] It might be also possible to consider the option of seasonally unit root and cointegration such as Franses and McAleer, (1998).

structural breaks and the alternative as a trend stationary process with possible structural change[6]. The general results of all these tests are certainly far from obvious. The empirical evidence gathered from these tests allows the researcher to argue in favor of different possible options. That is, in general, the ADF, the PP and the KPSS tests indicate that output and nominal interest rate are I(1) (Table 16.1). Alternatively the ADF indicates that the monetary aggregate and prices are I(2) meanwhile the PP tests confirms the same results for M_t but suggest that P_t is I(1) and the KPSS does not give clear cut evidence either (Table 16.1). Additionally, the joint ADF-KPSS tests indicate that yt is I(1), R_t is I(0) and M_t and P_t are I(1) (Table 16.2). Finally, Perron (1997) for the price index and the nominal interest rate yields that P_t and R_t are both I(1) considering possible structural changes in the series (Table 16.3.a, 16.3.b, 16.3.c and 16.3.d). In this sense, most of these tests suggest that y_t, R_t, M_t and P_t are all I(1) but there is also contradictory evidence. We will in any case proceed on this basis despite the theoretical doubts about the nominal interest rate and prices.

Variable	ADF			PP(3)				KPSS
	A	B	C	A	B	C	n_{ij}	n_t
Y_t	-3.088 (8)	0.462 (8)	-1.840 (8)	**-4.283**	-0.002	3.067	0.986	0.181
δy_t	**-3.855 (8)**	-3.456 (8)	-2.545 (8)	**-21.911**	**-20.574**	**-19.273**	0.118	0.057
R_t	-2.760 (0)	-0.856 (2)	-0.699 (2)	-2.667	-1.279	0.679	0.596	0.097
δR_t	**-8.371 (1)**	**-8.205 (1)**	**-8.220 (1)**	**-9.625**	**-9.508**	**-9.549**	0.232	0.078
$M2_t$	-1.406 (8)	-1.994 (8)	0.455 (8)	-0.139	**-4.108**	6.682	0.982	0.249
$\delta M2_t$	-2.039 (7)	-1.245 (7)	-1.228 (7)	**-6.712**	**-5.092**	-2.149	0.604	0.085
$\delta\delta M2_t$	**-5.336 (6)**	**-5.385 (6)**	**-5.358 (6)**	**-21.775**	**-21.087**	**-21.969**	0.076	0.068
P_t	-1.599 (3)	-2.893 (3)	-0.435 (3)	-0.237	**-3.771**	1.182	0.951	0.248
δp_t	-3.113 (2)	-1.958 (2)	-1.342 (2)	-3.447	-2.435	-1.594	0.591	0.078
$\delta\delta p_t$	**-8.705 (1)**	**-8.725 (1)**	**-8.774 (1)**	**-8.504**	**-8.541**	**-8.619**	0.088	0.065

Notes: Test statistics in bold indicates a rejection of null hypothesis. Critical Value at 5% significance level for the Augumented Dickey-Fuller and Philips-Perron tests for a size T = 100 are -3.45 including constant and trend (model A). -2.89 including constant (model B) and -1.95 without constant and trend (model C). (Maddala and Kim. 1998, p. 64) nij and nt is the KPSS test for null hypothesis of stationarity around a level and deterministic linear trend, respectively.
Both Test are calculated with a lag window size equal to 5. The 5% critical value for the two tests are 0.463 and 0.146, respectively (kwiatkowsky et. al. 1992 p.166) Period 2000:01 - 2003:05

Table 16.1. Unit root tests (ADF, PP and KPSS)

Under these circumstances, there is a mismatch between economic theory and the unit root tests (Hoover, 2000) because the econometric analysis does not represent a neat answer on the behaviour of the series. It is then possible to observe that in applied work the order of integration of the series depends on the time period of analysis, on the selection of tests and their specification and even on the personal choice of the researcher. This problem is stronger in the case of the Mexican economy with series that show important structural changes that depend on the sample period. This can be illustrated by the working papers of the Mexican Central Bank. In effect, Garcés (2002) argues that the price index and the monetary aggregates in Mexico are I(2) meanwhile

---

[6] In this matter there are new tests available such as using the KPSS in the presence of structural break (Lee and Strazictch, 2001).

Variable	Joint ADF and KPSS test for $\rho = 0.75$	
	Z_0	Z_k
Y_t	0.462 (8)	0.986
Δy_t	-3.546 (8)	**0.118**
R_t	-0.856 (2)	**0.596**
ΔR_t	**-8.205 (1)**	**0.232**
$M2_t$	-1.994 (8)	0.982
$\Delta M2_t$	-1.245 (7)	**0.604**
$\Delta\Delta M2_t$	-5.385 (6)	**0.076**
P_t	-2.893 (3)	0.951
Δp_t	-1.958 (2)	**0.591**
$\Delta\Delta p_t$	**-8.725 (1)**	**0.088**

Notes: Test statistics in bold indicates a rejection of null hypothesis. Critical Value at 5% significance level for the Augumented Dickey-Fuller and Philips-Perron tests for a size T = 100 are -3.45 including constant and trend (model A). -2.89 including constant (model B) and -1.95 without constant and trend (model C).
(Maddala and Kim. 1998, p. 64) n_{ij} and n_t is the KPSS test for null hypothesis of stationarity around a level and deterministic linear trend, respectively.
Both Tests are calculated with a lag window size equal to 5. The 5% critical value for the two tests are 0.463 and 0.146, respectively (kwiatkowsky et. al. 1992 p.166)
Period 2000:01 - 2003:05

Table 16.2. Joint ADF and KPSS test for $\rho = 0.75$

LP (method =STUDABS)					
Model	Break Date	K(t-sig)	$t_{(\alpha-1)}$	t_α	T
IO1	1987:04	5	3.54115	-5.04	80
IO2	1986:04	3	-4.77306	-5.33	70
AO	1989:02	3	-4.01910	-4.67	100
LP (method = UR)					
Model	Break Date	K(t-sig)	$t_{(\alpha-1)}$	t_α	T
IO1	1985:02	3	-3.51983	-5.09	80
IO2	1986:02	1	-4.93855	-5.59	70
AO	1989:03	3	-4.32965	-4.83	100

Notes: STUDABS: Max of the absolute value of the t-statistic on the parameter associated with the change in either the intercept (if IO1) or the slope (if IO2).
UR : Min the t-statistic for testing alpha = 1. IO1 innovational outlier with a change in the intercept: IO2 innovational outlier with a change in the intercept and in the slope and AO additive outlier with a charge in the slope but both segments of the trend function are joined at the time break. t critical value at 5% significance level for each model (Perron 1997, table 1). Test statistics in bold indicates the rejection of the null hypothesis.

Table 16.3.a. Test for broken stationary: Perron (1997)

LP (method =STUDABS)					
Model	Break Date	K(t-sig)	$t_{(\alpha-1)}$	t_α	T
IO1	1987:04	2	-4.67754	-5.04	80
IO2	1987:04	4	**-6.80615**	-5.33	70
AO	1983:03	1	-4.11325	-4.67	100
LP (method = UR)					
Model	Break Date	K(t-sig)	$t_{(\alpha-1)}$	t_α	T
IO1	1987:03	2	**-5.11664**	-5.09	80
IO2	1987:04	4	**-6.80615**	-5.59	70
AO	1982:01	1	-4.19065	-4.83	100

Notes: STUDABS: Max of the absolute value of the t-statistic on the parameter associated with the change in either the intercept (if IO1) or the slope (if IO2). UR : Min the t-statistic for testing alpha = 1. IO1 innovational outlier with a change in the intercept: IO2 innovational outlier with a change in the intercept and in the slope and AO additive outlier with a charge in the slope but both segments of the trend function are joined at the time break. t critical value at 5% significance level for each model (Perron 1997, table 1). Test statistics in bold indicates the rejection of the null hypothesis.

Table 16.3.b. Test for broken stationary: Perron (1997)

LP (method =STUDABS)					
Model	Break Date	K(t-sig)	$t_{(\alpha-1)}$	t_α	T
IO1	1993:04	2	-2.59292	-5.04	80
IO2	1987:04	0	-3.28738	-5.33	70
AO	1985:02	0	-3.05064	-4.67	100
LP (method = UR)					
Model	Break Date	K(t-sig)	$t_{(\alpha-1)}$	t_α	T
IO1	1987:03	0	-3.37227	-5.09	80
IO2	1987:03	0	-4.07819	-5.59	70
AO	1985:01	0	-3.05195	-4.83	100

Notes: STUDABS: Max of the absolute value of the t-statistic on the parameter associated with the change in either the intercept (if IO1) or the slope (if IO2). UR : Min the t-statistic for testing alpha = 1. IO1 innovational outlier with a change in the intercept: IO2 innovational outlier with a change in the intercept and in the slope and AO additive outlier with a charge in the slope but both segments of the trend function are joined at the time break. t critical value at 5% significance level for each model (Perron 1997, table 1). Test statistics in bold indicates the rejection of the null hypothesis.

Table 16.3.c. Test for broken stationary: Perron (1997)

LP (method =STUDABS)					
Model	Break Date	K(t-sig)	$t_{(\alpha-1)}$	t_α	T
IO1	1993:04	1	-8.83483	-5.04	80
IO2	1988:03	0	**-9.81008**	-5.33	70
AO	1988:01	1	-8.61170	-4.67	100
LP (method = UR)					
Model	Break Date	K(t-sig)	$t_{(\alpha-1)}$	t_α	T
IO1	1987:03	2	**-10.08786**	-5.09	80
IO2	1987:04	4	**-11.02957**	-5.59	70
AO	1982:01	1	-8.61170	-4.83	100

Notes: STUDABS: Max of the absolute value of the t-statistic on the parameter associated with the change in either the intercept (if IO1) or the slope (if IO2). UR : Min the t-statistic for testing alpha = 1. IO1 innovational outlier with a change in the intercept: IO2 innovational outlier with a change in the intercept and in the slope and AO additive outlier with a charge in the slope but both segments of the trend function are joined at the time break. t critical value at 5% significance level for each model (Perron 1997, table 1). Test statistics in bold indicates the rejection of the null hypothesis.

Table 16.3.d. Test for broken stationary: Perron (1997)

Copeland and Werner (1997) and Bailliu, Garcés, Kruger and Messmacher (2003) considers that these variables are I(1) (Table 16.4)

Authors	Tests	P_t	$M2_t$
Copelman and Werner (1997)	ADF		I(1)
Baillu, Garcès, Kruger and Messmacher (2003)	ADF		
Garcès (2002)	ADF	I(2)	I(2)
	PP	I(1)	I(1)

Table 16.4 Unit Root tests results.

16.2.2 VAR Analysis and Cointegration

The analysis of economic relationships in a modern econometric framework normally includes the estimation of a vector autoregressive model either in levels, first differences or in an error correction format if cointegration is present among the series. The VAR model can be represented as follows (Johansen, 1995):

$$X_t = \pi_1 X_{t-1} + + \pi_2 X_{t-2} + \phi D_t + u_t \qquad (16.1)$$

Where X_t is a vector that includes all variables, Dt can contain a constant, a trend seasonal and intervention dummies and ut is the error term with Gaussian errors and cero mean and constant variance. Under cointegration, the VAR model can be represented as in the Johansen (1998 and 1995) procedure as:

$$\Delta X_t = \pi X_{t-1} + \Delta \pi_2 X_{t-1} + \phi D_t + u_t \tag{16.2}$$

The use of any of these two representations of the VAR model is not unique and includes at least several alternatives that the applied researcher has to take into account in order to carry out the analysis. That is, the use of VAR models either in levels or first differences or in an ECM format implies the selection of lags (p) of the VAR, the appropriate specification of the VAR which includes the selected variables and the inclusion of constant and trend in the Johansen (1988) procedure if necessary, the correct rank identification and the appropriate order of integration of the variables (Johansen, 1995).

However, the selection of the lags, the rank and the trend or constant in the cointegrating vector, are not independent procedures (Mills, 1998 and Bahmani- Oskooee and Brooks, 2003). Moreover, the selection of these options is not straightforward and in practice is even more difficult in cases where the economic series show features of structural instability, high volatility and arguably a high order of integration. We consider each of these problems separately.

16.2.3 Mis-specification Tests

Econometric theory argues that it is only possible to analyze the crucial properties of macroeconomic systems by means of a correct and identified model (McCallum, 2000) with appropriate statistical properties (Spanos, 1986). However, there is a tendency in applied work that considers the econometric model to be just an extension of the theoretical model[7]. Furthermore, before making inferences there is normally not enough attention spent on building an appropriate econometric model that includes economic theory and empirical information. In this sense, it is indispensable to initially elaborate a general model that does not contain any misspecification problems (Spanos 1986). This implies essentially that the general model, approximated trough a VAR model, does not show evidence of autocorrelation, heteroscedasticity or non normality in the error terms. The lag selection should also be complemented by some information criteria (i.e. Akaike) (Mills, 1998 and Hatemi, 2003) and the goodness of fit especially in the case of several cointegrating vectors (Bahmani-Oskooee and Brooks, 2003).

This specification process includes the roles of the constant, trend and seasonal and interventional dummies. Franses (2001) argues that the best option is to initially not restrict the VAR model with a constant and a trend equal

---

[7] In the extreme case, the econometric model is the theoretical model plus the addition of an error term with stochastic properties.

to zero and then testing down the final specification. In particular, Pesaran and Smith (1998) argue that the relevant cases for the applied econometrician include restricted intercept and no trend or unrestricted intercept and restricted trend. Also Pesaran and Smith (1998) and Doornick, Hendry and Nielsen (1998) sustain that the inclusion of an unrestricted trend is problematic because it reduces the power of the tests and implies the existence of quadratic trends on the variables in levels. In general, Doornick, Hendry and Nielsen (1998) support an initial specification including an unrestricted constant and a restricted trend because the case of the unrestricted trend normally leads to excessive rejection of the null.

Additionally, the presence of structural breaks in the series, which is a common situation in the Mexican variables, can lead to spurious rejections in the cointegration tests (Leybourne and Newbold, 2003). That is, the exclusion of appropriate treatment of the structural change in the series might imply that the stochastic component is more important than the trend in the ADF statistics but while the residuals look trend stationary, they are in fact still stationary different (Leybourne and Newbold, 2003 and Baffes and Valle, 2003). This situation is particularly relevant in the case of a break at the beginning of the series[8] (Leybourne and Newbold, 2003). Therefore appropriate treatment of the structural breaks can be handled by the use of interventional dummies. The inclusion of interventional dummies can change the asymptotic distribution of the cointegrating tests which have been calculated by Johansen and Nielsen, (1993). However, if the dummy considered only includes ones in a few points and zeros otherwise then, despite a persistent shock in the non stationary series, the asymptotic effect is negligible (Doornick, Hendry and Nielsen, 1998). It is then normally advisable to use unrestricted dummies in the cointegrating space (Doornick, Hendry and Nielsen, 1998).

The evaluation of the causal relations between the variables in the context of a VAR model requires a careful identification of the possible endogeneity of the variables and possible autocorrelation problems since the asymptotical efficiency of the estimators depends on that condition (Caporale and Pittis, 1999). The analysis of the causality among the variables is normally based on the weak exogeneity and non Granger causality tests.

The weak exogeneity tests indicate that a multiple estimation is a valid option unless weak exogeneity conditions are present in the series. However, this cannot be imposed a priori and must be evaluated on empirical grounds. Additionally, econometric theory (Caporale and Pittis, 1999) argues that the best option, under cointegration, is to use a VAR model in an ECM form. The main reason is that in the ECM format the limited distribution conditions are standard which is only true in the VAR model under strong cointegration

---

[8] This partially explains the phenomenon that reversing the roles of the variables in the cointegrating regression also implies a change in the rejection of the null hypothesis.

properties[9] (Caporale and Pittis, 1999). However, this process implies that sufficient rank conditions of the matrix are well established which is normally not easy.

Recent empirical research is normally subject to the criticism of the identification problem across the Atlantic (Favero, 2000 pp. 225). That is, American research tends to use more vector autoregressive models without considering the order of integration of the series. In Europe, on the contrary, it is common practice to consider the unit root issue and then proceed to generate a small structural model possible in an ECM format.

Recent research indicates that both options (the VAR in levels and the ECM format) have advantages and disadvantages. That is, under cointegration, the error correction representation is only a reparametrization of the VAR model in levels (Favero, 2001). Furthermore, Sims, Stock y Watson (1990) argue that, under cointegration, the VAR model in levels is over-parametrizated which leads to inefficient but consistent estimations and Ohanian (1988) claims that the likelihood ratios tests on block exogeneity too often reject the null hypothesis. On the contrary, imposing restrictions trough the cointegration vector in the VAR can lead to the wrong long term equilibrium mechanisms (Favero, 2001). An option, proposed by Toda and Yamamoto (1995) is augmenting the VAR by the maximal order of integration for the process being examined (Caporale and Pittis, 1999, p. 3) and then using Wald tests to analyze the statistical significance of the variables (Dolado and Lutkepohl, 1996). In particular, Mills (1998) argue to augmented by one the lag order to the VAR and then use a conventional lag statistic considering that these changes allow standard asymptotic inference. The lag selection must consider the misspecification tests and the information criteria[10] because an over fitting VAR generates an increase of the mean square error while under fitting generates possible autocorrelation patterns (Ozcicek and McMillan, 1999). Naka and Tufte (1997) argue that for short run horizons it is possible to use the VAR in levels, in particular for impulse response functions and variance decomposition.

Therefore, this discussion is also a matter of research selection depending on the particular objective of the investigation. For example, there is a tendency to use VAR in levels to evaluate monetary policies in the short run while the restrictions in the cointegrating mechanism deal more with long run issues. Nevertheless, it is worth noticing that the limitations in both alternatives must be considered before making economic inferences and also that in any case it is necessary to elaborate a VAR model with the adequate statistical properties (Spanos, 1986). Under other circumstances the results are for limited use for practical purposes in monetary policy.

---

[9] For example more than one cointegrating vector.

[10] Ozcicek and McMillan (1999) argue in favor of the Akaike information criteria because it has better properties in small samples especially in the case of the use of impulse response analysis.

Additionally, it is worth noticing that the uncertainty about the unit root tests generates doubts about the exact order of integration of the series and therefore the exact specification to consider. This is for example the case for prices that can be considered as an I(2) or I(1) process (Juselius, 2000). However, recent literature suggests that the use of appropriate Wald test might solve this problem. For example, Toda and Yamamoto (1995) shows that the Wald statistic converges to a χ^2 no matter if the variables are I(0), I(1) or I(2) in particular when there are strong cointegration properties (i.e there are multiple cointegrating vectors) (Toda and Phillips, 1993). This result is very important in the Mexican case because it is possible to test for causality using series with different order of integration or where the order of integration is uncertain. It is worth noticing that the impulse response analysis with variables in levels that contain unit roots tend to converge to non-zero means instead of dying out and also show intervals too wide (Mills, 1998). Arguably, some of the economic analysis elaborated to discuss the effects of the monetary policy in Mexico, is subject to the problems that are mentioned in this paper. For example, Copeland and Werner (1996) and Schwartz, Tijerina and Torre (2002) use a VAR in levels to test for non Granger causality with variables I(1). The first authors argue in favor of this solution because the series are cointegrated and therefore this VAR is a representation of a restricted ECM. However, this analysis is clearly subject to strong criticism considering that the non Granger causality tests are subject to the criticism of spurious regression with I(1) variables and there is also no evidence of the lag selection criteria that might also generate misleading results. On the contrary, Castellanos (2000), with a more careful analysis of the residuals than the previous work, uses variables I(1) in first differences in a VAR framework to test for non Granger causality. Additionally, Díaz de León and Greenham (2001) use, while considering the effects of the el corto on interest rates, unit root tests with constant and trend. Nevertheless, these variables are afterwards not included in the VAR analysis and moreover, despite the presence of cointegration, the impulse response functions are only considered in levels without discussing the implications. With similar problems, Gunther and Moore (1993) combine variables of different order of integration in the VAR with non Granger causality tests. The potential problems of the VAR analysis that yields to alternative results can be well illustrated using the money neutrality hypothesis. One way to tests this hypothesis is considering the possible presence of cointegration between money and income[11]. In fact, Carstens and Reynoso (1997) use this type of analysis as evidence in favor of the neutrality hypothesis due the rejection of cointegration. However, this analysis includes at least three weak points:

1. The unit root tests indicate that money and income have, possibly, a different order of integration and therefore this equation is not balanced.

---

[11] Cointegration between money and income is considered empirical evidence against the neutrality hypothesis.

Furthermore, the difference of order of integration might be the cause of the rejection of cointegration properties in small samples especially in the case of the use of impulse response analysis. because the I(2) variable dominates the whole process[12]. In this case, an adequate specification should include variables of the same order of integration and open the discussion on the disadvantages of partial analysis against a more general approach. A partial analysis reduces the interactions and feedbacks between the variables limiting the ability of the model to simulate the real data.

2. The analysis of the money neutrality hypothesis might include as a general framework the quantity identity hypothesis. In this case, the uncertainty of the exact order of integration of the variables might be compensated but an identification problem might appear (Favero, 2001). The Johansen procedure using income, money, prices and interest rates indicates the presence of a cointegrating relationship (Table 16.6). This result implies the rejection of the neutrality hypothesis but also leads to an identification problem.

3. A general model must show trough a battery of misspecification tests, that all systematic information is included in the right hand side and therefore that the error term is white noise (Spanos, 1986). Otherwise, the inferences from the model might be misleading. The model including income, money, prices, the interest rate and a dummy for the structural change of 1995 reduces the problems in the error term (Table 16.5). This evidence indicates the presence of structural breaks in the series. This situation shows that the particular way to deal with the structural change is crucial for the final result and then for the rejection or not of the neutrality hypothesis in the Mexican economy. Moreover, it is necessary to incorporate options in the analysis to allow us to deal with different kinds of structural changes in the context of cointegration and proper identification.

Initially, a VAR framework including output, prices, nominal interest rates and a monetary aggregate can be specified. The estimated VAR, with the appropriate number of lags, is chosen considering a battery of misspecification tests. Therefore, the VAR does not show evidence of autocorrelation or heterocedasticty and the error terms do not reject the null hypothesis of normality (Table 16.5). Under these conditions it is possible to argue that all systematic information is contained in the explained variables.

Table 16.6 includes the maximum eigenvalue and the trace tests for cointegration including a constant and dummy variable for 95 in the cointegrating space. Both tests suggest the presence of at least three cointegrationg vectors suggesting strong cointegration.

---

[12] For example, with an I(2) and I(1) variables, then an I(1) linear combination which already reduces the order of integration will not reject the null hypothesis in the Johansen (1988) procedure.

	LM(4)	Arch(4)	J-B
y_t	F(4,38)=2.458 [0.062]*	F(4,34)=0.886 [0.482]*	$\chi^2(2)$=2.449 [0.294]*
p_t	F(4,38)=1.378 [0.259]*	F(4,34)=0.393 [0.812]*	$\chi^2(2)$=0.864 [0.649]*
$m2_t$	F(4,38)=1.607 [0.192]*	F(4,34)=0.453 [0.769]*	$\chi^2(2)$=1.413 [0.493]*
r_t	F(4,38)=0.494 [0.739]*	F(4,34)=0.359 [0.836]*	$\chi^2(2)$=3.946 [0.139]*

Notes: (*) indicate a rejection of the null hypothesis. Period 1980:01 - 2002:04

Table 16.5 Misspecification tests of the VaR

H_0	Trace	95%	λ-max	95%
r=0	44.76*	28.1	104.2*	53.1
$r \leq 1$	29.01*	22.0	59.49*	34.9
$r \leq 2$	17.6*	15.7	30.48*	20.0
$r \leq 3$	12.87*	9.20	12.87*	9.20

Notes: (*) significant at the 5% level,
λ-max = maximum eigenvalue test:
Trace = trace test:
r = number of co-integrating vectors.
Number of lags = 4. Period 1980:01 - 2002:04

Table 16.6 Statistics of the Johansen procedure

$$y_t = \beta_0 + \beta_1 * p_t + \beta_2 * m2_t + \beta_3 * r_t$$

Normalizing these cointegrating vectors as an output, price and demand for money equation yields coefficients with reasonable values in accordance with economic theory. That is, equation (1) indicates that output has a negative relationship with interest rate and the price index and a positive relationship with money. The coefficients of prices and money are similar but with opposite signs. This result suggests that real financial wealth has a positive effect on economic growth. Equation (16.2) is fundamentally a P* model (Hallman, Porter and Small, 1991) where the monetary aggregate has a positive effect on the price index and income and the nominal interest rate which is associated with the velocity of circulation has negative impact. Finally, equation (16.5) represents a demand for money function with a positive effect from income and a negative impact of the opportunity costs associated with the evolution of the nominal interest rate; additionally, the price index has a coefficient relatively close to one suggesting the option of a demand for money in real terms. Therefore, these equations suggest a general restriction on the coefficients with economic sense that might lead to a correct identification process.

$$y_t = 8.7506 + 0.90308 * p_t + 0.80871 * m2_t - 0.49374 * r_t - 1.2798 * d95\text{(16.3)}$$

$$p_t = 22.918 - 1.8576 y_t + 0.97804 * m2_t - 0.086787 * r_t - 0.21147 * d95\text{(16.4)}$$

$$m2_t = -6.4648 + 1.1140 * y_t + 1.322 * p_t - 0.24541 * r_t - 0.014355 * d95\text{(16.5)}$$

$$r_t = -44.706 + 4.6242 * y_t + 3.0098 * p_t - 2.9812 * m2_t - 0.77243 * d95\text{(16.6)}$$

The VAR, in an error correction form (VECM), allows to distinguish between the short run and long run causality and, in this context, to test for weak exogeneity and non Granger causality. In this sense, the weak exogeneity tests are performed over the α_I parameters of the Johansen (1988) procedure. This test implies that any deviation from the variables in the cointegrating vector will induce a feed back adjustment in the change of the dependent variable. Then the value of the coefficient of the error correction mechanism represents the proportion by which the long-run disequilibrium is corrected by the endogenous variable in each period (Ericsson and Irons, 1994 and Engle, Hendry, and Richard, 1983). On the other side, the statistical significance of the changes in the variables captures the existence of short run causality that is normally related with the non Granger causality tests (Granger, 1986 and Engle and Granger, 1987). Table 16.7 includes the weak exogeneity tests indicating that all variables are endogenous. Therefore, it is necessary to model all these variables together because of the existence of multiple feedback. Meanwhile the non Granger causality tests are presented in Table 16.6. Essentially, these tests indicate the existence of a feedback process between income and money. This result suggests that the relationship between income and money is much more complex than the suggestion of the neutrality hypothesis and that this hypothesis is rejected by the Mexican data.

$H_0 : \alpha_0 = 0$	$\chi_2(6) = 35.946 \ [0.000]^*$
$H_0 : \alpha_1 = 0$	$\chi_2(6) = 18.007 \ [0.006]^*$
$H_0 : \alpha_2 = 0$	$\chi_2(6) = 35.575 \ [0.000]^*$
$H_0 : \alpha_3 = 0$	$\chi_2(6) = 15.173 \ [0.019]^*$
Notes: (*) indicate a rejection of the null hypothesis at the 5% level. H_0: $\alpha_0 = 0$ (α_0 is related with GNP, α_1 is related with price α_2 is related with m2 and α_3 is related with interest rate). Period 1980:01 - 2002:04	

Table 16.7 Weak exogeneity test

These results indicate that the correct use of the variety of econometric techniques requires an explicit recognition of their limitations. Also, applied economic researchers should always consider that empirical models are a combination of economic theory with empirical information. In this sense, different econometric techniques do not represent a substitute for economic reasoning. Econometric models should simulate the main features of the real economy with a proper use of the different econometric methods. The uncertainty and lack of power of the different econometric options do not lead to their abolishment. On the contrary, the main results of this work show that an appropriate use of a battery of tests is the best way to obtain a real understanding of the empirical evidence. Therefore, it is even advisable, at this stage, to combine different techniques and through them to try to identify the main features of the real data.

Null hypothesis H_0	$F=(10,37)$	$\chi^2(10)$	Null hypothesis H_0	$F=(10,37)$	$\chi^2(10)$
Δp does not cause Δy	0.348 (0.945)	3.844 (0.954)	Δy does not cause Δp	0.738 (0.684)	7.382 (0.688)
Δp does not cause $\Delta m2$	1.747 (0.106)	17.471 (0.064)	$\Delta m2$ does not cause Δp	2.545 (0.019)*	25.451 (0.004)*
Δp does not cause Δr	1.019 (0.446)	10.195 (0.423)	Δr does not cause Δp	1.495 (0.181)	14.919 (0.135)
Δy does not cause $\Delta m2$	2.872 (0.009)*	28.728 (0.001)*	$\Delta m2$ does not cause Δy	1.861 (0.083)	18.614 (0.045)*
Δy does not cause Δr	0.390 (0.942)	3.902 (0.952)	Δr does not cause Δy	1.496 (0.179)	14.963 (0.133)
$\Delta m2$ does not cause Δr	1.116 (0.376)	1.168 (0.344)	Δr does not cause $\Delta m2$	1.233 (0.303)	12.330 (0.263)

Notes: (*) indicate a rejection of the null hypothesis. Period 1980:01 - 2002:04

Table 16.8 Non Granger causality using Wald test

16.3 Conclusions and General Coments

Applied monetary work requires the use of different econometric and statistical techniques. However, each approach produces very different results and conclusions because each economic time series has different statistical properties. Under these circumstances it is recommended that applied econometrics should be based on a probability approach and therefore, different econometric methods only represent an approximation of the data generation process (Spanos, 1986).

Therefore, different techniques show distinctive features of the real data that should still be validated with the rest of the empirical evidence. In this sense, the researcher should be conscious of the features and limitations of each econometric method. The problems of the modern econometric methods are stronger and more evident in the case of economic series with structural changes and high variability. In this case, optimal solutions derived from econometric theory are not necessary the best option.

In effect, econometric theory is mostly elaborated for the case of non stationary series, with stable parameters and correct specification. Unfortunately, real data in under developing countries are far from that. Therefore, applied econometricians should try to obtain maximum information by correctly applying different techniques without forgetting to keep the relevance of economic reasoning in mind.

Contemporary, applied econometricians should try to find a reasonable combination of the theoretical arguments of the real business cycle with the empirical insight of the time series analysis including unit root tests, cointegration and vector autoregressive models. But empirical research should also include a better combination of the use of the VAR models in levels with the

VAR, cointegration and identification procedures. This also implies the use of dummy variables to model changes in policy regimes[13] or new techniques that are particularly designed to deal with different kinds of structural change (Chang-Jin and Nelson, 2000).

These results indicate that despite the weakness and lack of power of most of the econometric techniques a decision to not use them does not solve the problem. On the contrary, their adequate use, even when they do not provide the perfect solution, does represent the best option. In this sense, despite its skeptical view of each technique this essay also shows that a reasonable combination yields results that are very insightful. Finally, we have to agree with Pesaran and Smith (1998) that the feedback between statistical and economic criteria is still a matter of judgment.

References

1. Badillo, R., J. Belaire-France, and D. Contreras, (2002), Spurious rejection of the stationary hypothesis in the presence of a break point, Applied Economics, 34, 1917-1923.
2. Baffes, J. and J.C. Valle, (2003), Unit root versus trend stationary in growth rate estimation, Applied Economic Letters, 10, 9-14.
3. Bahmani-Oskooee, M. and T.J. Brooks, (2003), A new criteria for selecting the optimum lags in Johansens cointegration techniques, Applied Economics, 35, 875-880.
4. Bailliu, J., Garcés, D. and M. Messmacher, (2003), Explicación y descripción de la inflación en mercados emergentes: el caso de México, Bank of Canada y Banco de México, Documentos de Investigación, No. 2000-3, February.
5. Banco de México (1996), La conducción de la política monetaria del Banco de México a través del régimen de saldos acumulados, Informe Anual del Banco de México, Anexo 4.
6. Caporale, G.M. and N. Pittis, (1999), Efficient estimation of cointegrating vectors and testing for causality in vector autoregressions, Journal of Economic Surveys, vol. 13, No. 1, 1-35.
7. Carstens A. and A. Reynoso (1997), Alcances de la política monetaria: marco teórico y regularidades empíricas en la experiencia mexicana, Documento de Investigación, no. 9705, Banco de México, pp. 45.
8. Castellanos, S.G. (2000), El efecto del corto sobre la estructura de tasas de interes, Documento de Investigación No. 2000-1, Dirección General de Investigación económica, Banco de México.
9. Chang-Jin, K. and C.R. Nelson (2000), State space models with regime switching, MIT Press.
10. Charemza, W.W. and E.M. Syczewska (1998), Joint application of the Dickey-Fuller and KPSS tests, Economic Letters, 61, 17-21.
11. Chen, M.Y. (2002), Testing stationary against unit roots and structural changes, Applied Economic Letters, 9, 459-464.

---

[13] Mixing two policy regimes is like having a structural change.

12. Copeland, M. and A.M. Werner, (1997), El mecanismo de la transmisión monetaria en México, Trimestre Económico, 75-104.
13. Díaz de León, A. and L. Greenham, (2001), Política monetaria y tasas de interés: experiencia reciente para el caso de México, Economía Mexicana. Nueva poca, vol. X, no. 2, segundo semestre, 213-208.
14. De Jong, D.N. and C.H. Whiteman (1991), Reconsidering trends and random walks in macroeconomic time series, Journal of Monetary Economics, 28, pp. 221-254.
15. Dickey, D. A. and W.A. Fuller, (1981), Likelihood ratio statistics for autoregressive time series with a unit root, Econometrica, 49, pp.1057-1072.
16. Dolado, J.J. and H. Lutkepohl, (1996), Making Wald tests work for cointegrated VAR systems, Econometric Reviews, 15, 4, 369-386.
17. Doornick, J.A., D.F. Hendry, and B. Nielsen, (1998), Inference in cointegrating models: UK M1 revisited, Journal of Economic Surveys, vol. 12, no. 5, 533- 572.
18. Engle, R.F. and C.W.J. Granger (1987), "Cointegration and error correction" representation, estimation and testing", Econometrica, 55, 251-276
19. Engle, R.F. D.F. Hendry and J. Richard (1983), "Exogeneity", Econometrica, 51,
20. Ericsson, N.R. and J.S. Irons (eds) (1994) Testsing exogeneity, Oxford University Press.
21. Favero, C.A. (2001), Applied Macroeconometrics, Oxford University Press.
22. Favero, C.A. (2000), New econometric techniques and macroeconomics, in R. Backhouse (ed.), Macroeconomics and the real world, Oxford University Press, pp. 225-236.
23. Franses, P.H. (2001), How to deal with intercept and trend in practical cointegration analysis, Applied Economics, 33, 577-579.
24. Frances, P.H. and M. McAller, (1998), Cointegration analysis of seasonal time series, Journal of Economic Surveys, vol. 12, no. 5, 651-678.
25. Galindo, L.M. and M.E.Cardero, (1997), Un modelo econométrico de vectores autoregresivos y cointegración de la economía mexicana, 1980-1996, Economía Mexicana, Nueva Época, vol. VI, no. 2, segundo semestre, pp. 223-247.
26. Garcés, D.G. (2002), Agregados monetarios, inflación y actividad económica, Documento de investigación, No. 2002-07, Banco de México.
27. Gil-Díaz, F. (1997), La política monetaria y sus canales de transmisión en México,
28. Gaceta de Economía, suplemento, ao 3, no. 5, 79-102. 25
29. Granger, C.W.J. and P. Newbold (1974), Spurious regression in econometrics, Journal of Econometrics, 2, pp. 111-120.
30. Granger, C.W.J. (1983), "Developments in the study of cointegrated economic variables", Oxford Bullettin of Economics and statistics, 48, 213-228
31. Gunter, J.W. and R.R. Moore, (1993), Credito y actividad económica en México, Economía Mexicana, Nueva Época, vol. II, no. 2, julio-diciembre, 415-428.
32. Hallman, J.J., R.D. Porter, and D.H. Small (1991), Is the price level tied to the M2 monetary aggregate in the long run?, American Economic Review, vol. 81, No. 4, 841-858.
33. Hatemi, A. (2003), A new method to choose optimal lag order in stable and unstable VAR models, Applied Economics Letters, 10, 135-137.
34. Hendry, D.F. and M.P. Clements (1999), Forecasting non-stationary economic time series, M.I.T. Press.

35. Hoover, K.D. (2000), Models all the way down: comments on Smith and Juselius, in in R. Backhouse (ed.), Macroeconomics and the real world, Oxford University Press, pp. 219-223.

36. Ibarra, C. (1998), Exchange rate policy credibility in Mexico, 1991-1994, Economía Mexicana, Nueva Época, vol. VII, segundo semestre, 229-266.

37. Johansen, S. (1995), Likelihood-based inference in cointegrated vector autoregressive models, Oxford University Press.

38. Johansen, S. (1988), Statistical analysis of co-integrating vector, Journal of Economics, Dynamics and Control, 12, pp. 231-54.

39. Johansen, S. and B.G. Nielsen (1993) Asymptotics for the Cointegrqation Rank Tests in the Presence of Intervention Dummies. Manual for the Simulation Program DisCo. Working Paper, University of Copenaghen.

40. Juseluis, K, (2000), Models and relations in economics and econometrics, in R. Backhouse (ed.), Macroeconomics and the real world, Oxford University Press, pp. 168-197.

41. Kwiatkowsky, D., P.C.B. Phillips, P. Schmidt and Y. Shin, (1992), Testing the null hypothesis of stationary against the alternative of a unit root, Journal of Econometrics, 54, pp. 159-178.

42. Lee, J. and M. Strazicich, (2001), Testing the null of stationary in the presence of a structural break, Applied Economic Letters, 8, 377-382.

43. Leybourne, S.J. and P. Newbold, (2003), Spurious rejections by cointegration tests induced by structural breaks, Applied Economics, 35, 1117-1121.

44. Maddala, G.S. and I-M. Kim (1998), Unit roots, cointegration and structural change, Cambridge University Press.

45. Martínez, L., O. Sánchez, and A. Werner, (2001), Consideraciones sobre la conducción de la política monetaria y el mecanismo de transmisión en México, Banco de México, Documento de Investigacion no. 2001-02, marzo.

46. McCallum, B.T. (2000), Recent developments in monetary policy analysis: the roles of theory and policy, in R. Backhouse (ed.), Macroeconomics and the real world, Oxford University Press, pp. 115-1139.

47. Mills, T, (1998), "Recent developments in modeling nonstationary vectors autoregressions", Journal of Economic Surveys, 12, 3, 279-311

48. Naka, A. and D. Tufte, (1997), Examining impulse response functions in cointegrated systems, Applied Economics, 29, 1593-1603.

49. Ng, S. and P. Perron, (1995), Unit root tests in ARMA models with data depend methods for the selection of the truncation lag, Journal of the American Statistical Association, 90, pp. 268-281.

50. Nelson, C.R. and C.I. Plosser (1982), Trends versus random walks in economic time series. Some evidence and implications, Journal of Monetary Economics, 10, pp. 139-162.

51. Ohanian, L.E. (1988), The spurious effects of unit roots on vector autoregressions, Journal of Econometrics, 39, 251-266.

52. Patterson, K. and S. Heravi, (2003), Weighted symmetric tests for a unit root: response functions, power, test dependence and test conflict, Applied Economics, 35, 779-790.

53. Perron P.C.B. (1997), Further evidence on breaking trend functions in macroeconomic variables, Journal of Econometrics, 80, pp. 355-385.

54. Phillips, P.C.B. and P. Perron, (1988), Testing for unit roots in time series regression, Biometrika, 75, pp. 335-346.

55. Phillips, P.C.B. and Z. Xiao, (1998), A primer on unit root testing, Journal of economic Surveys, vol. 12, No. 5, 423-469.
56. Obstfeld, M. y K. Rogoff (1996), Foundations of international macroeconomics, The M.I.T. Press.
57. Ozcicek, O. and D. McMillan, (1999). "Lag length selection in vector autoregressive models: symmetric and asymmetric lags", Applied Economics, 31, 517-524
58. Pesaran, M.H. and R.P. Smith, (1998), Structural analysis of cointegrating VARs, Journal of Economic Surveys, Vol. 12, No. 5, 471-505.
59. Schwartz, M.J., Tijerina, A. and Torre, L. (2002), Volatilidad del tipo de cambio y tasas de interes en México: 1996-2001, Economía Mexicana, Nueva Época, vol. XI, no. 2, Segundo semester, 299-331.
60. Sims, C., J. Stock and M. Watson (1990), Inference in linear time series models with some unit roots, Econometrica, 58, 1, pp. 113-144.
61. Spanos, A. (1986), Statistical foundations of econometric modeling, Cambridge University Press.
62. Toda, H.Y. and P.C.B. Phillips, (1993.a), Vector autoregressions and causality, Econometrica, 61, 6, 1367-1393.
63. Toda, H.Y. and P.C.B. Phillips (1993), The spurious effect of unit roots on vector autoregressions, Journal of Econometrics, 59, 229-255.
64. Toda, H.Y. and T. Yamamoto, (1995), Statistical inference in vector autoregressions with possible integrated process, Journal of Econometrics, 66, 225-250.
65. Walsh, C.E. (2000), Monetary theory and policy, The M.I.T. Press.
66. Weber, C. E. (2001) " F-test for lag length selection in Argumented Dickey-Fuller regression: some Monte Carlo Evidence", Applied Economics Letters, 8, pp. 455-458

Appendix A

Data sources:

- P_t = Mexican consumers price index (base 1994=100), Bank of Mexico.
- M_t = Monetary aggregate M2 in millions of Mexican pesos, Bank of Mexico.
- Y_t = Real gross domestic product in millions of 1993, INEGI
- R_t = CETES of 90 days of the average of the last month of the quarter, Bank of Mexico.

Themes of Growth and Development

Part III

Theories of Growth and Development

Environmental Policy Options in the Multi-Regimes Framework[*]

Salvatore Bimonte[1] and Lionello F. Punzo[2]

[1] Dipartimento di Economia Politica, Università degli Studi di Siena
bimonte@unisi.it
[2] Dipartimento di Economia Politica, Università degli Studi di Siena
punzo@unisi.it

Summary. In this paper we extend the multi-regime framework to variables involved in the debate on economic growth and environmental quality, starting from a reexamination of the so-called Environmental Kuznets Curve. The aim is to discuss the double convergence hypothesis that implicitly stems from a recent line of research. According to it, some stylized facts would support the almost paradoxical hypothesis that economic growth produce not only cross-countries or regions convergence in per capita output, but also in (the demand of) environmental quality.

Factual analysis seems to reject the hypothesis of convergence in output or income levels. Available evidence, rather, seems to point out that there is no such a thing as a unique avenue to sustainable development while the convergence predicted in more conventional analyses, in particular within the framework of the so called Environmental Kuznets Curve, is far away from being demonstrated. Actual growth processes do differ from each other in a deep qualitative sense, to the effect of profoundly influencing final outcomes as well as the unfolding of the processes themselves. This reflects differences in initial conditions, of course, but also the different sectoral or integrated policies that have been implemented along the way.

Therefore, in contrast to the *double convergence hypothesis*, in our contribution we argue that *growth is a necessary but not sufficient condition for the required* change in the individuals preferences needed to shift social preferences away from private to public goods and that, moreover, the relationship between growth and environmental quality depends crucially upon the countrys growth model. Therefore, more than the quantitative it is the qualitative aspects that matters. The theoretical context that seems to lend itself to the analysis of such issues falls within the boundaries of the theories of endogenous growth. We argue that sustainable development, if it emerges at all, is the result of investment in immaterial capital (research, education and the like) more than the reflection of the exogenous forces (technological progress and demographic) of the neoclassical theory. In the analysis

[*] This paper was read at the Third Latin American-European Workshop, October 3-5, 2002, Guanajuato, México. We are indebted to the two referees for their useful and stimulating comments and suggestions. All remaining errors are ours. Research for this paper has been carried out as part of the research project funded by PAR, University of Siena

of such issues, the environment offered by the multiregime approach proves useful as it highlights the qualitative properties of the dynamic processes, instead of focusing upon quantitative estimation of some special asymptotic states whose existence is often all but to be demonstrated.

Key words: : Environmental Kuznets curve; Growth; Regimes; Framework Space; Sustainable development.

17.1 Introduction

This paper extends the multi-regime framework (see e.g. Boehm and Punzo, 1994, 2001) to variables involved in the debate on the relation between environmental quality and economic growth. In this light, it reexamines the interpretation of the so-called Environmental Kuznets Curve (from now on EKC). The adapted framework can account for one fundamental finding, which does not find a place in the relevant literature: namely, the diversity across countries and regions of the development experience in terms of both growth performance *and* evolution of environmental quality. On the other hand, the paper reviews the proposition of the associated potential conflict between these two targets, and the presence of a trade-off between them. The issues captured by this simpler notion of tradeoff are essentially of a qualitative nature. We propose hereafter a formal way to think about these issues. The bonus is a framework that seems more appropriate for designing integrated policy plans apt to guide an economy along the difficult *traverse* between two different growth mixes.

Our argument can be introduced in the following way. The key issue traditionally associated with the notion of sustainable development *sustainable development* has, for a long time, been how to reconcile growth with environmental preservation. Sustainable development involves much more complex aspects revolving around problems of social, economic and cultural relevance. Often, objectives and aims of social and environmental nature are predicated as constraints to the growth of GDP, e.g. as objectives whose realization can be attained only at the detriment of growth. In recent years, however, we have witnessed the unfolding of a new line of research whereby some stylized facts have been identified supporting the almost paradoxical hypothesis that environmental quality represents the joint product of economic growth (a reference is to e.g. Beckerman, 1992)[3]. Economic growth is, therefore, seen to produce not only cross-country convergence in per capita output, but also in the levels of environmental quality: *a double convergence hypothesis.*

---

[3] These aspects are critically discussed in Torras and Boyce (1998) and Grossman and Krueger (1995).

According to this set of studies, generated by the interpretation of the EKC[4], there is a quadratic relation (a U-shaped curve) linking environmental quality, generally proxied by some index of emission, and *per capita* GDP. The current interpretation maintains that, in the initial phase of a growth process, environmental quality exhibits a tendency to deteriorate, while, once passed some threshold value of income, its further growth will generate improvement in the environmental quality. The hypothesis is that higher income level automatically boosts the demand for environmental protection and quality (income elasticity). In contrast to such strong hypothesis, our contribution argues that:

- *growth is a necessary but not sufficient condition* to produce the required change in the individuals preferences needed to shift social preferences away from private to public goods;
- the relationship between growth and environmental quality depends crucially upon the countrys growth model. In other words, more than the quantitative it is the qualitative aspect of growth that matter here, an aspect captured by the complex notion of *model of growth*.

Abundant factual evidence is known, which casts doubts on the *classical* hypothesis of convergence in output or income levels. It also seems to indicate that there is no such thing as a unique avenue to sustainable development. At the same time, double convergence implicitly predicted in conventional analyses, in particular within the framework of the so-called EKC, is far away from being demonstrated. Looked at in a multivariable framework, actual growth processes seem often to differ from each other also in some deeper qualitative sense, to the effect of profoundly influencing final outcomes, as well as the time unfolding of the processes themselves. This would reflect differences in initial conditions, so called idiosyncratic shocks, of course, but also the different sectoral or integrated policy actions that have been implemented along the way. The dependence upon initial conditions and implemented policies will be the focus for our analysis, hereafter.

The theoretical context that seems to lend itself to the analysis of such issues falls within the boundaries of the theories of endogenous growth. We argue that, if it emerges at all, sustainable development is bound to be the result of investment in immaterial capital (research, education and the like), more than the exogenous forces (technological progress and demographic growth) of the neoclassical theory. Accordingly, for its very nature, sustainability demands an approach based upon the design and implementation of integrated (non sectoral) policies. Only with these can a traverse path be initialized that may take low income/low environmental quality economies directly across to high income levels without having to pay for this with a phase of environmental degradation. The length of the latter is one of the uncertain elements in

---

[4] The seminal work of Simon Kuznets (1955) evidenced a quadratic relationship between equality and economic growth, i.e. an U-shaped curve.

this cost calculation: perhaps long or infinitely long, there being the risk for one such economy to get stuck in it.

In the analysis of such issues, the environment offered by the multiregime approach proves useful as it highlights the qualitative properties of the dynamic processes, instead of focussing upon quantitative estimation of some special asymptotic states whose existence is often all but to be demonstrated. This is a natural consequence of the fact that, in the standard studies, growth is generally described as a sequence of practically predetermined phases, each being characterized by peculiar structural features. That such phases can be changing over time, hence inducing changes not only on the levels of certain variables, but also and more importantly in the ways they dynamically interact, makes it natural to think in terms of an ever expanding portfolio of dynamic regimes and of a dynamics across them. This allows, among other things, a useful comparison of structurally different economies on a qualitative basis.

Although there is an ample choice in the multiregime literature, the approach introduced by Boehm and Punzo (2001) will be adopted and adapted to the treatment of environment as one of the variables defining regimes.

17.2 The Environmental Kuznets Curve: a Short Review

Studies on the relation between economic growth and environmental quality are synthesized by the debate around the EKC and in fact are based upon a dynamic re-interpretation of such curve. They all seem to accept a suggestion from conventional growth theory, whose scope is extended to include some environmental variable and whose predictions are articulated into phases chained together in the transition to the long run path the theory is really concerned with. The EKC relation, taken as a dynamical law, indicates that environmental quality would deteriorate in the initial phases of the growth process, therefore affecting the level of total welfare to an extent that may be difficult to anticipate, while, once surpassed some given threshold value of income level, further growth will go along with its improvement[5]. However, it is worth noting that empirical results depend crucially on the index of environmental quality used (World Bank, 1992), as we will see in the following sections, as well as the type of analysis, whether cross country or of the time series type, and that they also vary greatly across equipes of researchers (Ekins, 1997).

Why the relation takes up this shape, there is no consensus. The majority of studies has explained it uniquely in terms of income elasticity demand. Behind this set of studies lies the fundamental assumption (that should be tested rather than be taken for granted) according to which environment is an

---

[5] However, Pezzey (1989) argued that, at least in the long run, the inverted-U relationship may not hold. More likely is a so-called N-shaped curve.

income-elastic commodity (*luxury good*). According to the income elasticity hypothesis, in the advanced phases of the development process, environmental quality improves because people become more environmentally conscious, and can afford to build up political pressure for the enforcement of environmental regulations and protection. In other words, studies on the EKC implicitly assume that economic growth, because of the income elasticity of demand for amenities, and the greater information accessibility it produces, directly spurs an increase of demand for policies devoted to environmental protection and related environmental expenditures (Selden and Song, 1994).

More recently, the focus has shifted onto the role played by policies in shaping up the upward branch of the curve (Grossman and Krueger, 1995; Panayotou, 1995; Torras and Boyce, 1998). Other researchers took the road of trying to estimate empirically the relevance that some structural factors may have, together with income, in explaining the curve taken to be a stylized fact. Grossman and Krueger (1995, 1996) have identified three channels through which economic growth can exert an influence upon environmental quality: a scale effect, which would tend to prevail in the early phase of development; a composition effect modifying the productive structure of an economy; and finally, a technological effect, linked to the introduction of new and more efficient production techniques.

However, it is worth to point out that with a few notable exceptions, no *direct* measure of expenditure and/or of environmental policies has ever been tested for econometric relevance.[6] Actually, as a proxy for environmental quality of a country normally some index of emission is deployed, whose change, we will see, not necessarily must be imputed directly or solely to the impact of adopted environmental policies[7]. Moreover, this sort of indices neglect the stock effect typically associated with pollution emissions. In fact, long run sustainability does not only depend upon the annual rate of emissions; it crucially depends on the past levels of pollution, due to the cumulative effects of emissions and to the delayed effect of past accumulations of pollutants, and the capacity of the environment of absorbing it (Kaufmann and Cleveland, 1995).

17.3 The EKC and the Theory of Growth: Reconsidering the Convergence Issue

The literature on the EKC has gone along to a generalization of the convergence result associated with the traditional exogenous growth theory. According to the latter, whatever their initial conditions, countries will be converging

---

[6] Magnani (2000) constructs a test for the existence of the EKC using R&D expenditure for environmental protection.

[7] List and Gallet (1999), for instance, carried on an analysis of emissions of SO2 and NOx for the States of the US between 1929 and 1994.

to a long run path, tagged by a level of output per capita and the corresponding rate of growth. Under the known strict assumptions as to the production of innovation and new technologies, like their accessibility and the properties of production technologies being implemented, such long run is unique, implying the same level of output per capita or production, and the same rate of steady state growth, the latter being zero. Hence, the long run is a stationary state. The original model, of course, does not take into account the presence of resources and or the pollution of environment. The current interpretation of the EKC can be seen as an extension so as to fit in the latter problem.

In this enlarged framework, prediction is a little more sophisticated: one has to explain the two branches of the curve, which have opposite properties. The standard interpretation assumes there is a dynamical process behind the curve, so that points scattered around it do represent basically states on a trajectory. Accordingly, the downward sloping branch has been interpreted as the set of trade off equilibria where higher (lower) levels of one variable are consistent with lower (higher) levels of the other. In other words, it is a trade off relation, in conception similar to the one implied in the Phillips curve. At one point entered the interpretation of the Phillips curve that points along the curve could only be seen as short run equilibria, in other words, in a full dynamics as transients towards a long run equilibrium pair acting as the global attractor[8].

Similarly, for EKC, points on the left will eventually fly away towards the right or upward sloping branch of the curve.

There, the trade off relation between the two variables disappears, and both "move" together, tending towards some long run equilibrium values, which evidence cannot show but its apparent monotonicity suggests. It also seems to suggest the existence of some attractor, somewhere, to be imputed to external or internal constraints on the generating model. In the current interpretation, the Solows value of the growth rate of productivity will be married together with a corresponding value of the environmental index.

17.4 Modeling Income Growth and Environmental Protection

The multi-faceted nature of the environmental issue in relation to growth does not make it easy to construct synthetic indices. For want of such indices, as stated above, results tend to be dependent upon the kind of indicator our choice falls for, much more than one would like (World Bank, 1992). Although we are aware of this inherent limit, in this section we review the relationship using a typical index of emission, i.e. the index of CO_2 emission[9].

---

[8] In that case the natural rate of unemployment, associated with any level of the inflation rate. Notice the similarity with our treatment of the consequences of the three dimensional EKC.

[9] Data source World Bank (1999) and World Bank (2000).

The analysis deals with a sample of countries that is more or less homogenous from the point of view of economic performance (in growth terms), as they have all concluded the first phase of development and, all but Albania, display a composition of VA typical of advanced countries (where the service sector has the greatest share, representing no less than 45% of total VA, while agriculture is the least relevant)[10]. Figures 1 shows that with the C index one cannot find any Ushaped inverse relationship. Being countries just exited from the initial phases of their development, the first branch of such a curve is missing. Its no surprise that Albania finds itself in an anomalous situation, its V.A. composition being totally different from that of other countries. The upward sloping branch disappears also as a result of having considered only countries with a V.A. composition typical of the advanced world. This in turn has the effect of reducing the re-allocative effect of production. The more advanced countries show a tendency to relocate their dirty or technologically backward production lines into the lesser developed countries, and this automatically positions the latter along the upward sloping branch of the curve (Suri and Chapman, 1998; Musu, 2000). The same re-allocative process can account for the explanation offered by Vincent (1997), according to which the EKC would really be the result of the superposition of two distinct relations, a negative one for the less developed and a positive for the more developed countries.

Besides this aspect, it is to be noticed that the result graphed in Fig.1 is not necessarily to be attributed to more restrictive environmental policies, partially in contradiction with what is said in the example of the previous section. Energy efficiency may be un-related or independent of the degree of environmental awareness of a country, as said above. A large contribution to it comes from the level of energy price and the countries energy self-sufficiency. The explanation of the inverted-U form of the curve would lie in the price sensitivity of the curve, because of its energy foundations[11]. By implication the transition from the polluting phases to the one where environmental quality marries growth can be brought about by an external shock, as it has been the case of the petrol crisis at the beginning of the seventies, rather than of intentional policies. Unruh e Moomaw (1998) question the income determinism of the EKC. They maintain that it has been the petrol shock instead of the reaching of a high level of income, to have brought about the qualitative change in the growth-environment relationship[12]. Put differently, the

[10] de Bruyn et al. (1997) are a good reference for an analysis of the relation between growth and de-materialization of production processes.

[11] The majority of the studies on the EKC use as a proxy for the environmental quality pollutant emissions that originate almost entirely from fossil fuel burning. This depends on the energetic paradigm. Contrary to what usually believed, not only energy matters for environment. Matter matters too (just think of fertile soil).

[12] They find that the transition is not best correlated to a specific income level but to historic events common to the 16-country set they analysed, that is the

transition would have been the outcome of a Hicksian induced innovation (an endogenously triggered innovation), more than by a change in the social preference ordering over private and public goods. However, the essentially

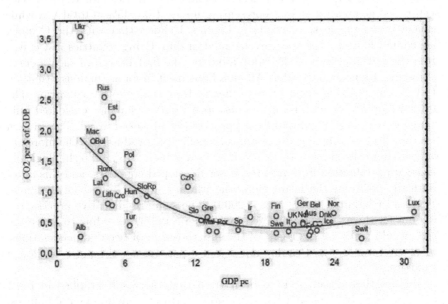

Fig. 17.1. Relationship between pc income and CO_2 per $ of GDP

economic nature of the relation finds strong support in the analysis of figure 2. There, per capita emission replaces the index of energy efficiency in the role of proxy for environmental quality. Despite a high variability, the observed relation indicates the prevalence of the scale effect, embedded into the level of *per capita* GDP, over efficiency enhancing effects of technological and substitution nature. This normally makes the total amount of emission to grow. Due to the public nature of the pollution (*public bad*) and the sensitivity of environment to the level of emission, this is what really matters.

17.5 The Environmental Kutznets Relationship: a Curve or a Surface?

In order to avoid some of the above highlighted ambiguities, in this section we test the EKC hypothesis introducing a different environmental indicator:

---

oil price shocks of the 1970s. Price shocks rather than income level appears to provide a sufficient incentive for new policy initiatives to overcome the political and economic inertia that maintained the previous trajectory or attractor (Unruh and Moomaw, 1998, p. 227).

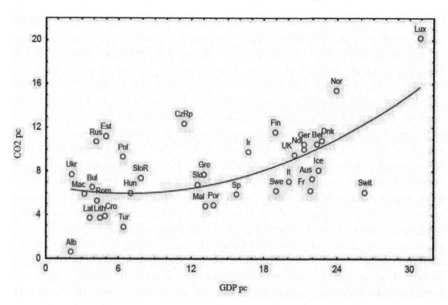

Fig. 17.2. Relationship between pc income and pc CO_2 emission

namely, the percentage of national territory that has been set aside and protected, i.e. destined to Parks and wild life reserves. Such variables, in fact, can
be taken as a direct measure of environmental policy of a country, as it is not
influenced by factors of any other nature (e.g. prices of fuels). According to
the IUCN *World Commission on Protected Areas* (1994), *a protected area is
defined to be a terrestrial or maritime area specifically destined to protection
and to the preservation of biological diversities, natural and cultural resources,
and is to be managed through the usage of appropriate tools dictated by laws
or of any other nature"*. Representing the effect of an active policy, the chosen
index can be considered to be a good proxy of the preferences of a country as
regards to such goods as natural resources and biodiversity. To a large extent,
the decision of protecting a given piece of national territory is unrelated with
the level of national wealth, though evidence shows that the richer countries
are those protecting more of their land. Vice versa, all indices of pollution
emission fundamentally depend upon the kind of technology being adopted
and implemented. The possibility of accessing certain technologies is often
precluded to poor countries lacking financial and human capital.

Figure 3 shows that, using the percentage of protected territory, the relationship reveals a quadratic form, although the upward branch appears to be
prevailing as our data refers only to advanced countries. The relation in the
EKC may induce to hastened conclusions, such as that growth automatically
generates greater environment protection. As we said, by increasing income
and therefore opportunities, in principle growth is a necessary element of this
nice story. That logical conclusion is not granted in reality, though.

To revise this set of issues without the temptation of falling into simplistic deductions, we have to take up and evaluate the whole analysis, and for this we better go back to the beginnings, to the way it was born in the mind of Kuznets in 1955. At a closer look, it is possible to notice that many of the studies on the EKC depart from the original curve by Kuznets himself, for it deletes a crucial variable in this latter, income distribution. This implicitly conveys the idea that in the KC, as well as in the EKC, per capita income is *the* explanatory variable and no room is left for mutual influences between variables. At any rate, beyond the history of ideas, it is this assumption that forces us to consider also income distribution. This also depends on the usual Income Elasticity Assumption (IEA) underlying the EKC hypothesis. In fact, growth in average income does not imply growth of income for the median individual. According to the IEA, if income growth goes along with increasing inequality (concentrated growth) growth can bring about a reduction or depression in environmental demand rather than increasing it, even though this demand is elastic to income (Magnani, 2000). Simply, a strongly concentrated income distribution discriminates opportunities across the citizens. Actually, as stressed by Kuznets, increasing inequality accompanying the initial growth phases determines the fall in environmental quality. Taking this extra variable into account, it is more appropriate to talk of a *Kuznets environmental surface* in three-dimensional space instead of a simpler curve.

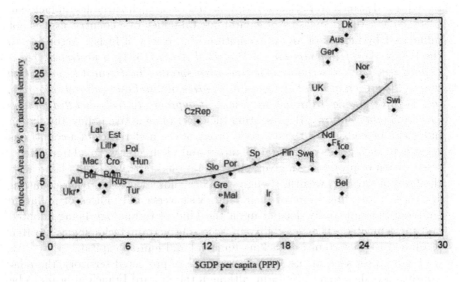

Fig. 17.3. Relationship between environmental quallity and income level Environmental Kutznets Curve

Figure 4 seems to show the net worth of this working hypothesis. As easily checked by inspecting sections of the surface, there is a whole *family of EKC*, parameterized by the Gini index.

In other words, the *position* of the representative EKC varies at the varying of the value of the Gini index, this showing the demand for environmental protection together with the increase in income equality just as the effect of the broader capacity of effectively choosing (and/or increase in opportunities) afforded by greater income equality. The worst famines were determined by entitlements issues, more than by overall food shortage, as pointed out by Sen (1981).

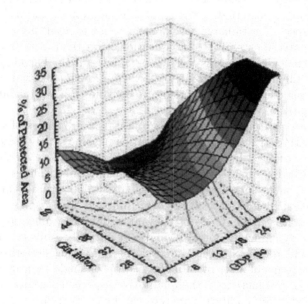

Fig. 17.4. Relationship between income level, equalilty and environemntal quality

Therefore, the standard version of the EKC isolating the relation between per capita income and pollution emission level, disregards an important aspect inbuilt into Kuznets own treatment. Evidence points out a key fact, that at the same income level (hence, level of development) environmental quality does improve parallel to the reduction in income concentration. A more equal distribution of income has the effect of speeding up the passage to growth paths that are more environmentally conscious, and thus have the effect of shifting upwards the conventional EKC.

This confirms that environmental demand is indeed influenced by the relative levels of income and social position, as well as by their absolute levels[13]. Then, a participated and diffused growth does represent one of the necessary conditions to generate the sought traverse process. A greater income equality increases individual readiness to pay for the environmental good, thus shifting upwards the minimal level of environmental quality a community or country is prepared to accept during the growth process (Bimonte, 2002). Such property proves to be crucial whenever there are threshold values to environmental damage beyond which it becomes irreversible.

Evidence, therefore, points out the relevance of thinking in terms of and designing integrated policies. The class of theories of endogenous growth offers the theoretical context that seems to best accommodate the task of dealing with these aspects. Sustainable development seems to result more from investment in immaterial capital (R&D, education, etc.) than from the exogenous forces of technological progress and demographic growth. Therefore, it is easy to introduce the multiregime framework where growth is basically seen as an endogenous phenomenon and therefore it does not follow standardized patterns or shapes.

This approach focuses upon certain qualitative properties of growth processes rather than the quantitative estimation of some steady state values for the involved variables. In economic analyses like the one we are dealing with here, growth is described as a sequence of predetermined phases, each being characterized in its turn by more or less specific structural features. That such component phases may change over time and across space, thus bringing about changes in the levels of relevant variables as much as in the way the economy dynamically operates, makes it natural to invoke such notions as regimes and regime switch. Moreover, in comparing performance across economies in terms of a set of variables instead of single indicators (e.g., the growth rate of productivity), the qualitative aspects naturally come to the fore and, often, only qualitative analysis can be carried on.

17.6 The EKC as a Dynamic Metaphor and Convergence

From a theoretical point of view, we can reconstruct the reasoning as follows. Confronted with a scatter plot in a plane of two variables, the ordinary attitude reacts by estimating a best fitting curve, via one of the ordinary methods econometrics teaches us. We get a curve, and the EKC is like any other curve fitting data for a *population of countries*. Now, what is crucial is the interpretation of the population. If its *internal structure* and *relative dynamics* are deemed not to be important, the curve is taken to tell us something about "average dynamics of the population". A country is taken to be

---

[13] This hypothesis seems to be also verified by the Easterlin paradox (Easterlin, 1974), according to which in spite of an overall wealth increase, people perceive a reduction in their own well-being.

representative of a state on a path; it does not count as an individual country. It is following this common interpretation, that we derive *dynamical* laws of tendency from distribution data at one point of time. We can break down a single curve, with separate interpretation of the two branches, the downward sloping with a trade-off relation between the two variables and the upwards piece with consistent behavior of variables, inferring basically a short run dynamics in two phases that will eventually land onto a monotonic approach to the *implied long run*. (A similar argument rationalizes the implied dynamics in other popular trade-off curves in the macroeconomic literature.) Thus, in the conventional interpretation, the EKC is understood as a law giving a tendency: all countries will eventually converge in the long run to a given common path, characterized by a double feature, i.e. the same level of GDP per capita (predicted by the exogenous growth theory) and a corresponding level of environmental quality. The latter is unique only if the curve holds true, otherwise it would be a whole (possibly dense) set. Adding the distribution variable has the effect of generating an interval of values of the index of quality corresponding to different values of the Gini coefficient. It would be hard to falsify the proposition of the existence of such a thing as an *implied long run*, to which all countries would approach had it not been for shocks and the like systematically shifting the target along the way. Still, in this paper we try out a new working hypothesis[14], according to which the distribution of countries in the plane of growth paths intended in this extended way, may embody a distribution across different qualitative features, summarized into the notion of the set of regimes. Therefore, the countries distribution is in a space of behaviors and not just paths and, if reducible to a finite set of internally homogenous models, these behaviors likely reflect mechanisms endogenous to the various economies. Thus, the distribution should really be understood and treated as the allocation of members of a population across a theoretical territory, rather than as a set of sampled individuals on transient paths, running towards some well defined final goal, an attractor or a distribution around its equivalent.

The logical consequence of this view is in that the points/countries, in the scatter diagram along the EKC curve, need not move together, actually they would be expected to move about where they are, if the cell of state space to which they belong is the *support of some regime* with some stability property. This is the idea we are going to formalize in the next section.

17.7 Growth as an Option set of Regimes

Theory behind the EKC assumes implicitly that different economies converge towards one another and therefore to a unique common path, and that the

---

[14] Consistent with the endogenous growth inspiration of the multiregime dynamics framework.

end state of such a process is independent of the initial conditions. What has been seen so far seems to suggest, on the contrary, the existence of a variety of regimes in the dynamics of sustainable development as shown by the sample of countries under our scrutiny, at the same time highlighting the key influence exerted by initial conditions.

It is now time to convert the previous analysis into the framework of the multiregime dynamics, and to construct the heuristic device called the Framework Space (FS). The former can be defined, in short, as a formal environment in which distinct countries (or generically *systems*) can follow different models of growth, depending upon where in their state space they happen to be. The definition implies therefore a way to capture dependence of growth performance and other dynamical properties upon initial conditions. This may play a relevant role in explaining why countries do not seem to comply to a common pattern, this in its turn raising a whole set of issues that span from the interpretation of the EKC to the discussion of why growth rates differ.

As explained elsewhere (Boehm, B., and Punzo (2001) and Brida and Punzo (2003)), behind a multiregime approach lies the hypothesis that quantitative differences among economies or sectors may sometimes be better explained by the existence of different models of behaviours, so that an economys history can be seen as a *choice* of which model to adhere to. This choice might generally be unconscious or even forced by external or domestic shocks, but more often than not these factors interact with the conscious implementation via active policies[15]. Two sorts of *ideal* histories may thus be encountered: at one extreme those exhibiting uncertain pattern with very frequent changes of the adopted models; at the other extreme, there are economies with a very high degree of "stability" with respect to the chosen model, up to the point that they never seem to depart from it. Of course, reality is generally somewhere in the middle. It is therefore *prudent* policy to work with a framework capable of accommodating the phenomena associated with this sort of qualitative variability (on top of the obvious quantitative variability, with which standard techniques are concerned)[16]. This is what the multiregime approach tries to do[17].

As constructed in Boehm and Punzo (2001), the multiregime approach was born to account for some generally accepted stylized facts of growth empirics, and to introduce structural change as discontinuous change embedded into observed dynamics. In order to deal with this twofold issue, we need an articulated dynamical framework, where multi-regime dynamics generates a chart of dynamical behaviours, the FS. *The latter becomes a space of growth paths*

---

[15] This is our point hereafter.

[16] An argument made clear and formal in the notion of entropy, see Brida and Punzo (2003)

[17] And of course not everything can be formalised, which justifies the prevalence of formal techniques devised to tame variability.

where trajectories are generally traverses from one path to another, some of them implying also crossing the border into a different growth model. A *regime*, in such a space, is defined to be a pair: a growth model (in the sense of a class of models generating the same prediction) and the *slice* of state space to which that class applies, its *supporting slice* or *domain*. There is a finite number of regimes, and therefore the partition is finite[18]. In a sense, behind dynamics observed in the FS there is a *dual* space of *generating models* with the explicatory variables, these being stochastic and deterministic factors, mechanisms of behaviour and growth, finally policy choices. Due to the way they are defined, regimes deserve the qualification of dynamic. Nothing prevents to use the notion of multi-regime framework with other different variables and to produce the corresponding version of the FS. This is what we are going to do hereafter. In fact, a FS is nothing but any n-dimensional space with a regime partition superimposed on it. Of course, traverses and episodes of structural change (i.e. *regime switches*) can only be fully represented if we have dynamical data at hands. We do not have this kind of data for the present application. Still, the idea of defining *qualitative* behaviours in terms of the variables chosen holds good, and via a reinterpretation we will be able to keep thinking of it as a dynamical framework, though a little special.

To represent and to classify homogeneously such different behaviors we need an adequate space representing all the variables believed to be crucial. Of course, there will be different constructions depending upon the choice of the variables and these latter will also determine *dimension*. The FS appropriate for the exercise here can be obtained in the following way. The foregoing discussion of the dynamics implicit along the EKC suggests that there are at least three variables involved in addition to *per capita* GDP, namely our two alternative indices of environmental impact (the first standing for pollution and the second for protection) and an index of income distribution. So far, the two environmental indices have been plotted (figs 1, 3 and 4) against *per capita* GDP as the measure of relative welfare in order to derive variants of the EKC. But we have already argued that we should consider the curve as really a *section* of the *true* relationship (as put forward by Kuznets): a surface in a four-dimensional space where distribution of income (together with its level) plays a key role as explanatory variable and indices are also plotted. The standard EKC, therefore, should be regarded as a section taken for some given distribution of income (and against one environmental index only). We can pick and choose among other sections, though. If we let income distribution inequality vary at the same time as a chosen environmental index (or else at the same time as *per capita* GDP), we get that different combinations of the values of the two *explanatory* variables would be consistent with the same level of *per capita* GDP (the same level of environmental quality, respectively). What can be seen from this sort of section? A given high level of protection

---

[18] The finesse of the partition, hence the number of recognised regimes, depends upon the criterion adopted to induce equivalence classes.

could turn up to be consistent with high *per capita* GDP and high inequality, or else low per capita GDP with a high equality. There would appear, in other words, a trade-off also between welfare and distribution. This relation would not be different from the classical EKC, though, still it would demonstrate that the world of choices over curves is greater than expected. If on the other hand, we take, sectioning for given *per capita* GDP and given value of the Gini index, our two environmental indices and plot them one against the other we obtain a full plane where a regime classification can be introduced. Every point in this plane is a chosen pair of values for the two pollution indices, and therefore they may be taken to reflect the result of a chosen integrated policy plan. The plane so obtained is the plane of (the effects of) implementable integrated policies. With a slight modification to our standard procedure[19], we can introduce the notion of regime by means of equivalence classes over a space of policy plans. Thus, a regime is (associated with) a whole set of integrated policy actions (implying choice of the corresponding mixes of the two environmental variables) driving a given model of growth[20], and four regimes are identified in the graphs below[21].

To each such policy option or choice corresponds at least a level of welfare and a value of income concentration: in other words, if we consider the values of the latter as equilibrium values, there is a correspondence between this version of the FS and a space of dynamical paths. The environmental policy FS is the space of values indexing (sets of) dynamical paths of an economy, which can therefore follow different behaviours depending upon the policy implemented. The original FS, as discussed in e.g. Boehm and Punzo (2001), is constructed in the state space of growth paths. The environmental FS is constructed in the dual space of the generating models and the explanatory variables, by introducing the hypothesis that environmental policies (perhaps together with other variables) can drive an economys dynamic path. Given their strong relation, we can still talk of traverses from one path to another in the economys own state space as the result of a policy choice represented in the present framework. In figure 5 the two indices of environmental quality are used: tons of CO_2 per million dollars of GDP produced and percentage of protected territory. Moreover, countries with income levels above the sample average have been identified. By plotting data for the different countries, after normalizing them on the basis of average values of the two indicators, we obtain the landscape of the models of growth followed by the various countries. These can be interpreted on the basis of the standard categories of substitutability of the models of sustainable growth proposed by Solow af-

---

[19] Where regimes are introduced in the framework space: the space of dynamical paths. Here, instead, they are defined in a space of policy plans: vectors of paths for policy targets inducing certain dynamic behaviours.

[20] Recall that a model of growth is a qualitative prediction of a set of growth paths.

[21] It is clear that our definition of sustainability is different from the one more commonly used in the literature, hence the qualification of relative.

ter Dasgupta[22]. With the latter classifying criterion, the picture shows four identifiable regimes: i.e. the regime of the relative sustainability, of relative un-sustainability, of technical and natural capital, respectively[23].

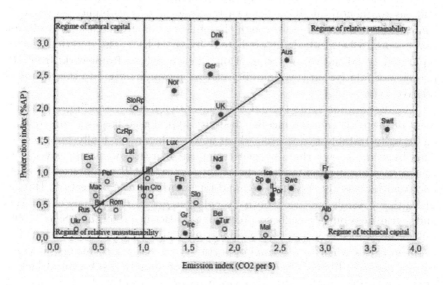

Fig. 17.5. Models of integrated policies

Notwithstanding the simplicity of the tool and without pretending to make universal statements, it is still possible to put forward a key to read the evidence so assembled. Countries tend to follow their own ways to environmental protection. The latter reflects to a large extent the relative availability of resources as well as the levels of social participation in the allocative process. Thus, apart from the countries in regime 1, of relative sustainability, which seem to implement adequate *integrated* measures to cope with the problem, all other countries still rely upon a purely sectoral approach. In particular, the few countries with a relatively lower density of inhabitants and a higher geographical concentration tend to implement a policy of conservation and to concentrate in regime 2 (the one defined of the natural capital). On the

---

[22] Just for ease of exposition, here the index of polluting emissions have been rendered as the inverse of the ratio of the emission rate of a given country to average rate. This has the effect that shifts rightwards along the horizontal axis correspond to abatements in the levels of polluting emission , or else to increased energetic efficiency.

[23] We would let the reader note that the term "sustainability" is here used in a comparative sense. Therefore, path of countries whose indices are above the sample average are defined sustainable. The same applies to define other : regime n. 2 is defined of natural capital because sustainability, as defined above, is pursued by protecting, whereas in regime n. 4 sustainability is pursued by innovating.

other hand, countries with a greater energetic dependence seem to concentrate on the objective of increasing energetic efficiency, thus locating themselves in the regime 4 of technical capital. The issue at this point becomes the following: is it possible to activate a process such that economies with low income and environmental quality levels can traverse directly to higher levels of income without having to go through the phases of environmental degradation that seems to be implied (or predicted) by the EKC? The question can be rephrased to advantage in terms of our chosen regime framework: from the regime 3 of relative un-sustainability can we jump directly into the regime of relative sustainability?

The assumption lying behind the conventional understanding and treatment of the EKC (i.e. the income elasticity hypothesis) has as a consequence an actually testable hypothesis, which however has never really been tested. This hypothesis says that, by generating an increase of income levels and therefore enlarging individuals choice sets, growth will automatically bring about an increase in the their demand for environmental protection. But, to make this possible, we need both an enlargement of individuals choice sets and a real possibility to choose. According to Berlin (1969) and Sens (1999) influential works, freedom is the opportunity to act, not action itself. The problem arises when to a formal opportunity to act, which means no violation of negative liberty (freedom "from"), corresponds a substantial lack of opportunity for action, that is, a violation of the positive liberty (freedom "to do"). If we accept this hypothesis, the answer to the previous question becomes obviously to integrate social, economic, environmental policies into a single coordinated, multi-valued action plan.

Figure 6 shows data on indices of equality, accessibility to information and education in a set of countries. Data have being normalized on the bases of average values[24]; countries with a protection level higher than the average have been marked for easier identification. Use of the three variables as coordinates makes it possible to identify four regimes[25], differing on the basis of the relative degree of participation (i.e. the level of literacy, information access, and equality) in the growth processes, degrees that can be interpreted by recalling categories typical of the theoretical models of sustainability associated with works of Sen and Schumpeter.

Sen (1999) focuses upon the "capability to function", i.e. what a person can really do or be, and defines development as improvement in such function This in its turn is seen as the primary goal for and a means to enhance development policies. In a Schumpeterian view, on the other hand, the focus is on intangible resources, that is on institutional features (social facts) that determine how

---

[24] It can be noticed that we are dealing with a smaller sample than before, this being due to the fact that for many countries data on the Gini index and annual sales of newspapers (this being our proxy for education) are unavailable.

[25] Regimes are defined according to the same principles used above for figure 5. Moreover, according to Bimonte (2002), inclusive (exclusive) stands for growth with high (low) participation.

effective an event is in generating growth. In our words, social participation increases "environmental returns", by modifying dynamical behavior.

Although our data set seems to indicate that it is hard to see any cross country convergence process, and that again this can be attributed to the relevance of initial conditions[26], on the whole the likelihood that environment be effectively protected increases with income equality and with the accessibility to information. This on its turn stresses once again the relevance of social policies and the necessity of abandoning a purely sectoral approach in favor of an integrated approach to the issues.

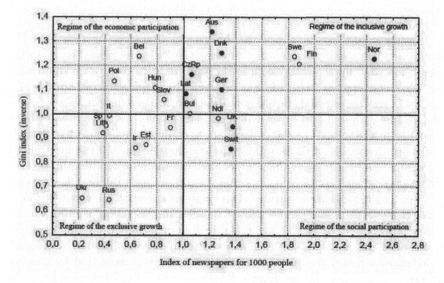

Fig. 17.6. Environmental quality and levels of partecipation

17.8 Some Conclusions

Functional relations, whether in terms of trade off or not, figure prominently in the macroeconomic literature, providing the hopeful reader with a menu *a la carte* for policy design and intervention. In practice, here it has been shown that around the scattered points of these curves there is a much richer dynamics than expected. The image of the flow of points transiting together towards some well defined state, or ergodic distribution, appears, at a closer look, marred by local cyclicity, irregular behaviors, roles interchanges, and the like. That much desirable long run predictable state has often emerged more to be a logical and statistical construction, or as the tendency implied by

---

[26] They only appear in the debate on β-convergence.

certain a priori assumptions. Here we tried to exit this situation treating each point state as a path of its own, along its own trajectory, at least in principle, till it can be "proved" that the hopes of theories turn up well founded.

On the basis of the argument above, we maintained that there is no unique formula to sustainable development and that the convergence implied and expected by all the traditional studies of the EKC as extensions of growth theory is far from being demonstrated. Growth processes, on the contrary, exhibit characteristics that are patently, dramatically different in some and many qualitative senses, and these do produce the outcomes that we can observe. Contrary to some well-established theorems, these outcomes carry the deep marks imprinted by their initial conditions, as well as those left by economic policies implemented on the way.

Initial conditions do appear in the growth literature, in that they determine the transitional dynamics towards the long run steady state. Initial conditions determine in the sample the sign and generally the value of the speed of convergence. They do not intervene in shaping up the growth process. On the other hand, initial conditions are expected to have no role in determining what the long run will look like. This is the twofold point here: just like in the models in the endogenous growth, we maintain that they do influence the values of the long run equilibrium. Our idea of the multiplicity of regimes as qualitatively different behaviors, implies something more: that there may be clusters of behaviors that on top of quantitative differences reveal different qualitative features. Our analysis is consistent with the vistas offered by certain endogenous growth models in the Schumpeterian family, where two factors of growth are taken to be crucial: accumulation of physical capital and innovation as it generates accumulation of immaterial capital. This paper stressed the relevance of the latter. We have seen that for a development path to become sustainable a balanced mix of technological progress, carefully designed environmental policies and social participation are necessary ingredients, and the latter has to be understood in the twofold aspect of participation in the choice process as well as in the division of the wealth so created.

As it brings about improvements in capabilities to function and/or modification in institutional features (as societys culture and attitudes), participation allows us to give an endogenous explanation of the shape of the EKC and the underlying dynamic behavior. This permits also to deal with some of the problems not accounted for by the traditional EKC. There, the major contributions to growth in environmental demand is left unexplained, set outside the model, by resorting to the income elasticity hypothesis or a structural explanation. The endogenous interpretation presented here links the overall increase in demand for environmental protection to the "increasing returns" that social participation produces through modifying dynamics. Therefore, in the spirit of some of the models of endogenous growth, social policies not only do not represent obstacle or constraint to growth; they actually increase demand, acceptability and even the efficacy of environmental policies, increasing its "return of scale"[27].

It is the available evidence that strongly suggests the need of exiting the deterministic income-driven approach. There is more than a single, uniquely determined avenue to sustainable development and, on the other hand, the special double convergence hypothesis implicit in studies on the EKC is far from being supported.

References

1. Beckerman, W. (1992), Economic growth and the environment. Whose growth? Whose environment? World Development, 20, 481-496.
2. Berlin, I., (1969), "Two concept of liberty", in: I. Berlin, Four essays on liberty, Oxford.
3. Bimonte, S., (2002), Information access, income distribution, and the Environmental Kuznets Curve, Ecological Economics, 41, pp. 145-156.
4. Boehm, B., and Punzo, L.F., (1994), Dynamics of industrial sectors and structural change in the Austrian and Italian economies, 1970-1989, in Boehm, B., and Punzo, L.F., (eds.), Economic performance. A look at Austria and Italy, Physica Verlag, Heidelberg.
5. Boehm, B., and Punzo, L.F., (2001), Productivity-investment fluctuations and structural change, in Punzo, L.F., (eds.), Cycles, growth and structural change: theories and empirical evidence, Routledge, London and New York.
6. Brida, G., and Punzo, L.F., (2003), Symbolic Time Series Analysis and Dynamic Regimes, Structural Change and Economic Dynamics, forthcoming
7. de Bruyn, S., Van der Bergh, J. And Opschoor, H., (1997), Structural change, growth, and dematerialization: an empirical analysis, in Van den Bergh, J.C.J.M. and Van der Straaten, J., (eds.), Economy and ecosystems in change: analytical and historical approaches, ISEE, Edgar Elgar.
8. Easterlin, R.A. (1974), Does economic growth improve the human lot?, in David, P., and Weber, R., (eds.), Nations and households in economic growth, Academic Press, New York.
9. Ekins, P. (1997), The Kuznets curve for the environment and economic growth: examining the evidence, Environmental Planning, A 29, 805-830.
10. Grossman, G.M. and Krueger, A.B. (1995), Economic growth and the environment, Quarterly Journal of Economics, 110, 353-378.
11. Grossman, G.M. and Krueger, A.B. (1996), The inverted U: what does it mean? Environmental Development Economics, 1, 119-122.
12. IUCN, (1994), 1993 United Nations list of national parks and protected areas, IUCN, Gland, Switzerland and Cambridge, UK.

---

[27] Some of the models of endogenous growth show that, in presence of imperfect capital markets limiting accessibility to the lower income brackets, income redistribution tends to create investment opportunities precisely for those categories where social marginal returns are greatest. In such conditions income equalization becomes one of the growth enhancing engines, rather than representing an obstacle to it.

13. Kaufmann, R.K. and Cleveland, C.J. (1995), Measuring sustainability: needed an interdisciplinary approach to an interdisciplinary concept, Ecological Economics, 15, 109-112.
14. Kuznets, S. (1955), Economic growth and income inequality, American Economic Review, 45, 1-28.
15. List, J.A. and Gallet, C.A., (1999), The environmental Kuznets curve: does one size fit all?, Ecological Economics, vol. XXXI, pp. 409-423.
16. Magnani, E. (2000), The Environmental Kuznets Curve, environmental protection policy and income distribution, Ecological Economics (32)3, pp. 431 443.
17. Musu, I, (2000), Introduzione alleconomia dellambiente, Il Mulino, Bologna.
18. Panayotou, T. (1995), Environmental degradation at different stages of economic development, in I. Ahmed and J.A. Doeleman (eds.), Beyond Rio. The environment crisis and sustainable livelihoods in the third world, Macmillan Press Ltd.
19. Pezzey, J., (1989), Economic analysis of sustainable growth and sustainable development, World Bank Environment Department Working Paper, no15, Washington DC.
20. Selden, T.M. and Song, D. (1994), Environmental quality and development: is there a Kuznets curve for air pollution emissions? Environmental Economic Management, 27, 147-162.
21. Sen, A., (1981), Poverty and famines, Clarendon press, Oxford.
22. Sen, A., (1999), Development as freedom, Oxford University Press, Oxford.
23. Suri, V. and Chapman, D. (1998), Economic growth, trade and energy: implications for the environmental Kuznets curve, Ecological Economics (25) 2, 195-208
24. Torras, M. and Boyce, J.K. (1998), Income, inequality, and pollution: a reassessment of the environmental Kuznets curve, Ecological Economics, 25, 147-160.
25. Unruh, G.C. and Moomaw, W.R. (1998), An alternative analysis of apparent EKC-type transitions, Ecological Economics (25) 2, 221-229
26. Vincent, J.R. (1997), Testing for environmental Kuznets curves within a developing country, Environmental and Development Economics, vol. 2, 417-431.
27. World Bank (1992), World development report 1992: development and the environment, Oxford University Press, New York.
28. World Bank (1999), World development report, 1998/1999, Oxford University Press, New York.
29. World Bank (2000), World development report, 1999/2000, Oxford University Press, New York.

An Empirical Analysis of Growth Volatility: A Markov Chain Approach[*]

Davide Fiaschi[1] and Andrea Mario Lavezzi[2]

[1] University of Pisa, Dipartimento di Scienze Economiche dfiaschi@ec.unipi.it
[2] University of Pisa, Dipartimento di Scienze Economiche lavezzi@ec.unipi.it

Summary. This paper studies the determinants of growth rate volatility, focusing on the effect of level of GDP, structural change and the size of economy. First we provide a graphical analysis based on nonparametric techniques, then a quantitative analysis which follows the distribution dynamics approach. Growth volatility appears to (i) decrease with per capita GDP, (ii) increase with the share of the agricultural sector on GDP and, (iii) decrease with the size of the economy, measured by a combination of total GDP and trade openness. However, we show that the explanatory power of per capita GDP tends to vanish when we control for the size of the economy.

Key words: growth volatility, Markov transition matrix, structural change, nonparametric methods.

18.1 Introduction

The relationship between income level and the volatility of growth rate of income (GRV henceforth) has received little attention up to now. Contributions can be divided into two main groups. The first highlights that economic development is accompanied by a sharp reduction in GRV (see [8]), while the second refers to a negative relationship between the *size* of an economy and GRV (see [3]).[3] The aim of the paper is to identify the main determinants of growth volatility of a country. This can be relevant for a better understanding of the development process, especially of low income countries, as well as for the design of economic policies aiming at stabilizing the growth path in underdeveloped economies.

---

[*] We are very grateful to Carlo Bianchi, Eugene Cleur, three referees and seminar participants at Siena and Guanajuato for their comments. The usual disclaimers apply.

[3] Here we are not interested in the link between GRV and long-run growth as in [11]

Economic development is generally intended as an increase in per capita GDP. Therefore a first possible empirical investigation regards the relationship between GRV (measured by the volatility of annual per capita GDP growth rate) and per capita GDP. We also analyze structural change, a typical phenomenon associated to development. In fact, a possible explanation of the inverse relationship between the level of development and GRV resides in the decreasing weight of sectors with a more volatile output, like agriculture and primary sectors, with respect to sectors with less volatile output, like manufacturing and services, as an economy develops. However, so far the literature on structural change has not paid attention to this issue (see e.g. [7]). Differently, the increase in the number of sectors (or productive units) associated to a growing size of the economy is the most common explanation of the second relationship mentioned above. In fact, a reduction in aggregate GRV may derive from averaging an increasing number of sectoral growth rates, since idiosyncratic shocks to each sector would tend to cancel out by the law of large numbers.[4]

In this paper we test for the existence of these relationships in a large sample of countries from [6]'s dataset. In particular we focus on the effect on GRV of three variables: (i) the level of per capita GDP (GDP henceforth), as proxy of the level of development, (ii) the share of agriculture on GDP (AS henceforth), as proxy of structural change and (iii) an interaction variable between total GDP and trade openness (GT henceforth), as proxy of the size of the economy.

When we study the individual effects of these variables we find an inverse relationship between GRV and both GDP and GT, and a positive relationship between GRV and AS as we expected, although some nonlinearities appear in the latter case. However, we show that the effect of GDP on GRV vanishes when it is considered jointly with GT. These results suggest that the explanation of growth volatility of a country can be found in structural change and, especially, in the extent of its economy (as defined above).

As a preliminary step in our empirical analysis, we follow the [3]'s approach, where all observations are pooled and then partitioned in classes. We measure GRV for each class of GDP, AS and GT as the standard deviation of growth rates associated to the observations belonging to that class. We estimate by nonparametric methods the relationship between GRV and our three explanatory variables, (this is a crucial difference with respect to [3]).

However, we argue that a drawback of this procedure is to ignore the relevant information on the dynamics of individual countries. Therefore we propose a new statistical methodology based on Markov transition matrices. In particular we propose some *growth volatility indices* derived from mobility

---

[4] This result holds if the strength of interactions, given for instance by the correlation of sectoral outputs, is weak or absent (see [4]). The estimation of these effects may be however difficult for the type of data needed for a cross-country analysis.

indices. By applying these indices to our sample, we find a confirmation of the previous findings. In particular, we propose the estimation of *conditioned* transition matrices showing that *GDP* is not informative, once we control for *GT*.

From the theoretical point of view, we relate this evidence to the framework of a multisector economy. Hence, our work is close to papers such as [12] and [5], which study the emergence of aggregate fluctuations from local shocks. None of them is however explicitly concerned with structural change. On the applied side, the statistical indices proposed in the paper are related to the literature on mobility (see, e.g. [1]). We reinterpret a set of indices generally utilized to measure intergenerational mobility as measures of volatility, and propose two new indices.

The paper is organized as follows. Section 18.2 contains a preliminary graphical analysis based on nonparametric data analysis; Section 18.3 introduces the growth volatility indices; Section 18.4 presents and discusses the results of the calculation of these indices; Section 18.5 concludes.

18.2 Graphical Analysis

We use data on *GDP* from [6]'s database for the period 1950–1998, and data on agriculture and trade from the World Bank's *World Development Indicators 2002.*[5] As noted, we proxy for the structure of the economy by the share of the agricultural sector in aggregate value added (*AS*), and measure the *effective* size of the economy, related to the number of sectors, by the interaction of *total* GDP *and* trade openness. That is, we create a variable by multiplying the total GDP (normalized with respect to the the maximum value in the sample) by the ratio of the sum of imports and exports on GDP (*GT*).

To evaluate the relation between *GRV* and level of development first we pool all observations for each variable. Then we separate these pooled observations on *GDP* into 196 classes with a similar number of observations (approximately 30), while to evaluate the relation between *GRV* and structural change we separate all pooled observations on *AS* into 109 classes. Finally, for the relation between *GRV* and *GT* we have 125 classes of observations. For every observation of *GDP* in year t we calculate the growth rate from t to $t + 1$.[6]

In Figs. 18.1, 18.2 and 18.3 we plot the standard deviation of growth rates (*STD*) relative to the observations in a class against, respectively, the (log of) average *GDP*, *AS* and *GT* in that class, and run a nonparametric estimation

---

[5] The Maddison database contains 122 countries. Data on *GDP* are in 1990 international dollars. When considering agriculture and trade the database is restricted to 119 countries for 38 years (1960–1998). Not all observations were available for each country for all years. See Appendix A for the country lists.

[6] For every data on *AS* and *GT* we consider the corresponding observation on *GDP* and calculate the associated growth rate.

of these relationships.[7] At first glance we see that STD tends to fall with (log

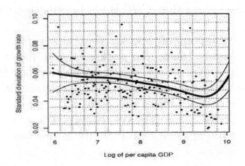

Fig. 18.1. *GRV* estimated by *STD* and log of *GDP*

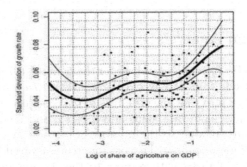

Fig. 18.2. *GRV* estimated by *STD* and log of *AS*

of) GDP. The increase in volatility at the highest and lowest (log of) GDP is associated with a wider variability band, meaning that there the estimate is not precise. Growth volatility appears to be increasing with (log of) AS. This relation is not monotonic, but the variability band is tighter where the upward sloping portion is steeper, indicating that the estimation is more precise where the curve is sharply increasing. The relationship is much clearer in Fig. 18.3: STD decreases with (log of) GT, and the reduction in volatility appears to

---

[7] The nonparametric estimate is obtained with the statistical package included in [2]. We used the standard settings suggested by the authors (i.e. optimal normal bandwidth). To test the robustness of this estimate, we ran an alternative nonparametric regression using the plug-in method to calculate the kernel bandwidth, and obtained a similar picture. We refer to [2] for more details. We report the variability bands, giving a measure of the statistical significance of the estimate (see [2, pp. 29–30] for details on variability bands). Data sets and codes used in the empirical analysis are available on the authors' websites (http://www-dse.ec.unipi.it/fiaschi and http://www-dse.ec.unipi.it/lavezzi).

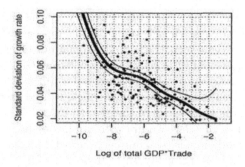

Fig. 18.3. *GRV* estimated by *STD* and log of *GT*

be exponential. Reference [3] presents a similar picture, but considers only the log of total GDP as a proxy for the size of an economy. They find that the log of *STD* decreases linearly with the log of GDP (that is, as a power law). However, they consider as an outlier the only observation they have for low GDP levels, which would generate a nonlinear relation. In our case the number of observations at low income levels is higher: results not presented here show that considering only total GDP would indeed produce a nonlinear relation, while the nonparametric estimation, particularly well-suited to detect the presence of nonlinearities, reveals that a nearly-linear shape (on a log-log scale) obtains when total GDP interacts with country openness.

This approach has the drawback of ignoring the information on the growth path of individual countries, being based on the pooling of observations and on the measurement of growth volatility by the standard deviation within a class. For example, consider two countries having a series of observation in the same *GDP* class, and with constant growth rates but at very different levels. If we compute the standard deviation of growth rates for that *GDP* class, we would obtain a high value, wrongly indicating high growth volatility. On the contrary, the method proposed in the next section, based on transition matrices, would correctly detect low volatility.[8]

18.3 Growth Volatility Indices

In this section we propose a set of synthetic indices to measure growth volatility. Their statistical properties are derived in [4]. In particular, the measurement of growth volatility requires first the estimation of a Markov transition matrix, whose states $S = \{1, 2, ..., n\}$ are defined in terms of growth rate classes. A transition matrix summarizes the information on the dynamics of growth rates (for more details see [9]), and represents a basis to calculate growth volatility indices.

---

[8] Reference [3] avoids this specific problem by detrending data, but the procedure is not immune from introducing spurious volatility.

Heuristically, the indices quantify volatility by the *intensity of switches across growth rate classes*. The advantage of the approach based on transition matrices is that we can keep track of the dynamics of individual countries in the sample. To evaluate the relationship between *GRV* and, for instance, *GDP* we calculate the values of these indices for different classes of *GDP*. A drawback of this methodology is that to calculate the transition matrix we need to partition the state space of the growth rates.[9]

To define indices of *GRV* we draw on studies of inter- and intragenerational mobility of individuals (see, among others, [1, pp. 24–30] and [13]), and propose two new indices. Basically, these indices are functions of the elements of a transition matrix. In a transition matrix high values on the principal diagonal indicate low mobility, while the values of off-diagonal elements refer to changes of state and, therefore, high values of the latter are associated to high mobility.

A simple mobility index is the following, proposed by [13]:

$$I^S(\mathbf{P}) = \frac{n - trace(\mathbf{P})}{n - 1}, \qquad (18.1)$$

where $\mathbf{P}$ is a transition matrix of dimension n. The range of the index is $[0, n/(n-1)]$ and a high value means high mobility. However, I^S is not well-suited to measure *GRV* because it is not affected by the value of off-diagonal elements, a key point for the present analysis, but we utilize it as a term of comparison with the other indices discussed below.

Reference [1, p. 28], proposes the following index which takes explicitly into account the *distance* covered by a transition from i to j, $(i, j \in S)$, when the states correspond to increasing or decreasing values of a variable:

$$I^B(\mathbf{P}) = \frac{1}{n-1} \sum_{i=1}^{n} \sum_{j=1}^{n} \pi_i p_{ij} |i - j|. \qquad (18.2)$$

In I^B, p_{ij} is an element of the transition matrix $\mathbf{P}$, while π_i is an element of the associated ergodic distribution.[10] The range of I^B is $[0, 1]$: a higher value means higher mobility.

In this case only the absolute value of the difference between i and j is taken into account. It is worth verifying the effect of increasing the weight attached to "longer" jumps, in order to better appreciate the magnitude of the fluctuations. Therefore we introduce the following index:

$$I^{BM}(\mathbf{P}) = \frac{1}{(n-1)^2} \sum_{i=1}^{n} \sum_{j=1}^{n} \pi_i p_{ij} (i - j)^2, \qquad (18.3)$$

---

[9] In [4] we present a graphical method of analysis which dispenses with the definition of growth rate classes. The drawback is the lack of a procedure to test the statistical significance of the results.

[10] The ergodic distribution represents the long-run distribution of the Markov process. For more details see [1] .

in which the distance of the transition enters in a quadratic form. As before $I^{BM} \in [0, 1]$ and a higher value means higher mobility/volatility.

Indices I^B and I^{BM} weight the transitions from growth rate class i by the corresponding mass in the long-run equilibrium (i.e. in the ergodic distribution). In other words, considering the elements of the ergodic distribution as weights amounts to measuring GRV in the long-run equilibrium. However, also the volatility along the transition path can reveal very interesting information. The following indices fill this gap:

$$I^{FL}(\mathbf{P}) = \frac{1}{A} \sum_{i=1}^{n} \sum_{j=1}^{n} p_{ij} |i - j| \, ; \tag{18.4}$$

$$I^{FLM}(\mathbf{P}) = \frac{1}{A^2} \sum_{i=1}^{n} \sum_{j=1}^{n} p_{ij} (i - j)^2 . \tag{18.5}$$

I^{FL} and I^{FLM} respectively correspond to I^B and I^{BM}, except for the absence of the elements of the ergodic distribution. The constant A normalizes both indices to the range $[0, 1]$.[11] A higher value still means higher mobility/volatility.

18.4 Empirical Results

In the following we study the relation between GRV, GDP, AS and GT by calculating the values of the indices described in Sect. 18.3. In particular: (i) we separate the observations on GDP, AS and GT in four classes for each variable, from "low" to "high" values; (ii) we calculate the transition matrix with five growth rate classes for each class, (iii) we compute indices (18.1), (18.2), (18.3), (18.4), (18.5) for every transition matrix and, finally, (iv) we make inference on these estimates.[12]

First we define five growth rate classes which will be used with the three variables. We set the central class to include the average growth rate of the sample, equal to 2%. In particular, we add and substract 1 percentage point. Then, we define the other classes symmetrically around this central class, maintaining the same class width (i.e. 3 percentage points).[13] With this criterion we obtain the state space:

$$S = \{[-\infty, -2\%), [-2\%, 1\%), [1\%, 3\%), [3\%, 6\%), [6\%, +\infty)\} . \tag{18.6}$$

[11] In particular:
$$A = \begin{cases} 2\sum_{i=\frac{n-1}{2}+1}^{n-1} i + \frac{n-1}{2} & \text{for } n \text{ odd;} \\ 2\sum_{i=\frac{n}{2}}^{n-1} i & \text{for } n \text{ even.} \end{cases}$$

[12] In the computation of all indices we assume that the assumptions of the Markov processes are always satisfied.

[13] We are assuming that there exists a common long-run trend and that the central class contains it. Results are not affected by slight changes of the classes' limits.

18.4.1 Per Capita GDP

We define four *GDP* classes which contain the same number of observations (≈ 30), obtaining the following:

$$I = [0, 1030), II = [1030, 2350), III = [2350, 5600), IV = [5600, +\infty).$$

For every *GDP* class we estimate a transition matrix relative to the state space S.[14] Table 18.1 contains the values of the indices calculated for each of the four transition matrices.[15] We observe that in all cases the value of the

Table 18.1. Growth volatility indices. Standard errors in parenthesis. *GDP*

Index\GDP class	I	II	III	IV
I^S	0.8086	0.7781	0.7733	0.7672
	(0.0162)	(0.0163)	(0.0162)	(0.0175)
I^B	0.2841	0.2733	0.2504	0.2309
	(0.0091)	(0.0082)	(0.0072)	(0.0067)
I^{BM}	0.1666	0.1569	0.1321	0.1108
	(0.0085)	(0.0076)	(0.0062)	(0.0054)
I^{FL}	0.3874	0.355	0.3249	0.3145
	(0.0109)	(0.0102)	(0.0093)	(0.0102)
I^{FLM}	0.286	0.2509	0.2116	0.1977
	(0.0122)	(0.0108)	(0.0096)	(0.0103)

index is strictly decreasing with respect to *GDP* class and that, in particular, the value of the index in the first *GDP* class is always higher than in the last. This result agrees with Fig. 18.1 in which volatility is measured by the standard deviation of growth rates.

We tested for the joint equality of the estimated value and strongly rejected the null-hypothesis.[16] Table 18.2 reports the p-values of tests of a null hypothesis of equality between the value of the index in the first *GDP* class versus its value in each of the other *GDP* classes, for all the indices.[17] Tests

---

[14] The four transition matrices and the four ergodic distributions (one for each *GDP* class) are available on request.

[15] For details on the computation of the standard errors see [4].

[16] Reference [4] show that these indices are asymptotically normally distributed. The test statistic:

$$F_{r-1, r(n-1)} = \frac{n \Sigma_{i=1}^r \left(I\left(\hat{P}^i\right) - I\overline{(P)} \right)^2 / (r-1)}{\Sigma_{i=1}^r \frac{\hat{\sigma}_{I_i}^2}{n} / r},$$

where r is the number of indices ($r = 4$), n is the number of observation used for the computation of every index $\hat{\sigma}_{I_i}^2 / n$, $i = 1, 2$, is the variance of the value of the index calculated from transition matrix $\hat{P}^i$ and $I\overline{(P)}$ is the average value of index, is distributed as a "'F'" distribution with $r - 1$, $r(n-1)$ degrees of freedom.

[17] The test statistic

Table 18.2. Test of equality between the *GRV* index of *GDP* class *I* versus its value in the other classes. * means rejection of the null hypothesis of equality at 5% confidence level.

Index\GDP class	*I* vs *II*	*I* vs *III*	*I* vs *IV*
I^S	0.092	0.062	0.041*
I^B	0.19	0.002*	0*
I^{BM}	0.20	0*	0*
I^{FL}	0.015*	0*	0*
I^{FLM}	0.016*	0*	0*

confirm that *GDP* class *I* generally has a statistically significant higher growth volatility. At a conventional 5% level, the null hypothesis is not rejected only in the comparison between the value of the index in the first and in the second *GDP* class for indices I^B and I^{BM}, and in the comparison of the value of the first *GDP* class with the second and the third for I^S (but it is rejected at 10%).

To check whether there is a monotonic decreasing relationship among the values of the indices at different *GDP* levels we also tested the following hypotheses of equality: (i) *GDP* class *II* vs *GDP* class *III*; (ii) *GDP* class *III* vs *IV*. In case (i), the hypothesis is strongly rejected for all indices but I^S; in case (ii) the hypothesis is rejected only for I^B and I^{BM}. Hence, according to indices I^B and I^{BM}, the decrease is statistically significant when we move from class *II* onwards, while with I^{FL} and I^{FLM} the decrease is statistically significant from class *I*, but it is less evident at higher *GDP* classes. According to I^S we do not find evidence of a monotonic decrease but, as noted, this index does not appear to be proper for our analysis.

The comparison between indices I^B (I^{BM}) and I^{FL} (I^{FLM}) highlights that transition dynamics may be relevant to evaluate the differences in growth volatility. It is not an easy task to disentangle the relative role of the two factors determining the values of the indices, that is the transition probabilities and the elements of the ergodic distribution. However, we obtain that in *GDP* class *I* the probabilities of transition to "distant" states (i.e. from $[-\infty, -2\%)$ to $[6\%, +\infty)$ and vice versa) take on a relatively high value, but I^B and I^{BM} weight these probabilities less, since they use as weights the elements of the ergodic distribution, 0.19 and 0.12, instead of the "fixed" weight 1. This may explain why the volatility in *GDP* class *I* is significantly higher than in the other classes, in particular in a comparison of *GDP* class *I* and *II*, when measured by I^{FL} (I^{FLM}) with respect to I^B (I^{BM}).

$$t = \left[I\left(\hat{\mathbf{P}}^1\right) - I\left(\hat{\mathbf{P}}^2\right) \right] / \sqrt{\frac{\hat{\sigma}_{I1}^2}{n} + \frac{\hat{\sigma}_{I2}^2}{n}}$$

converges towards a Gaussian distribution under the null hypothesis $I\left(\hat{\mathbf{P}}^1\right) = I\left(\hat{\mathbf{P}}^2\right)$. For details on these tests see [4].

Overall, the indices indicate the presence of a negative relationship between GRV and GDP. This is more apparent when the magnitude of a transition is taken into account, in particular for the indices I^{FL} and I^{FLM} (but notice that for these indices we cannot reject the null hypothesis of equality between their values in GDP classes III and IV). In any case, a comparison between the first and the last GDP classes always shows a significantly higher volatility in the former.

18.4.2 Structural Change

In this section we address the relationship between GRV and structural change proxied by AS. We first define four AS classes with the same number of observations. The resulting classes' limits are:

$$I = [0, 0.08), II = [0.08, 0.2), III = [0.2, 0.33), IV = [0.33, 1].$$

Table 18.3 contains the volatility indices calculated with this class definition.[18]

Table 18.3. Growth volatility indices. Standard errors in parenthesis. AS

Index\AS class	I	II	III	IV
I^S	0.7428	0.8152	0.7834	0.8756
	(0.0229)	(0.0214)	(0.0223)	(0.0207)
I^B	0.2297	0.2735	0.2576	0.3199
	(0.0088)	(0.0105)	(0.0102)	(0.0115)
I^{BM}	0.1105	0.1501	0.1377	0.1915
	(0.007)	(0.0094)	(0.0092)	(0.0114)
I^{FL}	0.0133	0.3611	0.3364	0.4119
	(0.3037)	(0.0132)	(0.0136)	(0.0138)
I^{FLM}	0.1905	0.2478	0.2249	0.2975
	(0.0134)	(0.0140)	(0.0142)	(0.0151)

Results seems to be in accordance with the pattern in Fig. 18.2. Moving from high levels of AS to low levels, that is following a typical development path, volatility decreases from class IV to class III, then increases in class II and decreases again in class I. However, tests of equality between the value of the indices in class III and in class II do not allow to reject the null hypothesis at conventional 5% level.[19] Finally, volatility is significantly higher in class IV than in class I: the hypothesis of equality between the value of the index in class I and in class IV is strongly rejected for all indices (we omit the details of the tests).

At this stage, we take this result as indicating the possible presence of a more complex behavior at intermediate levels of AS, which is in accordance

---

[18] See [4] for the transition matrices.

[19] The p-values of the tests for I^S, I^B, I^{BM}, I^{FL}, I^{FLM} are, respectively, $0.15, 0.14, 0.17, 0.10, 0.13$. As in the previous case we tested for the joint equality of the estimated value and strongly rejected the null hypothesis.

with the non–monotonic pattern of growth volatility found in Fig. 18.2. Notice that indices calculated for classes II and III are not significantly different at 5% level, but only at about 15%.

18.4.3 The Size of the Economy

In this section we repeat the exercise made with AS using the interaction variable GT. We define four GT classes with the same number of observations:

$$I = [0, 0.0005), II = [0.0005, 0.0017), III = [0.0017, 0.007), IV = [0.007, 1].$$

With this class definition, and with the same state space for growth rate classes, we obtain the volatility indices in Table 18.4. In this case we observe

Table 18.4. Growth volatility indices. Standard errors in parenthesis. GT

Index\GT class	I	II	III	IV
I^S	0.8437	0.8328	0.7996	0.7316
	(0.02)	(0.0203)	(0.0203)	(0.0229)
I^B	0.3161	0.2735	0.2565	0.2029
	(0.0113)	(0.0095)	(0.0094)	(0.0075)
I^{BM}	0.1958	0.1466	0.1355	0.0882
	(0.0112)	(0.0082)	(0.0084)	(0.0055)
I^{FL}	0.4038	0.3668	0.3402	0.2769
	(0.013)	(0.0128)	(0.0125)	(0.0123)
I^{FLM}	0.2988	0.2472	0.2263	0.1589
	(0.0141)	(0.0136)	(0.0133)	(0.0116)

a monotonic decrease in all the indices across the classes. The test of equality between the value of the indices in GT class I and the value in any other class is strongly rejected (with the exception of index I^S in a comparison with the its value in GT class II). Also, tests of equality to assess a monotonic decreasing relationship between the values of the indices are generally rejected at conventional significance levels. Exceptions are all indices (I^S, I^B, I^{BM}, I^{FL} and I^{FLM}) in a comparison of class II and III: the p-values are respectively: 0.12, 0.08, 0.17, 0.07, 0.14.[20] Hence we cannot reject the hypothesis of equality in a comparison of classes II and III. This partially agrees with Fig. 18.3, where the slope of the estimate appears less steep at intermediate GT levels.

18.4.4 On Conditioning

A natural question is how much of the growth volatility of a country can be imputed to its level of development, to structural change and to its size. Here we address this issue in the approach based on the Markov transition matrix.

---

[20] Also in this case the null hypothesis of joint equality of the estimated values is strongly rejected.

First, notice that the analysis in the previous section can be considered as an estimation of *conditioned* transition matrices. In fact, the basis for the calculation of each single index is a transition matrix, obtained from the transition matrix for growth rate classes only, indicating the probability of observing a transition to a certain growth rate class starting from a given growth rate class *and* a given e.g. GT class.[21]

For example, the formal definition of a transition probability from growth rate class S_i to growth rate class S_j, given that the observation is in GT class I is:

$$p\left(g_t \in S_j | g_{t-1} \in S_i, GT_{t-1} \in I\right) =$$
$$= p\left(g_t \in S_j | g_{t-1} \in S_i\right) \left[\frac{p\left(GT_{t-1} \in I | g_t \in S_j, g_{t-1} \in S_i\right)}{p\left(GT_{t-1} \in I | g_{t-1} \in S_i\right)} \right] \tag{18.7}$$

In Eq. 18.7 the term on the left–hand side is an element of the conditioned transition matrix from which we derived our indices relative to GT class I. The first term on the right-hand side is an element of the transition matrix for growth rate classes only; the second term is the probability that, given a transition from growth rate class S_i to S_j, the initial growth rate is associated to a GT in class I. If the conditioning variable GT is not relevant, $p\left(g_t \in S_j | g_{t-1} \in S_i, GT_{t-1} \in I\right) = p\left(g_t \in S_j | g_{t-1} \in S_i\right)$: any transition matrix calculated considering alternative values of GT would not be statistically different from the unconditioned transition matrix.[22] Therefore, GRV indices calculated from the former would not be statistically different from each other, and from that calculated from the unconditioned transition matrix. In the same manner, if we condition on two variables, e.g. GT and GDP, we have:

$$p\left(g_t \in S_j | g_{t-1} \in S_i, GT_{t-1} \in I, GDP_{t-1} \in I\right) =$$
$$= p\left(g_t \in S_j | g_{t-1} \in S_i, GT_{t-1} \in I\right) *$$
$$* \left[\frac{p\left(GDP_{t-1} \in I | g_t \in S_j, g_{t-1} \in S_i, GT_{t-1} \in I\right)}{p\left(GDP_{t-1} \in I | g_{t-1} \in S_i, GT_{t-1} \in I\right)} \right] \tag{18.8}$$

and the same reasoning for the relevance of GDP, given GT, applies.

Here we do not provide a complete discussion of this issue, but only some evidence on the irrelevance of GDP in explaining GRV once GT and AS (whose values are calculated on the basis of GT classes) are taken into account.[23] Namely, we compare the values of index I^B, computed for GT classes only, with the values obtainable when the transition matrix is calculated for

---

[21] An example of conditioned Markov chains is in [10].

[22] From the condition $p\left(g_t \in S_j | g_{t-1} \in S_i, GT_{t-1} \in I\right) = p\left(g_t \in S_j | g_{t-1} \in S_i\right)$ derives $p\left(GT_{t-1} \in I | g_t \in S_j, g_{t-1} \in S_i\right) = p\left(GT_{t-1} \in I | g_{t-1} \in S_i\right)$, i.e. the information on GT_{t-1} is irrelevant to know the state of g_t.

[23] The choice of GT as the "principal" variable is suggested from the results of Sect. 18.4.

every *GT* class conditioned to each class of *GDP* and *AS*. Tables 18.5 and 18.6 report the results (notice that we report the values for *AS* classes in descending order).

Table 18.5. Values of I^B for *GT* conditioned on *GDP*. Standard errors in parenthesis. * means rejection of the hypothesis of equality at 5% and ** at 10%.

	Non cond.	GDP(I)	GDP(II)	GDP(III)	GDP(IV)
GT(I)	0.3161 (0.0113)	0.3113 (0.0158)	0.3326 (0.0195)	0.2886 (0.0298)	
GT(II)	0.2735 (0.0095)	0.3272* (0.0255)	0.2766 (0.0179)	0.2483** (0.0134)	0.2568 (0.0308)
GT(III)	0.2565 (0.0094)	0.2923 (0.0760)	0.2545 (0.0262)	0.2512 (0.0158)	0.2591 (0.0140)
GT(IV)	0.2029 (0.0075)		0.2056 (0.0256)	0.2153 (0.0188)	0.1931 (0.0086)

Table 18.6. Values of I^B for *GT* conditioned on *AS*. Standard errors in parenthesis. * means rejection of the hypothesis of equality at 5% and ** at 10%.

	Non cond.	AS(IV)	AS(III)	AS(II)	AS(I)
GT(I)	0.3161 (0.0113)	0.3278 (0.0154)	0.3204 (0.0239)	0.2735 (0.0359)	0.1981* (0.0436)
GT(II)	0.2735 (0.0095)	0.3044 (0.0241)	0.2520 (0.0181)	0.2897 (0.0194)	0.2735 (0.0236)
GT(III)	0.2565 (0.0094)	0.2950 (0.0452)	0.2090* (0.0195)	0.2839** (0.0187)	0.2868** (0.0215)
GT(IV)	0.2029 (0.0075)	0.3929** (0.1421)	0.2320 (0.0227)	0.2216 (0.0201)	0.1943 (0.0105)

The first element of a line in these matrices is the value of I^B when we condition on the *GT* class only (see Table 18.4). In each line we report the value of the index and the estimated standard errors when we also condition on, respectively, *GDP* and *AS*. For example, in Table 18.5 0.3113 is the value of I^B calculated from a transition matrix based on observations belonging to *GT* class *I* and to *GDP* class *I*.

At first glance each row of Table 18.5 shows very similar values. In fact, tests of equality between unconditioned and conditioned indices are rejected in only two cases. In addition, there are two missing values, meaning that it is impossible to estimate the corresponding matrices due to lack of data. In Table 18.6 there appears more variability across each single row and, moreover, the table is full. Tests of equality are rejects five times in this case.

From these remarks we argue that, given the information on *GT*, the information on *GDP* is scarcely relevant. In fact, similar values in the same row indicates that *GDP* is not informative in explaining *GRV*.[24] On the contrary, *AS* adds information on the *GRV* of a country. Consider for example the case

---

[24] Missing values in the matrix reinforce this conclusion.

of the measurement of GRV in GT classes I and IV: in GT class I GRV is high ($I^B = 0.3161$), but when we control for AS, we observe that it is much lower when AS is low (in AS class I $I^B = 0.1981$ and it is statistical different from 0.3161). In other words, in a "small" economy growth volatility may be large because the number of sectors is low, but volatility decreases if the agricultural sector is small. Correspondingly, in GT class IV GRV is low ($I^B = 0.2029$), but when we control for AS we observe that it is much higher when AS is high (in AS class IV $I^B = 0.3929$ and iti is statistical different from 0.2029), meaning that in a "large" economy GRV is low, but it may substantially increase if the agricultural sector is large.[25]

18.5 Conclusions

This paper investigates the relation between growth volatility and: level of development, structural change and size of the economy. The two methods used to measure growth volatility, (i) the standard deviation of the growth rate and (ii) a set of indices inspired by the literature on social mobility, substantially lead to the same results: i) growth volatility appears to be negatively related to per capita GDP (proxy for the level of development), and to the size of the economy, measured by total GDP and integration in the world market; ii) growth volatility appears to be negatively related to the share of agriculture on GDP.

When we allow for the interaction of these explanatory variables, we are led to the conclusion that structural change and size of the economy play the most important role, while the information on per capita GDP becomes uninformative. In particular the analysis of conditioned transition matrices is revealing. A closer investigation of the relation between the size of the economy and structural change in the development process can be the subject for further research.

References

1. Bartholomew DJ (1982) Stochastic models for social processes, 3rd edition. John Wiley & Son, New York
2. Bowman AW, Azzalini A (1997) Applied smoothing techniques for data analysis. Clarendon Press, Oxford
3. Canning DL, Amaral YL, Meyer M, Stanley H (1998) Scaling the volatility of GDP growth rate. Economics Letters 60:335-341
4. Fiaschi D, Lavezzi AM (2003) Explaining cross-country growth volatility. Mimeo, University of Pisa

[25] These conclusions appear to be robust to different specification of growth rate and GT classes.

5. Horvarth M (1998) Cyclicality and sectoral linkages: aggregate fluctuations from independent sectoral shocks. Review of Economic Dynamics 1:781-808
6. Maddison A (2001) The world economy: a millenium perspective. OECD, Paris
7. Pasinetti L (1981) Structural change and economic growth: a theoretical essay on the dynamics of the wealth of nations. Cambridge University Press, Cambridge
8. Pritchett L (2000) Understanding patterns of economic growth: searching for hills among plateaus, mountains, and plains. TheWorld Bank Economic Review 14:221-250
9. Quah DT (1993) Empirical cross-section dynamics for economic growth. European Economic Review 37:426-434
10. Quah DT (1996) Regional convergence clusters across Europe. European Economic Review 40:951-958
11. Ramey G, Ramey VA (1995) Cross-country evidence on the link between volatility and growth. American Economic Review 85:1138-1151
12. Scheinkman J, Woodford M (1994) Self-organized criticality and economic fluctuations. American Economic Review 84:417-421
13. Shorrocks AF (1978) The measurement of mobility. Econometrica 46:1013-1024

A Country List

Table 18.7. Country list: the smaller sample used to analyze data on Agriculture and Trade contains all the countries with the exception of: Reunion, Taiwan and Burma

AFRICA	1 Algeria	2 Angola	3 Benin	4 Botswana
5 Cameroon	6 CapeVerde	7 Cent.Afr.Rep.	8 Chad	9 Comoros
10 Congo	11 Côte d'Ivoire	12 Djibouti	13 Egypt	14 Gabon
15 Gambia	16 Ghana	17 Kenya	18 Liberia	19 Madagascar
20 Mali	21 Mauritania	22 Mauritius	23 Morocco	24 Mozambique
25 Namibia	26 Niger	27 Nigeria	28 Reunion	29 Rwanda
30 Senegal	31 Seychelles	32 Sierra Leone	33 Somalia	34 South Africa
35 Sudan	36 Swaziland	37 Tanzania	38 Togo	39 Tunisia
40 Uganda	41 Zambia	42 Zimbabwe	LATIN AMERICA	43 Argentina
44 Brazil	45 Chile	46 Colombia	47 Mexico	48 Peru
49 Uruguay	50 Venezuela	51 Bolivia	52 Costa Rica	53 Cuba
54 Dom. Republic	55 Ecuador	56 ElSalvador	57 Guatemala	58 Haiti
59 Honduras	60 Jamaica	61 Nicaragua	62 Panama	63 Paraguay
64 PuertoRico	65 Trin. Tobago	OFF WESTERN	66 Australia	67 New Zealand
68 Canada	69 United States	WEST ASIA	70 Bahrain	71 Iran
72 Iraq	73 Israel	74 Jordan	75 Kuwait	76 Lebanon
77 Oman	78 Qatar	79 S. Arabia	80 Syria	81 Turkey
82 UAE	83 Yemen	84 W. Bank Gaza	EAST ASIA	85 China
86 India	87 Indonesia	88 Japan	89 Philippines	90 S. Korea
91 Thailand	92 Taiwan	93 Bangladesh	94 Burma	95 Hong Kong
96 Malaysia	97 Nepal	98 Pakistan	99 Singapore	100 Sri Lanka
101 Afghanistan	102 Cambodia	103 Laos	104 Mongolia	105 N. Korea
106 Vietnam	EUROPE	107 Austria	108 Belgium	109 Denmark
110 Finland	111 France	112 Germany	113 Italy	114 Netherlands
115 Norway	116 Sweden	117 Switzerland	118 Un. Kingdom	119 Ireland
120 Greece	121 Portugal	122 Spain		

New Measurement Tools of the External-Constrained Growth Model, with Applications for Latin America[*]

Juan Carlos Moreno-Brid[1] and Carlos Ricoy[2]

[1] Economic Commission for Latin America and the Caribbean (ECLAC) Mexico.
jcmoreno@un.org.mx
[2] Departamento de Fundamentos del Análisis Económico, Universidad de Santiago de Compostela, Spain. aericoy@usc.es

Summary. In the spirit of the New Tools perspective, this paper presents an overview that identifies the main changes in the measurement toolkit that have been used to test the empirical adequacy of the balance of payments constrained growth model (BPC-model) for comparative macroeconomic analysis as well as for the study of the long-term constraints on economic growth. It takes into account the BPC-model in its initial formulation, as put forward by A.P. Thirlwall nearly three decades ago, as well as in its recent developments introduced to account for alternative definitions of long-term equilibrium. It shows how the theoretical revisions of the BPC-model have been accompanied by a sophistication of its empirical testing techniques, better suited for the applied analysis of long-term economic relations. In addition, the toolkit currently used within this analytical perspective is illustrated with data for the Mexican economy, based on a version of the BPC-model that explicitly focuses on the relevance of the influence of interest payments on the economys long-term external equilibrium. The paper ends with some comments on the policy implications that may be derived from the application of the BPC-model, with special reference to the case of Latin America.

19.1 The Balance of Payments Constraint on Economic Growth: A Demand Based Approach

Nearly three decades ago, Anthony P. Thirlwall introduced the Balance of Payments Constrained Growth Model (BPC-model). Based on contributions by Sir Roy Harrod on the trade multiplier, and adopting the assumption that the current account deficit cannot be indefinitely financed, the BPC-model

[*] Comments by Julio López, Esteban Pérez, Martín Puchet, and Lionello Punzo are gratefully acknowledged. The third section of this paper is based on Moreno-Brid (2003). The opinions here expressed may not necessarily coincide with those of the United Nations.

identifies the evolution of net exports as a fundamental constraint on the
long rate of economic growth (Thirlwall, 1979). His seminal model, and the
revised and extended versions developed later, see long-term economic growth
as a "demand-induced" process. They thus belong to an analytical approach
opposed to the conventional framework that sees long-term economic growth
as fundamentally determined by the rate of expansion of labor, capital and
factor productivity.

By putting Harrod's foreign trade multiplier in a dynamic context, the
BPCmodel identifies the current account of the balance of payments as the
binding constraint on the long term rate of expansion of domestic output
(Thirlwall, 1979, 1980). Such identification is particularly relevant when assessing the possibility of sustaining a rate of expansion of output that would
allow the absorption of the increase in the supply of labor, given a socially
acceptable income distribution (cf. Singh, 1977). Situations in which the BPC-
growth rate is lower than this rate -denominated say as the "socially necessary" rate of growth- tends to be "self-perpetuating", placing the economy
in a vicious circle of economic slowdown, increased unemployment and social
tension. For the case of less developed countries, such scenario of long-term of
slow growth of domestic output will likely be associated with weak or insuffi-
cient structural change and low productivity growth. These in turn will tend
to curtail international competitiveness and, therefore, tighten the external
constraint on economic growth.

Clearly this view has important implications for the design of economic
policy. In particular it does suggest that policy makers must be particularly
careful in avoiding processes of sustained and persistent appreciation of the
exchange rate in real terms. It highlights, too, the need to monitor the evolution of the balance of payments current account deficit. In the next section
we identify the main algebraic formulations that have been put forward to
specify the BPC-rate of economic growth.

19.2 Balance-of-Payments Constrained Growth Models: An Overview of Key Theoretical Developments

19.2.1 The Canonical Expression

The balance-of-payments constraint model (BPCmodel) in its most parsimonious expression that assumes constant terms of trade, identifies the long-term
rate of economic growth (y_b) compatible with external equilibrium - as defined
by a zero deficit in its current account - as the ratio of the rate of growth of
real exports (dx/x) divided by the income elasticity of its imports (ξ)[3].

---

[3] For notation purposes "dx/x" represents the average rate of growth of the variable
"x".

$$y_b = \frac{dx/x}{\xi} \qquad (19.1)$$

If the terms-of-trade are not constant, or if the Marshall-Lerner condition is not fully satisfied, the BPC-model identifies the long-term rate of economic growth compatible with external equilibrium by a slightly different expression:

$$y_b = \frac{\pi dw/w + (\eta + \xi + 1)(dp/p - dp*/p*)}{\xi} \qquad (19.2)$$

where "p" stands for domestic prices, "p*" for foreign prices expressed in local currency, "w" for the worlds real income, $\eta < 0$ and $\pi > 0$ for the price and income elasticities of exports, while $\varphi < 0$ and $\xi > 0$ for the respective elasticities of imports, and the nominal exchange rate is assumed to be constant. In turn, the income and price elasticities of imports and exports are derived from the dynamic formulations of the standard demand functions for exports and for imports:

$$dx/x = \eta(dp/p - dp*/p*) + \pi dw/w$$
$$dm/m = \varphi(dp*/p* - dp/p) + \xi dy/y$$

where "x" and "m" represent total exports and total imports at constant prices expressed in local currency.

19.2.2 The Extended BPC-model with Foreign Capital Flows

In the early eighties Thirlwall in collaboration with N.Hussain extended the BPC-model in order to capture the influence of foreign capital flows on long-term economic growth. They take as starting point the standard accounting identity of the balance of payments expressed in nominal terms as:

$$M = X + F \qquad (19.3)$$

where M, X and F stand for nominal imports, nominal exports and net foreign capital flows, all measured in current US dollars. Now, to simplify the algebraic notation, the proportion of the import bill covered by export revenues is defined as:

$$\theta = \frac{px}{p*m} \qquad (19.4)$$

Substituting equations 1.3, 1.4 and 1.6 in the balance of payments identity they arrived to the following expression of the long-term rate of economic growth given by the ratio of, on the one hand, a weighted average of the rates of change of i) net foreign capital inflows, ii) world's real income and iii) the terms of trade and, on the other hand, the income elasticity of imports:

$$y_b = \theta\pi dw/w + \frac{(1-\theta)(df/f) + \theta(\eta + \varphi + 1)(dp/p - dp*/p*)}{\xi} \qquad (19.5)$$

where $\frac{df}{f}$ stands for the rate of change of net foreign capital inflows measured in constant local prices. This extended model, as pointed out in Moreno-Brid 1998-99, has the limitation that it imposed no restriction on the trajectory of foreign capital flows - or debt accumulation. In their words:

> "In the model ., balance of payments constrained growth must be interpreted to mean nothing more than the growth rate associated with the balance of payments balancing i.e. with all debit and credits summing to zero." (Thirlwall and Hussain, 1982, p.502)

Therefore the long-term rate of economic growth given by Equation (7), and supposedly consistent with external equilibrium, does not guarantee a sustainable pattern of foreign debt accumulation. This shortcoming may be relevant because the persistent accumulation of foreign debt tends to provoke balance of payments crises that, in turn, severely affect investment and reduce the potential for economic growth.

19.2.3 Stock-flow Equilibrium in the BPC Model

To overcome such limitation, Moreno-Brid (1998-99) and McCombie and Thirlwall (1999) put forward a further revised version of the BPC-model. Their version is based on an alternative notion of long-run equilibrium defined in terms of a trade deficit given as a constant proportion of national income. With this modification, their revised BPC-model identifies the longterm rate of economic growth compatible with a sustainable and non-zero path of foreign debt accumulation as:

$$y_b = \frac{\theta\pi dw/w + (\theta\varphi + 1)(dp/p - dp*/p*)}{\xi - (1-\theta)} \qquad (19.6)$$

If the terms-of-trade are constant, (8) leads to a modified version of Thirlwall's Law:

$$y_b = \frac{\theta\pi dx/x}{\xi - (1-\theta)} \qquad (19.7)$$

Recent additions to this theoretical perspective by Dutt (2002) and Moreno-Brid (2003) explicitly capture the influence of interest rate payments on the long-run rate of economic growth compatible with balance of payments equilibrium. Such modification is important for the analysis of the long-term growth path of developing economies whose net interest payments abroad are a large component of their current deficits. Following Moreno-Brid (2003), and defining net interest payments abroad as "r" -measured in real terms, we substitute Equation (1.6) by the following two identities:

$$\theta_1 = px/p * m$$
$$\theta_2 = pr/p * m$$

Thus $\theta_1 > 0$ is the proportion of the import bill covered by export earnings, and $\theta_2 > 0$ measures net interest payments abroad as a proportion of imports at current prices. Introducing (1.10) and (1.11), and the dynamic specification of the demand functions for exports and imports - (1.3) and (1.4) - in the standard balance of payments accounting identity that explicitly identifies interest payments in the current account, leads to the following expression for the long-term rate of economic growth:

$$y_b = \frac{\theta_1 \pi dw/w - \theta_2 dr/r + (\theta_1\eta + \varphi + 1)(dp/p - dp * /p*)}{\xi - (1 - \theta_1 + \theta_2)} \tag{19.8}$$

This expression captures the influence of interest payments and, at the same time, guarantees a sustainable long-run trajectory of external debt accumulation. When the terms-of-trade are relatively constant in the long-run, a more simple expression is obtained:

$$y_b = \frac{\theta_1 dx/x - \theta_2 dr/r}{\xi - (1 - \theta_1 + \theta_2)} \tag{19.9}$$

Accordingly, if the current account deficit is zero $(1-\xi_1+\xi_2 = 0)$, the following modified version of Thirlwalls Law is obtained:

$$y_b = \frac{\theta_1 dx/x + (1 - \theta_1)dr/r}{\xi} \tag{19.10}$$

19.3 Old and New Tools for the Empirical Analysis of the Balance of Payments Constrained Growth: A Concise Overview

The evaluation of the empirical relevance of the BPC-model has by now a considerable tradition. It has gone through various changes, characterized by the use of new statistical and econometric tools in the quest to better serve in the study of long-run economic phenomena. The early empirical tests of the BPC-model, pioneered by Thirlwall and Hussain relied mainly on non-parametric methods, *inter alia* the use of rank correlation analysis in comparative studies of growth performance. Such analysis examined, for a subset of countries, the association between the rates of economic growth actually observed for given periods and the corresponding rates of growth estimated with ordinary least squares regressions (Thirlwall and Hussain, 1982). Typically these regressions took the log-level of real GDP as the endogenous variable and the log-level of real exports as well as a time trend as exogenous variables. Their non-parametric analysis of the deviation between the actual and the thus estimated rate of economic growth for selected groups of economies was rather limited in statistical power. The next phase in the empirical studies of the adequacy of the BPC-model was characterized by the use of cross-country regressions between the actual level of real GDP (y) and its estimated level (y*) obtained through regression analysis:

$$y = \alpha + \beta y*$$ (19.11)

where the null hypothesis is $\alpha = 0$, $\beta = 1$.

The conclusions of such cross-country studies, however, were heavily conditioned by the particular sample of countries chosen in the comparative analysis. The presence or inclusion of "outliers" played a crucial role in their overall conclusion. The third wave of empirical studies within this tradition relied, not on cross country comparisons, but on the econometric analysis of individual case studies, of selected economies' long-term growth. These studies relied on log-linear regressions to estimate the association between real GDP, real exports and a time trend. The analyses used annual data, sometimes expressed in terms of moving averages.

Notwithstanding their merits, these testing procedures had important shortcomings, reflecting in many cases the weakness at the time in the econometric analysis of long-term phenomena. Indeed though the BPC-model is concerned with long-run economic growth, many empirical studies in this tradition failed to verify the non-stationary properties of the time-series used in the analysis. Neither did they check for the presence of cointegration vectors. In addition, in the econometric analysis of the import and export functions they simply assumed away (instead of verifying its non significance) the influence of relative prices. The currently most accepted methodological tool for the empirical study of the BPC-model was put forward by McCombie (1997). This tool gauges the empirical relevance of the BPC-model by testing whether the long-run income-elasticity of import demand "ξ" -as estimated by modern econometric methods- does not significantly differ from its hypothetical equilibrium "ξ_H". The equilibrium level of the income-elasticity of import demand ξ_H is in turn defined as the value of the income elasticity that would equate the actual growth rate of the economy "dy/y" with its BPC-growth rate "y_b" in the period under consideration. Clearly, "ξ" must be estimated using econometric tools explicitly tailored to study long-run properties of economic phenomena. This testing procedure boils down to a verification that there is no significant difference between ξ_H and ξ . If such is the case it is then concluded that the balance of payments poses a significant restriction on the country's long term rate of economic growth.

The method is contingent on the formulation of the BPC-growth rate "y_b". Its formulation, as shown in the previous sections, differs according to the assumptions made regarding the influence of the terms-of-trade, of foreign interest rate payments and -particularly- it depends too on the adopted notion of long-term external equilibrium. The following section of this paper illustrates this methodological tool by contrasting the relevance, for the case of the Mexican economy, of three alternative formulations of the BPC-model. The first one corresponds to the canonical version of Thirlwalls Law, the second to its revised version that guarantees a sustainable path of foreign indebtedness, and the third explicitly captures the influence of foreign interest payments.

19.4 Applying the new Methodological Tool to the BPC Model: the Mexican Case

The first task to test the BPC-model using Mc.Combie's methodological tools is to calculate the hypothetical equilibrium values of the income-elasticity of imports. For the canonical version of Thirlwalls Law, such value (ξ_T) is defined as the ratio of the actual growth rate of exports and of GDP:

$$\xi_T = (dx/x)/(dy/y) \tag{19.12}$$

For the revised version of the BPC-model consistent with a stock-flow notion of longrun equilibrium -defined in the first section- the hypothetical equilibrium elasticity may be derived from Equation (9) by substituting in it the actual average growth rate of real GDP "dy/y" for the BPC-growth rate "y_b", and then solving for ξ. The value thus obtained is here denoted as ξ_X:

$$\xi_x = (1 - \theta) + [(\theta dx/x)/dy/y] \tag{19.13}$$

Finally for the most recent version of the BPC-model commented above, the hypothetical equilibrium elasticity of imports, derived from Equation 1.13 is ξ_M

$$\xi_M = (1 - \theta_1 + \theta_2) + [(\theta_1 dx/x - \theta_2 dr/r)/dy/y] \tag{19.14}$$

All calculations of ξ_T , ξ_x and ξ_M assume that the terms-of-trade are not important determinants of the economys long-run growth rate in the period of analysis. The next step was to estimate Mexicos long-term import demand using Johansens cointegration method, explicitly allowing for the effects of non-tariff restrictions (a very popular policy in Mexico until the late 1980s). Though not here reported, Dickey Fuller tests were applied to check the stationarity properties of the data. The Akaike Information and Schwartz Bayesian criteria served to estimate the optimum lag for the unrestricted VAR system for import demand. And Lagrange Multiplier tests were conducted to check for residual serial correlation of the VARs individual equations. The results of Johansen tests, under the assumption of an unrestricted intercept, identified one cointegration vector for import demand. With an intercept restricted to the cointegrating space, the vector corresponding to the largest eigenvalue was chosen. In any case, the respective coefficients were very similar, reporting an estimated long-run income elasticity β_y around 1.8, a long-run price elasticity β_p close to -0.5 and an estimated parameter for the long-run effect of import permits β_q around -1.0. The results of Johansens tests assuming an unrestricted intercept identified one cointegrating vector among the log-levels of GDP and of imports and the index of non-tariff restrictions "q" (See Table 1, part B). The estimated long-run income elasticity of import demand was $\beta_y = 1.772$, practically the same as the corresponding estimate obtained using the larger VAR-system.

Having calculated the estimated long-term income elasticities of imports - via cointegration analysis- the final step is to check whether it is significantly different from its alternative hypothetical equilibrium values given by ξ_T , ξ_x or by ξ_M. Using official data of Mexicos real GDP and real exports, and measuring θ_1 and θ_2 as the ratios at the beginning of the period, the hypothetical equilibrium value of the income-elasticity of import demand during 1967-99 were: $\xi_T = 2.189, \xi_x = 1.991$ and $\xi_M = 1.913$.2 Though these three values are, apparently, not too distant from the estimated coefficient of 1.777 obtained as the long-run income elasticity of import demand via Johansens techniques (See Table 2) or the estimated $\xi = 1.772$ derived by the cointegration tests applied on the trivariate VAR-system excluding relative prices, the significance of such differences must be formally tested.

19.5 Conclusions

The tests on the over-identifying restriction Ho: $\xi = \xi_T$ imposed on the cointegrating vector for the full VAR-system (including relative prices) reject the null hypothesis at a 5% critical level of significance (See Table 2). Such result suggests that Thirlwalls Law, in its canonical formulation, does not offer an adequate interpretation of Mexicos longrun economic growth during 1967-99. For the alternative definition of the BPC-growth rate that allows for a long-run stock/flow equilibrium position, the results do not reject the hypothesis $\xi = \xi_x$ even at a 10% level of significance. The results were even stronger for the tests on the BPC-model that allowed for the influence of interest payments abroad. Such result should not be surprising given the enormous volume of interest payments that Mexico incurred during part of the period analyzed. Thus there is strong support to the modified versions of Thirlwalls Law as a relevant hypothesis for the Mexican case. These results may help to claim that the new generation models -and the new methodological tools- recently introduced may strengthen the empirical relevance of the theory of balance-of-payments constrained growth economies.

Table 1: Mexico's long-term import demand,1967-99.
(Estimated with Johansen's cointegration procedure) (a)

$$\ln(m) = \alpha + \rho_y \ln(y) + \rho_p \ln(p) + \rho_q q + v$$

Part A. Results for Var(1) system with three endogenous variables: ln(m), ln(y) and ln(p)

	Test on Max eigenvalues			Test on trace			Cointegration vector and chi test on the significance of ρ_p and ρ_q
	H_0	H_1	LRS	H_0	H_1	LRS	
unrestricted intercept ($\alpha = 0$)	r=0	r=1	48,7*	r=0	r$\geq$1	67,7*	$A_1.\ln(m) = 1.777 \ln(y) - 0.536 \ln(p) -1.044(q)$
	r$\leq$1	r=2	14,1	r$\leq$1	r$\geq$2	19	(0,12) (0,27) (0,16)
	r$\leq$2	r=3	4,9	r$\leq$2	r=3	4,9	$\chi^2[\rho_p = 0]p - value = 0.193$ $\chi^2[\rho_q = 0]p - value = 0.000$
restricted intercept ($\alpha \neq 0$)	r=0	r=1	72,2*	r=0	r$\geq$1	103,9*	$A_2.\ln(m) = 1.872 \ln(y) - 0.577 \ln(p) - 0,983 (q) - 6.52$ (b)
	r$\leq$1	r=2	20,8*	r$\leq$1	r$\geq$2	31.8*	(0,15) (0,28) (0,15) (2,16)
	r$\leq$2	r=3	11.0	r$\leq$2	r=3	11	$\chi^2[\rho_p = 0]p - value = 0.187$ $\chi^2[\rho_p = 0]p - value = 0.000$

Part B. Results for VAR(1) system excluding the relative price of imports ln(p)

	Test on Max eigenvalue			Test on trace			Cointegration vector
	H_0	H_1	LRS	H_0	H_1	LRS	
unrestricted intercept ($\alpha = 0$)	r=0	r=1	41.7*	r=0	r=1	49,9*	$B_1 . \ln(m) = 1,772 \ln(y) - 1,269\ q$
	r$\leq$1	r=2	8.3	r$\leq$1	r=2	8,3	(0,18) (0,19)
restricted intercept ($\alpha \neq 0$)	r=0	r=1	56,9*	r=0	r=1	75,2*	not available (b)
	r$\leq$1	r=2	18,3*	r$\leq$1	r=2	18,3*	

Notes: (a) Tests carried out assuming no deterministic trend and taking the coverage of import licence requirement (q) as an exogeneous I(1) process. (b) In part A, when two cointegrating vectors were identified, the one associated with the largest eigenvalue is here reported. In part B, since there are only two endogenous variables there can be at most one linearly independent cointegrating relation between them. The identification of two such vectors by Johansen tests may reflect specification error in the VAR system. H_0 = null hypothesis, H_1 = alternative hypothesis, r = number of cointegrating vectors, LRS = likelihood ratio statistics, y = real GDP, m = real imports, p=ratio of the implicit price deflators of imports relative to domestic output, q = production-weighted coverage of import licences. An asterisk(*) denotes significance with a 5% critical level. Asymtotic standard errors of the estimatedcointegration coefficients are reported in parenthesis. Source: Own calculation with Microfit.

Table 2
Test of the empirical relevance of Thirlwall's Law (original and extended versions) for the Mexican economy , 1967-99
(based on McCombie's procedure)

VAR-system for import demand	Income elasticity of import demand				LRS-tests of equality of the long-run income-elasticity and its hypothetical equilibrium values (p-values) [a]		
	Johansen's cointegration coefficient	Hypothetical equilibria consistent with Thirlwall's Law as expressed in the:			null hypothesis		
		original BPC-model[b]	extended BPC-models[c]				
	ξ	ξ_T	ξ_x	ξ_M	$\Xi = \xi_T$	$\xi = \xi_x$	$\xi = \xi_M$
A. With four variables ln(m), ln(y), ln(p) and q	1.777	2.189	1.991	1.913	0.048	0.177	0.337
B. With three variables [d] ln(m), ln(y) and q	1.772	2.189	1.991	1.913	0.072	0.282	0.468

Notes: (a) p-values of the χ^2 corresponding to the LRS to test the over-identifying restriction equalizing the cointegrating coefficient for the income-elasticity of import demand to its hypothetical equilibrium derived from three versions of Thirlwall's Law. (b) ξ_T is derived from Equation 1.1, (c) ξ_x is derived from Equation 1.2 taking θ (the export/import ratio) reported for the beginning of the period. ξ_M is derived from Equation 1.3 calculated with the values of θ_1 and θ_2 given by ratio of exports to imports and of interest payments abroad to imports reported at the beginning of the period ..."

References

1. Blecker R.A. (1999) Taming Global Finance, Washington: Economic Policy Institute.
2. Dutt, A. (2001), "Income elasticities of imports, north-south trade and uneven development", Department of Economics, University of Notre Dame, unpublished.
3. Enders, W. (1995), Applied Econometric Time Series, New York: John Wiley.
4. Kaldor, N. Further Essays on Economic Theory. London: Duckworth. 1978.
5. Kaldor, N. "Discussion", in D. Currie, R. Nobay, D. Peel (eds.) Macroeconomic Analysis. London: Croom Helm. 1981.
6. McCombie, J.S.L. (1998), "Harrod, economic growth and international trade", in G.Rampa et al (Eds.) Economic Dynamics, Trade and Growth, London: McMillan Press.
7. McCombie, J.S.L. (1997), "On the empirics of balance-of-payments-constrained growth", Journal of Post Keynesian Economics, Vol.19, pp.345-375.
8. McCombie, J.S.L. (1989), "Thirlwalls Law and balance-of-payments-constrained growth: a comment on the debate", Applied Economics, Vol.21, pp.611-629.
9. McCombie, J.S.L. and A.P.Thirlwall (1999), "Growth in an international context: a Post Keynesian view" in J.Deprez and J.T.Harvey (Eds.) Foundations of International Economics: Post Keynesian Perspectives, London: Routledge.
10. McCombie, J.S.L. and A.P.Thirlwall (1997), "Economic growth and the balance-ofpayments constraint revisited", in Arestis et al (Eds.) Markets,Unemployment and Economic Policy, New York: Routledge.
11. McCombie, J.S.L. and A. P.Thirlwall (1994), Economic Growth and the Balance of Payments Constraint, New York: St.Martins Press.
12. Moreno-Brid, J.C. (1998-99), "On capital flows and the balance-of- payments constrained growth model", Journal of Post Keynesian Economics, Vol.21, pp.283- 289.
13. Moreno-Brid, J.C. (2001), Essays on Economic Growth and the Balance of Payments Constraint; with special reference to the case of Mexico, Ph.D. dissertation, Faculty of Economics and Politics, University of Cambridge, United Kingdom.
14. Moreno-Brid, J.C. (2003), Capital Flows, Interest Payments and the Balance-of-Payments Constrained Growth Model; a theoretical and an empirical analysis, Metroeconomica, Vol.54, May/September, No.2 & 3.
15. Singh, A. "UK industry and the world economy: a case of de-industrialisation", Cambridge Journal of Economics, Vol., 1, pp. 113-136, 1977.
16. Thirlwall, A.P. (1998), "The balance of payments and growth: from mercantilism to Keynes to Harrod and beyond", in G. Rampa, et al (Eds.) Economic Dynamics,
17. Trade and Growth, London: McMillan Press.
18. Thirlwall, A. P. Balance of Payments Theory and the United Kingdom Experience. London: Macmillan. 1980.
19. Thirlwall, A.P. (1979), "The balance of payments constraint as an explanation of international growth rates differences", Banca Nazionale del Lavoro, Quarterly Review, Vol.128, pp.45-53.

20. Thirlwall, A.P. and M.N. Hussain (1982), ''The balance of payments constraint, capital flows and growth rates differences between developing countries'', Oxford Economic Papers, Vol.34, pp.498-509.

The Fractal Structure, Efficiency, and Structural Change: The Case of the Mexican Stock Market*

Guillermo Romero-Meléndez, Mauricio Barroso-Castorena, Jorge Huerta-González, Manuel Santigo-Bringas, and Carlos Alberto García-Valdéz

Universidad de las Américas Puebla,
Hda. Sta. Catarina Mártir,
72820, Puebla, México,
grome@mail.udlap.mx

Summary. In the first part of this paper we present experimental evidence that the Mexican stock market has a fractal structure. We obtained the experimental results by using the Matsushita-Ouchi method, the box-counting method, and the fractal image compression technique of M. Barnsley. The results obtained by applying the Matsushita-Ouchi technique to the returns of the Indice de Precios y Cotizaciones (IPC) of the Mexican stock market support the assertion that the returns of the IPC behave like a fractional brownian motion. The box-counting method allows us to calculate the dimension of the graphs of the IPC, and its Hurst exponent H. The application of the fractal image compression technique produces attractors which are closed to the graphs of the returns of the IPC, and has permited us to estimate H. In the second part of the paper, we present two applications: first we study the efficiency graphs of the Mexican stock market, by plotting the Hurst exponent H as a funtion of time, and then we localize its structural changes by making use of a fractal attractor located in the phase space: H-IPC. We show that the fall of the IPC, that occurred between 1994 and 1995, corresponded to a rise of the value of H. We localize approximately the structural changes of the Mexican economy between 1987 and 1996.

20.1 Introduction

Fractional brownian motion (fbm), also called Biased random walk or Fractal time series was studied by Hurst in the 1940's, and by Mandelbrot in the 1960's and 1970's. (See [Hurts] and [Mandelbrot & Van Ness]). Its characteristic exponent H, called Hurst exponent, has been calculated for several

---

* This paper was read in the Workshop: "New Tools of Qualitative Analysis in Economic Dynamics", held Siena, Italia, in 2000

international markets, including those of U. K., Japan and Germany. The applications of the fbm include the study of volatility, risk and market efficiency. (See [Peters, Chapter 8, Table 8.3]). In the first part of this paper, we present the results obtained when we applied a method of Matsushita and Ouchi to the returns of the Mexican stock market, in order to verify the selfaffinity of the graphs of their returns. (See [Matsushita & Ouchi]). We find that these returns behave like a fractional brownian motion, and we compute its Hurst exponent by different methods, including the Box-counting method and the fractal image compression image technique of Barnsley. (See [Barnsley & Sloan]). In the second part of this paper, we present an application to study the evolution of the efficiency of the Mexican stock market and to localize its structural changes. We use a fractal attractor situated in the phase-space: H-IPC, where H denotes the Hurst exponent, and IPC denotes the returns of the Mexican stock market. [2]

20.2 The Matsushita-Ouchi Method Applied to the Mexican Stock Market

Matsushita and Ouchi have developed a numerical method to analize the self-affinity of curves such as noise and topographical curves. (See [Matsushita and Ouchi]). The method for a curve in a two dimensional space, given by $y = f(x)$, is the following: Take the smallest scale a_0 and measure by this scale the length Na_0 between two arbitrary points A and B of the curve. Then calculate x- and y- variances X^2 and Y^2 for all measured points of the curve between A and B:

$$X^2 = \frac{1}{N} \sum_{i=1}^{N} (x_i - x_c)^2 , \qquad Y^2 = \frac{1}{N} \sum_{i=1}^{N} (y_i - y_c)^2 ,$$

with

$$x_c = \frac{1}{N} \sum_{i=1}^{N} x_i , \qquad y_c = \frac{1}{N} \sum_{i=1}^{N} y_i ,$$

where (x_i, y_i) is the coordinate of the i-th measured point on the curve. Repeat the procedures described above for many pairs of points on the curve. Then we have the following behavior:

---

[2] The first part of this paper was included in the thesis work of the third and the fourth authors. (See [Huerta & Santiago]). The second part is included in the thesis work of the second and the fifth authors. (See [Barroso] and [García]). These thesis works were realized under the advice of the first author, and were presented in the Universidad de las Américas-Puebla. The work of the first part was supported by the Consejo Nacional de Ciencia y Tecnología (CONACYT) ls of México. We are very grateful to Juan Antonio Navarro for his invaluable help to write this paper in Latex.

$$X \sim N^{\nu_x}, \qquad\qquad Y \sim N^{\nu_y}.$$

For self-similar fractal curves we have: $\qquad \nu_x = \nu_y.$

For self-affine fractal curves we have: $\qquad \nu_x \neq \nu_y.$

For fractional brownian motion we have: $\qquad \nu_x = 1$, and $H = \nu_y.$

We take here $x = t$ (time). We applied the Matsushita-Ouchi method for the returns of the Indice de Precios y Cotizaciones of the Mexican stock market (IPC), from January 2., 1987 to June 25., 1992, in the following four periods:

Period 1: from January 2., 1987 to May 31., 1988.
Period 2: from June 1., 1988 to October 23., 1989.
Period 3: from October 24., 1989 to March 27., 1991.
Period 4: from April 1., 1991 to June 25., 1992.

The results we obtained are:

	Period 1	Period 2	Period 3	Period 4
ν_y	0.6076	1.0210	0.5292	0.8542
ν_x	0.9662	1.0049	0.9591	1.0014

We see that the value of ν_x for the four periods is very near to 1. Thus we consider this fact as experimental evidence that the Mexican stock market behaves like a fractional brownian motion.

20.3 The Hurst Exponent H and the Fractional Brownian Motion

We can explain the fractional Brownian motion as a function $V_H(t)$, with the property:

$$\Delta V_H \propto \Delta t^H ,$$

with $\Delta t = t_2 - t_1$, $\Delta V_H = V_H(t_2) - V_H(t_1)$, and $0 < H < 1$. Here H is the Hurst exponent. For $H = \frac{1}{2}$ we obtain Brownian motion, for $H > \frac{1}{2}$ we have persistence, and for $H < \frac{1}{2}$ we have antipersistence. More exactly the fractional Brownian motion is defined as follows. (See [Falconer,16.2]):

Fractional Brownian motion is a random process $V : [0, \infty] \to \mathcal{R}$ on some probability space such that:

(i) With probability 1, $V(t)$ is continuous and $V(0) = 0$.
(ii) For any $t > 0$, and $h > 0$, the increment $V(t + h) - V(t)$ has the normal distribution with mean zero and variance h^{2H}.

The graph of a fractional brownian motion with Hurst exponent H has Hausdorff and box dimension $2 - H$. (See [Falconer, 16.7]).

20.4 Other Methods to Calculate H: The Box-counting Method, and the Fractal Image Compression Technique

We present here two more methods to calculate H. The first one is the box-counting method to calculate the box dimension, and the second one uses the fractal image compression technique.

20.4.1 Values of H Obtained by Box-counting Method

The box-counting method to calculate the fractal dimension of a curve is well known: Cover $\mathcal{R}^2$ or a square, containing the curve as a subset, by closed boxes of side lenght $1/2^n$. Let $\mathcal{N}_n$ denote the number of boxes which intersect the curve. Then the fractal dimension of the curve is:

$$D = \lim_{n \to \infty} \left\{ \frac{\ln(\mathcal{N}_n)}{\ln(2^n)} \right\}.$$

(See [Barnsley, 5.1, Theorem 1.2]). We applied this method to the graphs of the four studied periods of the IPC, and the corresponding values of H obtained by the relationship: $H = 2 - D$ are:

	Period 1	Period 2	Period 3	Period 4
H	0.910	0.891	0.886	0.810

20.4.2 The Fractal Image Compression Technique of Barnsley and Sloan

The fractal image compression technique of Michael Barnsley and Alan D. Sloan makes use of iterated function system codes and in the best cases can achieve a compression ratio of 10,000 to 1. (See [Barnsley & Sloan]). This method consists of the following steps:

1. Localize the parts I_i of the image I to be compressed, which are similar to I.
2. Find affine transformations $W_i : \mathcal{R}^2 \to \mathcal{R}^2$,

$$W_i \begin{pmatrix} x \\ y \end{pmatrix} = \begin{pmatrix} a_i & b_i \\ c_i & d_i \end{pmatrix} \begin{pmatrix} x \\ y \end{pmatrix} + \begin{pmatrix} e_i \\ f_i \end{pmatrix},$$

with the property: $W_i(I) = I_i$.
3. Associate a probability p_i with each transformation W_i determinig the relative size of I_i. This can be done as follows:

$$p_i = |a_i d_i - b_i c_i| / \sum_{i=1}^{n} |a_i d_i - b_i c_i| \, .$$

Here n is the number of affine transformations.

4. Choose one point (x_0, y_0) of I, and apply the following steps:
   i) Initialize $x = x_0$, $y = y_0$.
   ii) For $m = 1$ to 2500, do steps (iii) to (vii).
   iii) Choose k to be one of the numbers $1, 2, \ldots, n$ with probability p_k.
   iv) Apply the transformation W_k to (x, y) in orden to obtain (x', y').
   v) Set $(x, y) = (x', y')$.
   vi) If $m > 10$, plot (x, y).
   vii) Loop.

This method produces an attractor which is close to the given image. The theoretical foundation of this technique is the Collage Theorem. (See [Barnsley, 3.10, Theorem 1]). This theorem tell us that the more accurately the image I is described by the union of the I_i's, the more accurately the attractor will aproximate the image I. The collection of the transformations is called an iterated function system (IFS). The IFS code of an image I consists of the coefficients of the transformations W_i described above, together with its probabilities p_i.

We applied this method to the graphs of the four mentioned periods of the IPC. We show here only the graph of the first period (Graph 1), and the attractor produced by the fractal image compression technique (Graph 2):

The IFS code of the attractor shown in graph 2 is the following:

There is a method to calculate the fractal dimension of the attractor produced by an IFS like the one produced above. Theorem 1 in [Barnsley, 6.3] tell us that its fractal dimension is the unique real solution D of the equation:

$$\sum_{i=1}^{n} |d_i| \, a_i{}^{D-1} = 1 \, .$$

We know that for this kind of attractor the Hausdorff dimension and the fractal dimension are equal. (See [Barnsley, 5.4, Theorm 3]). Then we can obtain the value of H for the four studied periods of the Mexican stock market, using the dimension of the attractor produced by the method explained above. The results are:

We observe that these values of H are very close to the respective values obtained with the box-counting method. Nevertheless, the values of H (ν_y) obtained by the method of Matsushita and Ouchi are not close to the values obtained by the last two methods. We will use mainly the box-counting method in this work.

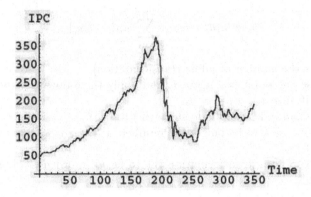

Fig. 20.1. GRAPH 1

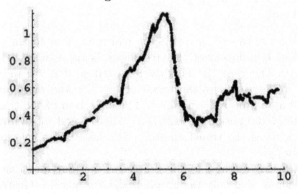

Fig. 20.2. GRAPH 2

Table 20.1. IFS CODE OF GRAPH 2

a	b	c	d	e	f	p
0.09	0.00	0.01	-0.02	0.00	0.15	0.02
0.07	0.00	0.01	-0.06	0.88	0.23	0.04
0.09	0.00	0.01	-0.04	1.58	.31	0.03
0.20	0.00	0.05	-0.19	2.45	0.45	0.36
0.08	0.00	0.03	0.10	4.38	0.79	0.08
0.03	0.00	-0.01	0.05	5.20	1.14	0.02
0.02	0.00	-0.03	-0.10	5.50	1.09	0.01
0.02	0.00	-0.01	-0.08	5.65	0.76	0.01
0.02	0.00	-0.01	-0.05	5.80	0.58	0.01
0.02	0.00	-0.01	-0.08	5.98	0.51	0.02
0.06	0.00	0.00	-0.08	6.20	0.41	0.04
0.10	0.00	0.03	-0.15	6.75	0.37	0.14
0.07	0.00	-0.01	0.15	7.70	0.53	0.10
0.04	0.00	-0.01	0.06	8.40	0.54	0.02
0.10	0.00	0.01	-0.09	8.80	0.54	0.09

	Period 1	Period 2	Period 3	Period 4
H	0.907	0.920	0.882	0.783

20.5 An Application: Efficiency Graphs and Structural Changes in the Mexican Economy

We present here an application of the fractional brownian motion to obtain efficiency graphs and to localize structural changes in the Mexican economy.

20.5.1 The Hurst Exponent and the Efficency of Stocks Markets

According to Peters, "An efficient market is one in which assets are fairly priced according to the information available, and neither buyers, nor sellers have an advantage". (See [Peters, p. 5]). The efficiency of a stock market is related with its randomness: If the market follows a random walk, then it is efficient (See [Peters, p. 14]).

Because of the value of the Hurst exponent indicates the randomness of a motion, we have the following relationship between the Hurst exponent and the efficiency of a market: The closer H is to $1/2$, the more efficient is the market. The closer is H to 1, the less efficient is the market. Thus the evolution of the Hurst exponent indicates the evolution of the market.

Graph 3 below show the evolution of the IPC of the Mexican stock market from 1987 to 1996, and graph 4 show the evolution of its efficiency. The second one was obtained plotting, for each time t, the value of H corresponding to the part of the first graph situated between t and six months before.

We can observe that the last fall of the IPC ocurred between 1994 and 1995 (time coordinate between 1800 and 1900) corresponds with a rise in the value of H.

20.5.2 Localizing Structural Changes in the Mexican Economy

We present here as an application, the localization of the structural changes of the Mexican economy occurred between 1987 and 1996. The method we used consisted in the following steps:

1. To graph the value of the IPC of the Mexican stock market against its Hurst exponent H. We obtained the following attractor in the phase space $H - IPC$:
   This attractor has fractal dimension 1.48, therefore it is a fractal.
2. To calculate the mass center of the attractor obtained in 1.- we got the point (1363, .8)- , and localize the dates with these H and IPC values. We localize four data with values close to the corresponding values of the mass point. Three of them are situated in October 1991, and one in December 1991. Graph 6 and graph 7 show such data with vertical lines.

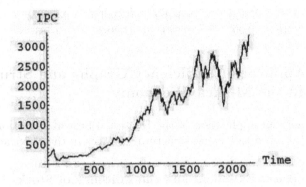

Fig. 20.3. GRAPH 3

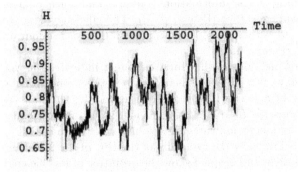

Fig. 20.4. GRAPH 4

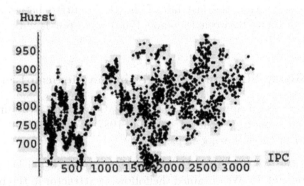

Fig. 20.5. GRAPH 5

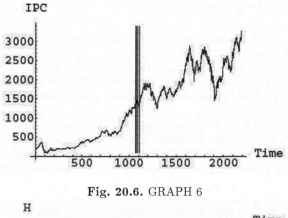

Fig. 20.6. GRAPH 6

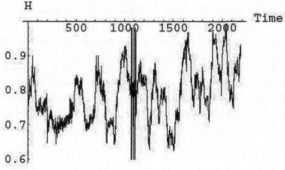

Fig. 20.7. GRAPH 7

Professor Gonzalo Castañeda, economist of the Universidad de las Américas - Puebla, told the first author, that between 1989 and 1991 several structural changes took place: Mexican banks were sold by the goverment, credit policy turned more liberal, and the bank assets were expanded. These facts validate our result. We recommend the use of this method, when there exists only one episode of structural changes in the studied period.

20.6 Conclusions

We obtained experimental evidences that the Mexican stock market has the fractal structure of a fractional brownian motion, using the method of Matsushita and Ouchi. We calculated the Hurst exponent of the Mexican stock market by three different method, in two of those methods we obtained very similar values. As an applications we obtained a graph of the efficiency of the Mexican market, and we localized approximately the structural changes of the Mexican economy between 1987 and 1996.

References

[Barnsley and Sloan] Barnsley, M. F. and Sloan A. D. *A Better Way to Compress Images*, Byte 215, pp. 215-223. 1988.

[Barnsley] Barnsley, M. F. *Fractals Everywhere*, Academic Press, 1988.

[Barroso] Mauricio Barroso-Castorena. *Atractores Fractales de las Gráficas del IPC en un Espacio Fase*, Thesis work. Licenciatura en Actuaría. Universidad de las Américas-Puebla, 2000.

[Falconer] Falconer, K. *Fractal Geometry*, John Wiley and Sons, 1990.

[García] Carlos Alberto García-Valdéz. *Algorítmos de Extrapolación para el Movimiento Browniano Fraccionario y sus Aplicaciones a la Economía*, Thesis work. Licenciatura en Actuaría. Universidad de las Américas-Puebla, 1996.

[Huerta & Santiago] Huerta-Gonzalez J. and Santiago-Bringas M. *Modelación Fractal del Mercado Accionario Mexicano y sus Implicaciones en la Teoría de la Eficiencia*. Thesis work. Maestría en Dirección y Administración de Empresas. Universidad de las Américas-Puebla, 1992.

[Hurst] Hurst, H. E. *Long-term Storage of Reservoirs*, Transactions of the American Mathematical Society of Civil Engineers 116, 1951.

[Mandelbrot & van Ness] Mandelbrot, B. and van Ness, J. *Fractional Brownian Motion, Fractional Noises and Applications*, SIAM Review 10, 1968.

[Matsushita & Ouchi] M. Matsushita and S. Ouchi. In *Dynamics of Fractal Surfaces*. Edited by F. Family and T. Vicsek, World Scientific 1991, pp. 66-71.

[Peters] Peters, E. *Chaos and Order in the Capital Markets*, John Wiley and Sons, 1996.

Processes of Evolutionary Self-Organization in High Inflation Experiences

Fernando Tohmé, Carlos Dabús, Silvia London[1]

CONICET Departamento de Economía, Universidad Nacional del Sur
12 de Octubre y San Juan,
8000 Bahía Blanca, Argentina ftohme,cdabus,slondon@criba.edu.ar

Summary. We study some features of the processes that have generated high inflation in Latin- American countries. The statistical evidence shows that these inflationary experiences are *fractional brownian* noises. Several authors showed that self-organized criticality (SOC) processes may constitute the best explanation of the origin of such noises. But this hypothesis requires that the underlying structure remains timeinvariant. We conjecture, instead, that the economic structures evolve in time being, at each stage of their evolution, self-organized structures. We find that such ESO (evolutionary self-organized) processes still generate fractional brownian noises. Thus, they seem to provide a better explanation for the economic phenomenon of high inflation.

21.1 Introduction

The Latin-American inflation episodes of the last decades showed a remarkable persistence, peaked by periods of extremely high inflation. This makes them interesting objects of study. Despite the fact that those levels of inflation seem byproducts of a past era, they still deserve a careful analysis. We presented elsewhere a hypothesis about the behavior of the notorious Argentinean inflation: that it was the result of a critically self-organized (SOC) process (Tohmé et al., 1995). The main phenomenological property of such a system is that it generates time series behaving like $\frac{1}{f}$ noises. The hypothesis is that the underlying economic structure that determines the price levels in the economy is in a permanent critical state, such that any shock can generate a non-correlated and rather unbounded inflationary response. We showed data that made this contention plausible. This is also intuitively plausible, since SOC processes exhibit "avalanche-like" behavior, and this is consistent with the evidence of hyperinflationary burst. From November 1988 to July 1989 the inflation rate in Argentina jumped from 5.7% to 196%.

Nevertheless, there are two main problems with this line of analysis. On one hand, it lacks a solid econometric support. On the other, to accept the SOC hypothesis involves to assume some properties which have a very low likelihood when interpreted in the context of real economies: SOC processes can generate fluctuations of all sizes in response to uniformly distributed exogenous shocks. If this were true, it would imply that an economy like Argentina's is prone, at any moment, to generate wild and unexpected bursts of inflation, contrary to the fact that prices remained stable for almost the entire decade of 1990. This is because of another feature of a SOC process, namely that a single process generates the entire series of its outcomes. Again, this seems counterintuitive in the case of a real world economy for a period of more than thirty years since it would mean, if true, that the institutional, political, social and economic rules that determine the behavior of inflation remain invariant during such a long period of time. Therefore, to accept the SOC hypothesis for inflation means to assume that the economy has a strong tendency to generate inflation. Moreover, if this is so, any level of inflation, no matter how high, should be expected.

In order to widen the domain of evidence we analyze two other Latin-American high inflation experiences, Chile and Colombia. The first issue to address is whether the inflation rates behave like $\frac{1}{f}$ noises, the key signature of SOC processes. We find that this kind of behavior is indeed a common trait of these economies. The common characteristic of being $\frac{1}{f}$ noises makes the different inflationary experiences very similar in a crucial aspect: the statistical features of their respective time series of observations. However, this seems odd, especially when we consider the remarkable historical, political and economic differences among the countries in this small sample. Hence, even if our database has been slightly expanded, we have to face a hard-to-believe set of conclusions. Moreover, we still lack a strong econometric tool to make sure that this is not just a statistical artifact. In other words, we face the necessity of justifying our methodology of analysis. In this sense, we follow a method of making conjectures and looking for refutations similar to Lakatos (1976). That is, we just make a claim without being "ontologically" committed to it and then we confront that conjecture with data and theoretical arguments.

The alternative hypothesis we will confront to the SOC hypothesis is based on a simple insight: it may be possible that the underlying economic structure that generates inflation consists of a sequence of SOC processes, instead of a single one. We call such a process an Evolutionary Self-Organized (ESO) system. In fact, $\frac{1}{f}$ noises are generated also in this case. Therefore, the ESO hypothesis is able to explain the same statistical information that can be concluded from the SOC hypothesis. Nonetheless, instead of obeying a single set of behavioral rules, an ESO process goes through an evolutionary process, characterized by sudden (but sporadic) changes of structure. This seems to be a better explanation of the behavior of the economies under study, since these have exhibited a history of erratic and unstable political and economic conditions. In other words, the ESO hypothesis supersedes the SOC conjecture since it explains equally well the same set of data, being at the same time far less vulnerable, because of its generality, to attacks based on political or institutional criteria.

In this paper we analyze the three inflationary series (Argentina, Chile and Colombia) under the alternative SOC and ESO hypotheses. We show that both are similar in their main predictions and that those are verified by the Latin-American inflations. The next step in our argument consists in applying a version of Occam's Razor, in order to retain only the less demanding of both hypotheses and to drop the other. This procedure lead us to a general explanation of inflationary processes, with less precision and predictive capabilities than any of the known theories of inflation in macroeconomics. This seems to go beyond the limits of economic analysis, and directly into the field of Systems Theory. In the next section we make precise the methodological foundations that allow us to support the idea that the high inflation processes in Latin-America have been ESO processes. Section 1.3 presents a quick refresher of the key notion of $\frac{1}{f}$ noise and its relation to SOC processes. Section 1.4 defines and characterizes what we call ESO processes, the core of our alternative hypothesis. In Section 1.5 we analyze and discuss the price data, the regimes of inflation of each economy and the overall behavior of the inflation series. We look for evidence that may favor either the SOC or ESO hypothesis. Finally, section 1.6 presents the conclusions. We find the data compatible with both SOC and ESO hypotheses. Since most of the inflationary processes under study have clearly suffered structural changes, the ESO hypothesis appears to be a better explanation for the statistical properties of the series.

There exists a wide consensus about the influence of money over inflation. Since Fisher's quantitative theory of money it is accepted that if money is pumped into the economy, the level of prices will tend to rise.

21.2 Methodological Foundations

Economics has long ago adopted a clear and precise *inductive* method for accepting or rejecting hypotheses. It begins with the formulation of a hypothesis, that should be made quantitatively precise (i.e. it must be a sentence about the magnitude of certain variable). Then, a sample from the population over which the hypothesis is predicated must be provided. The magnitude of the variable under analysis is measured for every element in the sample, so that a sample distribution of its values obtains. Then, the hypothetical value is placed in this distribution. If it lies more than a given number of standard deviations from the sample's average the hypothesis is rejected (in a probabilistic sense). On the other hand, the two traditional ways of making deductive inferences are *modus ponens* (from $A \rightarrow B$ and A it follows the validity of B) and modus tollens (from $A \rightarrow B$ and $\neg B$ it follows $\neg A$). If we define A to be the statement "the real value of variable X is a" and B, "the sample value of X is b", from the rejection of B and $A \rightarrow B$ it follows the rejection of A, while from the non-rejection of B nothing can be inferred about the validity of A. This explains why the main hypothesis to be tested is the one that the analyst is interested in rejecting.

This discussion shows implicitly that in order to test a hypothesis, we need first to state it clearly (which allows also to make clear its negation). Second, a sufficiently large sample is needed to ensure that the sample distribution approximates (asymptotically) the real distribution of the population. These requirements can be fulfilled in case the hypothesis, as said, is about the magnitude of a variable. But when the hypothesis is of a higher order, say about the structure underlying a set of observations, this is no longer so clear. In fact, a certain number of crucial assumptions about the distribution of values in the real population, which are assumed typically in empirical analyses, cannot be either assumed nor verified. Known examples of how very different hypotheses about the distribution of a single population can be all accepted, with the same degree of precision, show that this is not a minor problem when working with data (e.g. Tufte, 1983).

In that sense, if different hypotheses about the structure are all verified by the same database, then the method based on modus tollens, i.e. trying to reject the complement of the hypothesis of interest, is no longer tenable. This is because, unlike the magnitudes of a variable, a set of possible structures does not always match observable partitional information. To make this clear, consider just an example. The statement "variable X has value a" induces a partition on the domain of values of the measurable variable X in two sets: $\{a\}$ and $\{x \in Dom(X) : x \neq a\}$. Instead, a more complex statement, like "$f(x)$ is concave", on a population where only the first derivative at each point can be evaluated, generates a partition on its domain (a functional space F) : $\{f \in F : f \text{ is concave}\}$ and $\{f \in F : f \text{ is not concave}\}$. The measurability of the first derivative leads to the following partition of the domain of observable variables: $\{f \in F : f' < 0\}$ and $\{f \in F : f' \geq 0\}$, which is clearly not a refinement nor a coarsening of the partition induced by the hypothesis. Then, the rejection of $f' < 0$ does not allow any conclusion about the concavity of f.

Therefore, the statement of hypotheses involving theoretical entities without a straightforward real world counterpart does not allow the application of any test, except for just concluding that the hypothesis is consistent with the data. This does not mean that the hypothesis is accepted, but in Popper's (1992) approach it would mean that its degree of confirmation has increased. Of course, even this assertion is not quite certain, since a precondition for the definition of degrees of confirmation is the existence of a σ-algebra over the set of possible hypotheses. In the case of the hypothesis of concavity of a function, the existence of a σ-algebra is ensured, since such an algebra can be defined over the functional space F. For more elaborated hypotheses, instead, this is completely unclear.

In fact, except for a few exceptions, most discussions in economics involve hypotheses that are not directly translatable to observational terms. In several areas of economic analysis it is common to base explanations on the behavior of non-observable entities like preferences, beliefs, expectations, goals, etc. Great efforts have been made to measure them. But, since they are all prone to strategic manipulation, the possibility of testing hypotheses requires relying on the validity of approximations to measurable variables. Even so, empirical studies are performed and their conclusions are used to the design and evaluation of policies.

A possible justification for doing so is based on the following points:

- Any first-order sentence ϕ, involving theoretical concepts, can be translated into a proposition ϕ' involving only observational terms (Craig, 1953).
- Since ϕ' is also a first-order statement, it can be decomposed as the conjunction of binary and unary terms (i.e. as the conjunction of predicates involving either one or two observable variables) (Quine, 1954).
- Since the binary and unary terms are observable, they must be testable, then ϕ' and indirectly ϕ are also testable (the translation from the theoretical sentence to the observable is not, in general, one-to-one).
- If the tests indicate that ϕ' cannot be rejected, ϕ is accepted until a better hypothesis is proposed.

That is, all the non-rejected hypotheses must be deemed as temporarily accepted. On the other hand, if $\phi \to \lambda$ and λ is such that the observational λ' is identical to ϕ', the accepted hypothesis becomes λ. This is because from the two hypotheses, the more general one is preferable, since it is the least constrained. That is, it requires less extra assumptions and yields the same set of observable conclusions.

But, even if the previous description already characterizes what we will do in the next sections, there is an additional point that requires clarification. Assume that an hypothesis ϕ is temporarily accepted. To reject it we could argue that even as a theoretical construction ϕ is not feasible because $\phi \wedge \gamma \to \perp$, where γ is another accepted hypothesis. That is, both hypotheses cannot be true at the same time because their conjunction leads to inconsistencies. Then, one of them must be dropped. If there exists a plausibility ordering of them, i.e. $\phi \prec \gamma$, meaning that γ is more plausible than ϕ, it is clear that ϕ must be dropped. This ordering $\prec$ may be purely subjective, based on the appraisal of the specialist or it can be objective, say because γ is a more precise claim that has already been tested. For example, suppose that ϕ is "f is concave", and ϕ' is "$f'' > 0$". Then, if ϕ' is accepted, we assume temporarily ϕ. But then, if γ is "$f'' > 0$", and has been accepted, we have to keep γ and drop ϕ. Of course, what happened here is that the translation from ϕ into ϕ' was poor and so the acceptation of the later should not imply that of the former.

As Rescher (1976) prescribed, no matter how a plausibility ordering arises, it should be linear since it must allow the analyst to decide over every pair of incompatible hypotheses. So, once decided which one to drop, a replacement

for it must be provided. Although there does not exist a mechanical procedure that provides the alternative, some guidelines are rather obvious. The
new hypothesis θ should be such that $\gamma \to \theta$ i.e. it must be implied by the
validity of the defeating hypothesis, while at the same time, $\phi' \to \theta'$. That
is, its observational consequences should be implied by those that have been
already tested. In our example, θ could be "f is increasing for $x \in Dom(f)$,
$x \geq \overline{x}$", since it implies that $f' > 0$ for an upper segment of its domain (in
particular for the non-negative segment). In fact, this shows that an extralog-
ical requirement is the *minimality* of θ. This means that from all the elements
that verify the desired conditions we should choose the one that "differs the
less" from ϕ. This is certainly a very tricky point, but it means that $\phi \to \theta$.
That is, that ϕ is a particular instance of θ.

In the next sections we present a discussion about two possible hypotheses
that could explain the same statistical evidence for the inflation processes in
some Latin-American countries. Our argument for accepting one of them and
dropping the other is based on the methodological principles presented here.
In these terms we confront two candidate hypotheses, ϕ and θ. ϕ represents
that the inflation processes are SOC, while θ represents that they are ESO
processes. As said, there exist reasons to think that the SOC hypothesis is
inconsistent with the fact that all these inflation processes underwent structural changes, which will be represent by γ. Moreover, these reasons indicate
that γ is more plausible than ϕ, because the SOC hypothesis implies that a
single process explains the entire series. The rest of the paper is devoted to
make clear the distinction between the two hypotheses in dispute and to derive their approximation by means of empirical data. This will allow us to use
the method presented in this section to choose one of them as an explanation
for the inflationary processes under analysis.

21.3 SOC Processes

Given a time series $\{x_t\}_{t=0...T}$, two functions, derived from the series provide
information about the structure of the process that generates its values:

- S_x, the spectral density of the series. To each frequency it associates the
  average of the corresponding Fourier coefficients.
- P_x, the sample density of the series.

S_x indicates the degree of dependence among values of the series. P_x approximates the theoretical density function of the generating mechanism. A
taxonomy of possible cases is the following:

- If $S_X \approx \lambda^0$ (where $\lambda \in \left[\frac{1}{T}, \frac{2}{T}, \dots \frac{\frac{T}{2}}{T} \right]$ is the frequency) the series is a white noise[1].
- If $S_x \approx \lambda^{-2}$ the series is a brownian noise.
- If $S_x \approx \lambda^{-\alpha}$ with $0 < \alpha < 2$ the series is a $\frac{1}{f}$ noise.

The case of $\frac{1}{f}$ noises is the most interesting, because they cover the widest variety of situations. A white noise is a series in which observations are noncorrelated. A brownian noise, instead, is a series in which observations are highly correlated. A $\frac{1}{f}$ noise shows a certain correlation among data, but diminished by the effect of random movements.

Fractal behavior in time series is evidenced if P_x exhibits a variance that changes with the size of the sample (the infinite variance property). Among the distributions that verify this property, the most representative is the Pareto-Levy distribution, for which

$$P_x \approx | x |^{-\beta} \tag{21.1}$$

with $0 < \beta < 2$.

On the other hand, additional evidence for $\frac{1}{f}$ noise and fractal behavior in a time series can be detected by means of the Hurst coefficient. Given the average of the series $\bar{x} = \sum_{i=1}^{T} \frac{x_i}{T}$ the minimal and maximal "accumulated flux" of the series are, respectively:

- $M_x = {}^{min}_{t=0\dots T} \sum_{i=1}^{T} \frac{x_i - \bar{x}}{t}$
- $M^x = {}^{max}_{t=0\dots T} \sum_{i=1}^{T} \frac{x_i - \bar{x}}{t}$

Then, if $R_T = M^x - M_x$ and given the standard deviation of the series, $S_T = (\sum_{i=0}^{T} \frac{(x_i - \bar{x})^2}{T})^{\frac{1}{2}}$ the coefficient of Hurst H is defined as

$$H = log \frac{\frac{R_T}{S_T}}{log T} \tag{21.2}$$

This number yields a kind of degree of correlation among data in the time series. In particular, $H > 0.5$ indicates that an increase of magnitude in an earlier stage is associated to an increase at later stages. There exists a great number of processes in nature and society that show $\frac{1}{f}$, fractal and $H > 0.5$ behavior, which we characterize as fractional brownian noises, by an abuse of language[2]. Attempts to explain these phenomena have been highly ad-hoc. To our knowledge the most articulate framework that explains the emergence of fractional brownian noises is that of the Self- Organized Critical processes(SOC). The idea is that fractional brownian noises are generated by underlying complex processes, such that the magnitude of the connections

---

[1] The interval of frequencies chosen is intended to avoid the problem of aliasing, i.e. the spurious increse in the spectral weight of higher frequencies due to the superposition with cycles of lower frequency.

[2] See Mandelbrot and Van Ness (1968) and Voss (1988)

among components of the support system can be affected by exogenous shocks that modify its structure. The result is that any such system goes from one meta-stable configuration to another. In the case that the fluctuations are non-correlated with the magnitude of the shocks, it seems natural to assume that they have been endogenously generated by the system[3]. A clear representation that captures this property is given by a sandpile, in which a constant amount of sand is added, and there exists a critical value for the slope of the pile. Once the sandpile attains that critical state, avalanches of all sizes can be expected, even some of a size far greater than the amount of sand added. This is because, at the critical slope, any overflow due to local configurations can generate a cascade (Bak and Chen, 1991).

A property of these processes is that the generated series are $\frac{1}{f}$ noises, the density functions approximate the Pareto-Levy distribution and $H > 0.5$. SOC is the most systematic and reliable model explaining those behaviors. Thus, if evidence is given in form of series behaving like fractional brownian noises, the existence of an underlying self-organized critical system generating the series cannot be discarded. According to Krugman (1996), this does not mean that no other explanation may exist or that a general hypothesis (instead of a particular representation for each phenomenon) is a theoretical necessity. It just means that the SOC hypothesis is not implausible. No statistical test exists to detect its presence since there is no well-defined class of reference for processes generating fractional brownian noise.

21.4 ESO Processes

The shortcomings of SOC processes as mechanisms to explain phenomena that develop over long periods, are all related to their static features: the dimension of the underlying structures and the systems of rules remain unchanged. Moreover, the size of the shocks does not affect those parameters. But it is relatively easy to relax those stringent conditions and create a model in which shocks are able to transform the very nature of the system. London and Tohmé (1998) presents an alternative:

Definition 1. *A ESO system is a tuple* $\Sigma = \langle T, R, M \rangle$, *where* $T = T_0, T_1, \ldots$ *is a sequence of topologies,* $R = R_0, R_1, \ldots$, *a sequence of local uniform rules and* M *a set of meta-rules. In turn, a topology* $T_i = \langle S_i, C_i \rangle$, *where* S_i *is a finite set of sites and* C_i *a regular structure of connections among them.*

The notion of topology used here is that of the class of neighborhoods of sites. For example, a N-sites complete graph is a topology where the neighborhood of a site consists of the entire set of nodes. Instead, if we consider a 2-dimensional orthogonal arrangement of sites (a "chessboard"), the neighborhood of a site consists of the set of eight adjacent sites.

---

[3] See Bak et al. (1993) and Scheinkman and Woodford (1994).

Each $s \in S_i$ can be in a state $\mid s \mid$, a numerical value in a discrete interval $[0, 1, ..., n_s]$. Given a site s and the state of the sites in the neighborhood in time t, $\mid Neigh_s \mid_t$, the value of the site at $t + 1$ is the result of the application of the rule $R_i :\mid s \mid_{t+1} = R_i(\mid Neigh_s \mid_t, \mid s \mid_t)$. That is, the state of the site depends both on the previous state of the neighborhood and on the state of site. We add the requirement that each R_i should be a non-linear function, to avoid the trivial case in which the asymptotic result consists of all sites having the same (maximum) value.

A shock is a random variation of the state of one or more sites. Assume that a site s is in a state $\mid s \mid$ and the shock is σ_s. The result of the shock depends on the rule R_i. In the case of a sandpile, any excess over n_s is transferred to the neighborhood. In the Game of Life, for example, the shock generates a new state with values in the legal interval. Instead, if potential variations are not limited they can change the structure of the system. Variations in excess over the admissible value of a site can provoke that the meta-rules transform *both* the topology and the set of rules.

A shock is applied in time t, and it propagates until it settles down to a definite pattern. This pattern may be periodic or non-periodic. In any case, the result is recorded, and in time $t + 1$ a new shock is applied to the system. If shocks in time t are within the "stable" range, i.e.[4] $\mid s \mid +\sigma_s < n_s$ then $T_{t+1} = T_t$, and $R_{t+1} = R_t$. If not, any meta-rule $M_k \in M$ is selected at random, and $\langle T_{t+1}, R_{t+1} \rangle = M_k(T_t, R_t, \sigma_s - n_s)$. That is, the meta-rule generates a new topology and a new set of rules, depending on the previous structure and the magnitude of the shock over the limit value. We add a rule of parsimonious behavior: S_t and S_{t+1} cannot be disjoint (this makes the history of the system continuous).

According to Bak et al. (1989), if shocks are always within their normal range, the behavior of the system is analogous to the Game of Life, a cellular automaton that constitutes a paradigmatic example of a SOC system. The non-linearity of R provides for a version of the *birth-reproduction-death* rule of the Game of Life, and therefore, the propagation of shocks generates $\frac{1}{f}$ noise, fractal structures and universal computability (see Berlekamp et al., 1982). We use, in the following statements, the Game of Life as a canonical characterization of SOC systems:

Proposition 1. *If for every t, $\mid s \mid +\sigma_s < n_s$, then Σ is a SOC system.*

Proof. If for every t, $\mid s \mid +\sigma_s < n_s$, then $T_{t+1} = T_t$ and $R_{t+1} = R_t$. That means that Σ can be represented as $\langle T_0, R_0, M \rangle$ since the initial topology and rules are permanent. The set of meta-rules M is irrelevant, since these rules are never activated. As R_0 is non-linear, Σ is an extension of the Game of Life. It is equivalent to a sandpile model: each site's state depends non-linearly on the state of its neighbors and the range of values is discrete, with more than

---

[4] We abuse of notation denoting, by $\mid s \mid +\sigma_s$, the result on s of σ_s.

two possible values. Therefore, as a sandpile is a SOC system it follows that Σ is also a SOC.

When shocks are not in range, the whole system "mutates", adding new sites, new connections or new rules. The overall behavior of the system is no longer easy to describe, but we have this immediate result that characterizes an ESO system as a sequence of SOCs:

Theorem 1. *The sequence of average values of the sites in an ESO system is a $\frac{1}{f}$ noise.*

Proof. First, let us show that for any site s, the sequence of its values in time, $\mid s \mid_0, \mid s \mid_1, \ldots$ must be a $\frac{1}{f}$ noise. The Fourier transform $F(s)$ of the time series of this sequence, which yields the spectral density of the time series, must verify, for the SOC stage T of Σ, $F(s) \approx \lambda^{-\alpha^T}$ with $0 < \alpha^T < 2$, while it must verify $F(s) \approx \lambda^0$ or $F(s) \approx \lambda^{-2}$ for the transients between SOC stages . Since $F(\cdot)$ is a linear operator, the transform of the average of the entire sequence must be the average of all the stages. Then, we have that $0 < \frac{log F(\bar{s})}{log \lambda} < 2$ for $\bar{s}$, the average of the sequence (because an average always falls between the extreme values). That is, the sequence is a $\frac{1}{f}$ noise. Then, given the set of sites $\{s^1, s^2, \ldots\}$ we consider the corresponding set of their average values $\{\bar{s}^1, \bar{s}^2, \ldots\}$. Then, given the average of all these values, $\bar{s}$, $F(\bar{s})$ is the average of the transforms of the individual sites. Call α^H the highest value of the corresponding α parameter (for, say, the site i_H) while α_L is the lowest value among all the sites. Then

$$0 < \alpha_L \leq -\frac{log F(\bar{s})}{log \lambda} \leq \alpha^H < 2 \qquad (21.3)$$

This means that the sequence of average values constitutes an aggregated $\frac{1}{f}$ noise.

We can consider a measure of homogeneity of sites, for each period t and the corresponding average of values of all sites, $\bar{s}^t$: the standard deviation of values $\delta(\bar{s}^t)$, or δ^t in short. This measure allows us to understand the scope of $\frac{1}{f}$ noisiness in the behavior of an ESO system:

Proposition 2. *The sequence $\delta^0, \delta^1, \ldots$ is, in average, a $\frac{1}{f}$ noise.*

Proof. Each standard deviation δ^t is such that $-2\bar{s}^t < \delta^t < 2\bar{s}^t$ and according to Theorem 1 the sequence $\{\bar{s}^t\}_{t \leq 0}$ is a $\frac{1}{f}$ noise. Therefore the sequence of standard deviations, $\delta^0, \delta_1 \ldots$ must be also, in average, a $\frac{1}{f}$ noise.

Similar arguments can be presented to show that an ESO system must exhibit, in average, a spatial distribution that approximates a Pareto-Levy distribution. The identification of regions of meta-stability in the history of an ESO (a sequence of SOC processes, according to Theorem 1) provides regions of robustness for the relevant parameters. In particular for the signature parameter ϕ^t, such that w^t, the spectral weight obtained from $F(\bar{s}^t)$ verifies $w \approx \lambda^{-\phi^t}$. If so, the identification of meta-stable stages in the history of a ESO system is equivalent to the detection of stages of robustness (i.e. periods with little change) of the ϕ^t parameter. Besides, an interesting consequence of assuming that an economic system can be represented as an ESO system is the following:

Theorem 2. *A ESO system may never reach a stationary state.*

Proof. Trivial. Since the sequence of standard deviations $\delta^0, \delta_1, \ldots$ is a $\frac{1}{f}$ noise, there may not exist (except for a degenerate case) a t^* such that for each $t > t^*$, $\delta^{t+1} = \delta^t$, even if the number of sites remains fixed.

In short: a ESO system never settles down in a specific pattern of diversity. That is, there may exist a permanent variation of the set of rules.

21.5 Evidence from Inflationary Series

Inflation is, in general, a consequence of economic instability. Its characteristics change with its level. So, situations of more instability induce a higher inflation rate, its effects being in turn stronger on the inner workings of the economy. Therefore, the performance of the economy becomes affected by these processes.

To describe the actual behavior of inflation processes we use a version of Leijonhufvuds (1990) taxonomy of regimes. Each regime represents an economic context characterized by an inflation rate fluctuating in a certain range of values, and to which a degree of uncertainty and a system of expectations are associated. According to Leijonhufvud (1981) and Heymann and Leijonhufvud (1995), these features are related to the degree of coherence of expectations. The concept of inflationary regime represents, therefore, how economic agents respond to both the available information and to unexpected shocks. When the inflation is higher, the individual expectations are not uniform, leading to a diversity of decisions, and then to a dispersion of values. The four regimes we consider here are the following:

- Moderate inflation (less than 2 % of monthly inflation).
- High inflation (between 2 % and 10 %).
- Very high inflation (between 10 % and 50 %).
- Hyperinflation (over 50 %).

The definition of inflationary regime, therefore, can be stated in the framework of SOC and ESO processes, given that each regime may correspond to a pattern of connections in the economy. This pattern is a result of the previous performance of the system and also of exogenous shocks. Each regime corresponds to a state of the system, and each one generates inflation episodes of different magnitude, according to the past experience and the shocks of economic policy. In order to see if the series behave like fractional brownian noises, we analyze the inflation of three Latin-American countries: Argentina, Chile and Colombia. For the three countries we use inflation series derived from the Consumer Price Index (CPI). The periods are as follows[5]

- *Argentina*: January 1957-December 1993
- *Chile*: January 1934-September 1995
- *Colombia*: January 1954-December 1992

These countries exhibit different inflationary processes. Argentina experienced moderate inflation during the 1960s, high and very high inflation in the 1970s and 1980s, with hyperinflationary episodes in 1989 and 1990. Since 1991 the country returned to the regime of moderate inflation, until the economic crisis of 2001. In turn, Chile experienced moderate and high inflation during almost all the period, with very high inflation episodes in the 1970s. Colombia showed a pattern of chronic moderate inflation. Despite the differences among these inflationary histories, our results are quite homogeneous. We determine the values of S_x, P_x and Hurst's coefficient (H) for the three countries. In the case of the spectral density, it can be approximated by the function λ^α while the sample density can be approximated by the function $\mid x \mid^\beta$: [6]

Country	α	β	H
Argentina	−1.06	−1.5	0.74
Chile	−0.40	−1.1	0.73
Colombia	−0.80	−0.7	0.76

Even if the three series roughly fall in the category of $\frac{1}{f}$ noises, there are some differences. Chile, for example, seems closer to a white noise, while Argentina reveals a heavier participation of low frequencies in the explanation of the series. This means that long range correlation is more important in this case than for the other countries.

With respect to the sample density, it is typically close to a power law (i.e. with $\beta \sim 1$), with small differences among the countries. Finally, for the Hurst's coefficient the result is, again, that all the countries show long range correlation, all positive. In other words, an increase (decrease) of the

---

[5] Notice that for each country the number of data points exceeds by large 450, which is sufficient for the kind of tests we run over them.

[6] The coefficients obtain as the regression coefficients in the log-log transformations of both the sample density and the Fourier transform. The goodness-to-fit of these regressions is ensured by the high values of the adjusted R^2. For example, in the case of Argentina, this value is 0.71.

inflation level in a period leads to an increase (decrease) in its level in the future. In brief, all the evidence suggests that inflation in the four countries is a fractional brownian noise.

The causes of fractional brownian noises can be understood in terms of two kinds of processes presented in Sections 3 and 4: SOC and ESO. The latter are the more general and encompass the former as particular cases. If inflation is generated by a SOC system, the existence of meta-stable states (in- flationary regimes) may be a warning. Hence, the possibility of very high and hyperinflation would be always present, although with a very low probability. However, the SOC hypothesis seems to be hard to accept: the bottom line of this assumption is that inflation is a result of a process that remains unaltered during the entire period of analysis. In other words, the structure of the economy must be assumed invariant and the evolution of inflation must be explained by its interaction with non-correlated shocks. Despite the fact that certain underlying mechanisms may remain "constant", the variations of policy affecting inflation could not be considered as uniformly distributed external shocks. To incorporate the changes of economic policies (for example the application of different stabilization programs), one should consider the possibility that they may generate drastic structural changes. In this sense, while ESO and SOC processes exhibit the same statistical properties, the former allows to consider the changes of structure that are discarded in the SOC processes while preserving the self-organized nature of the phenomena. The evidence is consistent with such hypotheses, but economic intuition indicates that the first one should be chosen: it is more natural to assume that the social, political and economic rules operating in an economy evolve in time. In particular, the Latin-American economies under study have suffered political, social and economical shocks that changed suddenly their structures, or at least their rules of operation. For example, Dabús (2000) reports evidence of significative structural change in the relation between inflation and relative price variability for Argentina. In terms of the characterizations presented in Section 2, we have the following statements:

- $\phi = $ "The inflationary process is a SOC".
- $\phi' = $ "The inflation series is a fractional brownian noise".
- $\theta = $ "The inflationary process is an ESO".
- $\theta' = $ "The inflation series is a fractional brownian noise".
- $\gamma = $ "The inflationary process underwent structural changes".

We argued that $\phi \prec \gamma$, i.e. that the SOC hypothesis seems incompatible with the idea that the inflationary process suffered structural changes. This is supported both by evidence and by economic intuition. On the other hand, by definition we have that $\phi \to \theta$ and $\gamma \to \theta$. Finally, the evidence provided by the inflationary series indicates that $\phi' = \theta'$. In other words, the ESO hypothesis is more plausible than the SOC one.

21.6 Conclusions

This paper shows two results, one strong and the other weaker. The strong result is that several, rather different, Latin-American economies exhibit inflation processes that behave like fractional brownian noises. In other words, all of them present power laws in the domain of values (revealed by the sample density), in the frequencies domain (seen in the spectral density) and in the domain of "fluxes" (shown by H, Hurst's coefficient).

The weak result is that ESO processes could provide an explanation for the emergence of those fractional brownian noises. Moreover, they allow a natural explanation in terms of periods of relevant analysis. If economies in Latin America are structures in evolution, with sudden changes and jumps, they would be explained to an explanation in terms of ESO processes.

A natural step in this analysis is to detect the periods of meta-stability in the time series, each one associated to a SOC system. If our hypothesis is close to be true, those periods, should have endpoints characterized by sudden social, political or economic changes. Otherwise we have to look for other explanations. The inflation rates of other inflationary economies could also constitute $\frac{1}{f}$ noises. Interesting cases should be the European episodes of hyperinflation, which were followed by decades of price stability. It would be possible to distinguish periods of meta-stability in terms of both the inflation rate and the economic structure. Then, if a strong correlation could be found between them (say, using dummy variables for the structure) support for the hypothesis of ESO behavior would receive a strong empirical support.

References

1. ak 19961 Bak, Per: *How Nature Works: The Science of Self-organized Criticality* (Springer-Verlag, Berlin).
2. ak and Chen 19912 Bak, Per and Kan Chen: Self-Organized Criticality, *Scientific American*, January.
3. ak et al. 19893 Bak, Per, Kan Chen and Michael Creutz: Self-Organized Criticality in the Game of Life, *Nature* **342**:780–782.
4. ak et al. 19934 Bak, Per, Kan Chen, José Scheinkman and Michael Woodford: Aggregate Fluctuations from Independent Sectoral Shocks: Self-Organized Criticality in Model of Production and Inventory Dynamics, *Ricerche Economiche* **47**:3–30.
5. erlekamp et al. 19825 Berlekamp, Elwyn, John Conway and Richard Guy: *Ways for Your Mathematical Plays* **II**, (Academic Press, New York).
6. asti 19956 Casti, John: *Complexification*, (Abacus, London).
7. raig 19537 Craig, William: On Axiomatizability within a System, *Journal of Symbolic Logic* **18**:30–32.
8. abús 20008 Dabús, Carlos: Inflationary Regimes and Relative Price Variability: Evidence from Argentina, *Journal of Development Economic* **62**:535–547.
9. eymann and Leijonhufvud 19959 Heymann, Daniel and Axel Leijonhufvud: *High Inflation*, (Oxford University Press, Oxford).

10. rugman 199610 Krugman, Paul: *The Self-Organizing Economy*, (Blackwell Publishers, Cambridge).
11. akatos 197611 Lakatos, Imre: *Proofs and Refutations: the Logic of Mathematical Discovery*, (Cambridge University Press, Cambridge U.K.).
12. eijonhufvud 198112 Leijonhufvud, Axel: *Inflation and Coordination*, (Oxford University Press, Oxford).
13. eijonhufvud 199013 Leijonhufvud, Axel: *Extreme Monetary Instability:* High Inflation, Hohenheim University Lecture.
14. ondon et al. 199514 London, Silvia, Carlos Dabús and Fernando Tohmé: Self-Organized Criticality and Inflation: the Argentinean Case, *VII World Meeting of the Econometric Society*, Tokyo, Japan.
15. ondon and Tohmé 199823 London, Silvia and Fernando Tohmé: Evolutionary Self-organized Systems, *CEFES98: International Conference of The Society for Computational Economics*, Cambridge, UK.
16. andelbrot and Van Ness 196815 Mandelbrot, Benoit and John Van Ness: Fractional Brownian Motions, Fractional Noises and Applications, *SIAM Review* **10**:422–437.
17. eitgen and Sauppe 198816 Peitgen, Heinz-Otto and Dietmar Sauppe: *The Science of Fractal Images*, (Springer-Verlag, New York).
18. opper 199217 Popper, Karl: *Logic of Scientific Discovery*, (Routledge, London).
19. uine 195418 Quine, Willard V.O.: Reduction to a Dyadic Predicate, *Journal of Symbolic Logic* **19**: 180-182.
20. escher 197619 Rescher, Nicholas: *Plausible Reasoning: an Introduction to the Theory and Practice of Plausibilistic Inference*, (Van Gorcun, Amsterdam).
21. cheinkman and Woodford 199420 Scheinkman, José and Michael Woodford: Self-Organized Criticality and Economic Fluctuations, *American Economic Review* **84**:417-421.
22. ufte 198321 Tufte, Edward: *The Visual Display of Quantitative Information*, (Graphic Books, Cheshire, Connecticut).
23. oss 198822 Voss, Richard: Fractals in Nature : from Characterization to Simulation, In: Peitgen, Heinz-Otto, Sauppe, Dietmar (eds.) *The Science of Fractal Images*, (Springer-Verlag, New York).

Semi-Infinite Programming: Properties and Applications to Economics

Francisco Guerra Vázquez[1] and Jan-J. Rückmann[2]

[1] Universidad de las Américas, Escuela de Ciencias, Sta. Catarina Mártir, Cholula, 72820 Puebla, Mexico fguerra@mail.udlap.mx
[2] Universidad de las Américas, Departamento de Física y Matemáticas, Sta. Catarina Mártir, Cholula, 72820 Puebla, Mexico rueckman@mail.udlap.mx
CONACyT grant 44003

Summary. In recent years the semi-infinite programming became one of the most substantial research topics in the field of operations research. The goal of this paper is to make more familiar to a broader class of mathematicians, economists and engineers the idea of what is a semi-infinite programming problem and how it can be applied to the modelling and solution of real-life problems from economics, finance and engineering. Several examples illustrate the characteristic geometric features of this class of problems as well as their consequences for the use of solution methods. Then, the paper refers to applications which are modelled as semi-infinite programming problems: a control problem, a Stackelberg game, a portfolio problem, a robust optimization problem and a technical trading system for future contracts. Finally, conclusions and open questions are discussed and a special bibliography on applications of semi-infinite programming is presented.

Key words:(Generalized) semi-infinite programming, feasible set, topological properties, optimal control, Stackelberg games, portfolio, robust optimization, trading system, future contracts

22.1 Introduction and First Examples

The goal of this paper is to illustrate in a conclusive manner how one of the recently most substantial research topics in operations research and applied mathematics, the *semi-infinite programming*, can be applied successfully to the modelling and solution of real-life problems from economics, finance and engineering. Before giving a mathematical definition of a semi-infinite programming problem we will start with an example.

Example 1. **A technical trading system for future contracts**

In [33], the authors deal with a model of the behaviour of an asset price (e.g. the price of a collection of stocks, bonds, warrants or the value of a stock exchange index). Based on discretized information about the asset price in the past, the model predicts the future price where its uncertainty is described by using several parameters whose values may vary in known fixed intervals. The model uses a deterministic (i.e. non-stochastic) approach which can be described as a *semi-infinite programming problem*. In particular, the authors apply this model to the real asset of the DAX future (the German stock exchange index DAX represents the 30 best capitalized companies in Germany) and obtain a technical trading system that produces selling and buying signals. We will later come back to this example in Section 22.3, Example 10.

In this paper we consider selected examples of so-called (generalized) semi-infinite programming problems; examples which illustrate some characteristic peculiarities of this class of problems as well as their differences to the finite case and, more important, examples of successful applications of the calculus of semi-infinite programming to the modelling and solution of problems from economics, finance and engineering. The paper does not include any mathematical demonstration; it will make familiar to a broader class of mathematicians, economists and engineers the idea of what is a semi-infinite programming problem and how it can be applied.

We start with some necessary mathematical background on generalized semi-infinite programming problems (GSIP). *Semi-infinite* means that we have a set of *finitely* many decision variables $x_1, x_2, \ldots, x_n$ and the feasible set M is basically defined by *infinitely* many inequality constraints. Then, the problem under consideration is to minimize (or to maximize) an objective function f subject to the feasible set M:

$$(\text{GSIP}): \qquad \text{minimize } f(x) \text{ subject to } x \in M$$

where M is defined as

$$M := \{x \in \mathbb{R}^n \mid g(x, y) \leq 0, \ \forall y \in Y(x)\}.$$

For sake of simplicity this model only contains one inequality constraint of the type $g(x, y) \leq 0$; in the more general case the description of the feasible set M may contain finitely many inequality constraints of this type (cf. Section 22.3, Example 8) as well as, additionally, finitely many equality constraints which only depend on x.

Obviously, each index $y \in Y(x)$ represents an inequality constraint $g(\cdot, y) \leq 0$ and the set $Y(x) \subset \mathbb{R}^k$ is assumed to be compact (and *infinite*); it is given as

$$Y(x) := \{y \in \mathbb{R}^k \mid h_i(x, y) = 0, \ i = 1, \ldots, p, \ h_j(x, y) \leq 0, \ j = p+1, \ldots, r\}.$$

Therefore, a generalized semi-infinite programming problem (GSIP) is characterized by *finitely many variables* $x \in \mathbb{R}^n$, *infinitely many inequality constraints* $g(\cdot, y) \leq 0$ and *the index set* $Y(x)$ *depends on* $x \in \mathbb{R}^n$. If the index set $Y(x)$ does not depend on the decision vector $x \in \mathbb{R}^n$, i.e. if $Y(x) = Y$ is constant for all $x \in \mathbb{R}^n$, then the corresponding problem is called a *standard semi-infinite programming problem* (SIP):

$$\text{(SIP)}: \qquad \text{minimize } f(x) \text{ subject to } x \in M$$

where M is defined as

$$M := \{x \in \mathbb{R}^n \mid g(x, y) \leq 0, \ \forall y \in Y\}.$$

Obviously, (SIP) is characterized by *finitely many variables* $x \in \mathbb{R}^n$, *infinitely many inequality constraints* $g(\cdot, y) \leq 0$ and *the index set Y does not depend on $x \in \mathbb{R}^n$*. We also define a finite programming problem (FP) which is characterized by *finitely many variables* $x \in \mathbb{R}^n$ and *finitely many constraints*:

$$\text{(FP)}: \qquad \text{minimize } f(x) \text{ subject to } x \in M_{\text{FP}}$$

where M_{FP} is defined as

$$M_{\text{FP}} := \{x \in \mathbb{R}^n \mid g_i(x) = 0, \ i = 1, \ldots, p, \ g_j(x) \leq 0, \ j = p+1, \ldots, r\}.$$

Note that the index set $Y(x)$ of (GSIP) has the structure of the feasible set of a finite programming problem. Throughout the paper assume that all appearing functions are continuously differentiable.

In recent years, many problems from economics, finance and engineering have been modelled and solved by using the calculus of mathematical programming. In particular, the classes of (GSIP) and (SIP) have become a very active and successful research area in applied mathematics. Exemplarily, we refer to the survey papers [12] and [21], to the recent monographs [8] and [22] as well as to the recent books [9] and [24] which contain several tutorial papers on theory, numerics and applications in semi-infinite programming. For those who are interested in more details, we also refer to some recent publications on properties, e.g. optimality conditions and numerical solution methods of semi-infinite programming problems: [3, 4, 5, 14, 15, 17, 23, 25, 26, 30, 31, 32, 37].

A statement about an origin of the theory of semi-infinite programming is published in [1]: *"In March 1962 Charnes, Cooper and Kortanek developed the theory of semi-infinite programming which associates the minimization of a linear function of finitely many variables subject to an arbitrary number of linear inequalities in these variables with maximization of a linear function of infinitely many variables subject to a finite system of linear inequalities"*.

The classical problem of (Chebyshev) approximation of a function can also be considered as an origin of semi-infinite programming.

Example 2. **Approximation of a function**

Frequently, in several applications a real-valued function $F(y)$ with a "complicated" structure has to be approximated on a set Y by a real-valued parameter-dependent function $f(y, x)$ (x is the parameter) with a more "simple" structure; e.g. $f(y, x)$ may be a polynomial and its vector of coefficients $x \in I\!R^n$ is taken from a corresponding set $X \subset I\!R^n$. Then, one obtains the "best" approximation function $f(y, \bar{x})$ by minimizing the maximal deviation $|F(y) - f(y, x)|$ on the set Y:

$$(AP): \quad \operatorname*{minimize}_{x \in X} \operatorname*{maximize}_{y \in Y} |F(y) - f(y, x)|.$$

Obviously, the approximation problem (AP) can be reformulated equivalently as a *standard semi-infinite programming problem*:

$$\text{minimize} \quad q \quad \text{subject to} \quad (x, q) \in M$$

where
$$M = \{(x, q) \in I\!R^n \times I\!R \mid |F(y) - f(y, x)| \le q, \ \forall y \in Y\}.$$

The modern theory of semi-infinite programming was obtained as a generalization of this class of approximation problems by allowing a wider class of objective functions and feasible sets.

As already mentioned above, there exists a wide range of applications, which can be modelled as (SIP) or (GSIP). Exemplarily, we refer to

- Chebyshev and reverse Chebyshev approximation problems,
- maneuverability of a robot,
- environmental problems,
- design and design centering problems,
- minimax problems,
- defect minimization for operator equations,
- optimal layout of an assembly line,
- wavelets (as a special approximation problem),
- semidefinite programming

and many others [2, 7, 10, 11, 13, 16, 18, 19, 20, 27, 30, 33, 34, 37]. If the mathematical description of a real-life problem requires an *infinite number* of restrictions, then, obviously the standard methods of mathematical programming which only allow a *finite number* of restrictions cannot be applied. Note that the infinite index set $Y(x)$ of restrictions may represent e.g.

- a time interval: in time-dependent processes at each moment $y \in Y(x)$ an appropriate inequality constraint has to be fulfilled;
- or a geographic area: each vector of coordinates $y \in Y(x)$ of this area corresponds to a constraint, e.g. in environmental problems.

The paper is organized as follows. In the next section several examples illustrate specific geometric properties of a semi-infinite programming problem. In particular, some differences to the standard case of finite mathematical programming as well as the consequences of these features to the design and application of solution methods are discussed. These results are new and original. Section 22.3 contains selected examples from recent publications which show how the calculus of semi-infinite programming can be applied to the modelling and solution of problems from economics, finance and engineering. Finally, in Section 22.4 some conclusions and open questions are presented and in Section 22.5 a list of related articles is given which are dealing with possible applications of semi-infinite programming.

According to the motivation mentioned at the beginning of this section, the present paper illustrates a possible bridge between semi-infinite programming and its applications to the modelling and solution of real-life problems. A main reason for writing this paper was that, to our opinion, both the community of applied mathematicians and economists as well as people from industry and management need this type of *bridge articles* in order to get more knowledge and a better recognition that methods and algorithms from applied mathematics can be very useful for tackling real-life problems. It is our hope that this type of articles will help mathematicians, economists and engineers initiating further cooperations and investigations in the field of applying mathematical programming methods to problems from economics, finance and engineering.

22.2 Properties of the Feasible Set

In this section we present examples of the feasible set of a semi-infinite programming problem in order to become more familiar with the peculiarities of its geometric structure. The properties illustrated are characteristic for semi-infinite programming problems; in particular, they do not appear in case of a feasible set

$$M_{\text{FP}} = \{x \in I\!\!R^n \mid g_i(x) = 0,\ i = 1, \ldots, p,\ g_j(x) \leq 0,\ j = p + 1, \ldots, r\}$$

of a finite programming problem. Furthermore, we briefly discuss some consequences of these features when applying the calculus of solution methods known from finite programming.

Example 3. **A convex set as a feasible set of a linear semi-infinite programming problem**

We consider the convex set

$$M = \left\{ x \in I\!\!R^2 \ \middle| \ \frac{x_1^2}{a^2} + \frac{x_2^2}{b^2} \leq 1 \right\}$$

with $a, b \in I\!\!R$, $a \neq 0$, $b \neq 0$, whose boundary ∂M is a two-dimensional ellipsoid (see Fig. 22.1). For each boundary point (x_1^0, x_2^0) we can define the corresponding tangent

$$\frac{x_1 x_1^0}{a^2} + \frac{x_2 x_2^0}{b^2} = 1 \qquad \text{(cf. Fig. 22.1).}$$

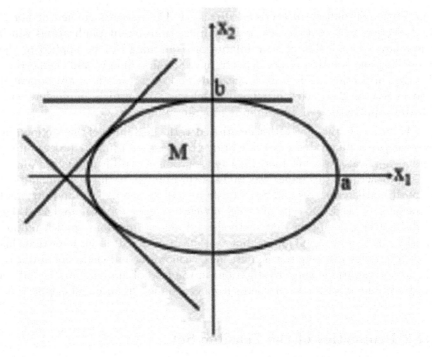

Fig. 22.1. The ellipsoid as an intersection of infinitely many closed halfspaces

Then, M can be described by *infinitely many linear inequality constraints*:

$$M = \left\{ x \in I\!\!R^2 \ \Big| \ \frac{x_1 y_1}{a^2} + \frac{x_2 y_2}{b^2} \leq 1, \ \forall y \in Y \right\},$$

where the index set Y consists of all boundary points

$$Y = \left\{ y \in I\!\!R^2 \ \Big| \ \frac{y_1^2}{a^2} + \frac{y_2^2}{b^2} = 1 \right\}.$$

Obviously, the latter description of M is that of the feasible set of a linear (SIP). It can easily be seen that, in general, *each convex set* can be described as a feasible set of a linear SIP: by defining at each boundary point the corresponding tangents we obtain infinitely many supporting halfspaces whose intersection is just the original convex set.

Problems and difference to the finite case: In general, the simplex method cannot be used for the solution of linear semi-infinite programs since the feasible set is not necessarily characterized by its vertices (in case of an ellipsoid there are no vertices). Furthermore, we cannot conclude as in the finite case that the set of the so-called active constraints in a neighbourhood of a feasible point $\bar{x}$ under consideration is a subset of the set of active constraints at $\bar{x}$. We will explain that in more details and consider the following finite programming problem in the two-dimensional space $I\!\!R^2$:

(FP) :   minimize $f(x_1, x_2) = -x_1$ subject to $x \in M_{FP}$

where

$$M_{FP} = \left\{ x \in \mathbb{R}^2 \;\middle|\; \begin{array}{l} g_1(x) = -x_1 \leq 0, \; g_2(x) = -x_2 \leq 0, \\ g_3(x) = x_1 + x_2 - 1 \leq 0 \end{array} \right\}.$$

Obviously, the solution point is $\bar{x} = (1\ 0)$ (note that the gradient of the objective function is $\nabla f = (-1\ 0)$, cf. Fig. 22.2).

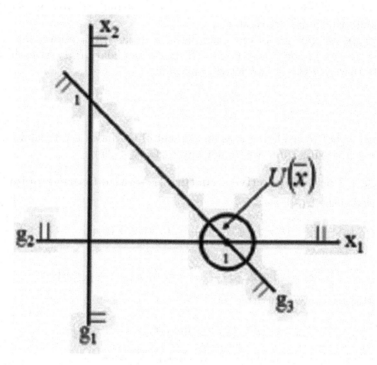

Fig. 22.2. The set M_{FP}

At the solution point $\bar{x} = (1\ 0)$ the constraints g_2 and g_3 are fulfilled as equalities:

$$g_1(1, 0) = -1 < 0,$$
$$g_2(1, 0) = 0,$$
$$g_3(1, 0) = 0,$$

i.e. g_2 and g_3 are *active* constraints. It is well-known that the optimality of a feasible point depends on its active constraints (and not on those which are not active). Generally, we define the set of active constraints at a point $\bar{x} \in M_{FP}$ as

$$J_0(\bar{x}) = \{j = \{1, \ldots, r\} \mid g_j(\bar{x}) = 0\}.$$

The continuity of finitely many constraints implies that there exists a neighbourhood $U(\bar{x})$ of $\bar{x} \in M_{FP}$ such that

$$J_0(x) \subset J_0(\bar{x}) \quad \text{for all } x \in U(\bar{x}), \tag{$*$}$$

i.e. at each point from this neighbourhood $U(\bar{x})$ there are only active constraints which are also active at $\bar{x}$. (In Fig. 22.2 we have $J_0(x) \subset \{2, 3\}$ for all $x \in U(\bar{x})$.) An important consequence of the latter fact for solution methods is that in a neighbourhood of a solution point $\bar{x}$ one only has to consider those constraints which are active at $\bar{x}$.

However, as we can see in our example of a linear (SIP) above, *the condition* $(*)$ *does not hold* at any point $\bar{x}$ from the boundary ∂M. At each point $\bar{x} \in \partial M$ there is exactly one active constraint:

$$\frac{x_1 \bar{x}_1}{a^2} + \frac{x_2 \bar{x}_2}{b^2} = 1,$$

i.e. $J_0(\bar{x}) = \{\bar{x}\}$ (considering $\bar{x}$ as an element of the index set Y) and for any point $x \in \partial M$ with $x \neq \bar{x}$ we do not have $J_0(x) \subset J_0(\bar{x})$.

Example 4. **Local disjunctive structure (re-entrant corner point)**

We consider the feasible set

$$M = \{x \in \mathbb{R}^2 \mid -y \leq 0, \forall y \in Y(x)\}$$

where

$$Y(x) = \{y \in \mathbb{R} \mid x_1 + 1 - y \leq 0, \ x_2 - 1 - y \leq 0\}.$$

The latter set can be rewritten as

$$Y(x) = \{y \in \mathbb{R} \mid y \geq \max\{x_1 + 1, x_2 - 1\}\}$$

and, therefore, M is the *union* of two closed halfspaces

$$M = \{x \in \mathbb{R}^2 \mid \max\{x_1 + 1, x_2 - 1\} \geq 0\}$$
$$= \{x \in \mathbb{R}^2 \mid x_2 \geq 1\} \cup \{x \in \mathbb{R}^2 \mid x_1 \geq -1\} \quad \text{(cf. Fig. 22.3)}.$$

This description using a max-function implies a so-called *disjunctive structure* of the shape of M. The point $(-1, 1)$ is also called *re-entrant corner point* (cf. e.g. [30]).

Problems and difference to the finite case: In contrast to the disjunctive structure of M which is (locally) characterized by the *union* of sets of type $\{x \in \mathbb{R}^n \mid g(x) \leq 0\}$ (where g is a continuously differentiable function), the feasible set M_{FP} of a finite programming problem is always the *intersection* of such sets (for sake of simplicity assume that $p = 0$):

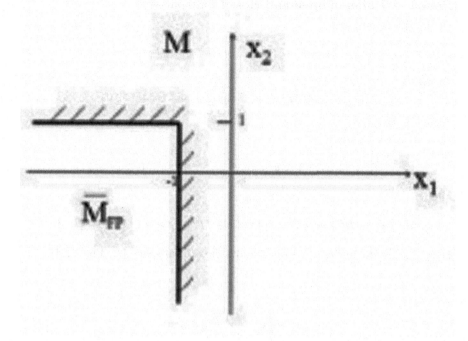

Fig. 22.3. M is the union of two closed halfspaces

$$M_{FP} = \{x \in I\!R^n \mid g_i(x) \leq 0, \ i = 1, \dots, r\}$$
$$= \{x \in I\!R^n \mid g_1(x) \leq 0\} \cap \dots \cap \{x \in I\!R^n \mid g_r(x) \leq 0\}.$$

Therefore, M can locally be considered as the complement set of the feasible set of a finite programming problem; in this example, M is the complement set of

$$\bar{M}_{FP} = \{x \in I\!R^2 \mid x_1 \leq -1\} \cap \{x \in I\!R^2 \mid x_2 \leq 1\} \quad \text{(cf. Fig. 22.3)}.$$

Hence, the set M need not be convex although all constraints may be convex functions; in this example all constraints are linear and M is not convex. The latter fact implies that also a linearization of the problem (that we obtain by replacing all functions by its first order Taylor expansion at the considered point) need not have a convex feasible set. Since all functions are linear in this example, the set M and its linearization are locally identic (and not convex). Altogether, the replacement of the feasible set by its linearization does not provide a convex set as in the finite case: in finite programming several solution algorithms solve in each iteration step the corresponding linearized problem (which has a convex feasible set) using the *calculus of convex programming*. In general, this is not possible for semi-infinite programming problems since the feasible sets of the corresponding linearized problems need not be convex.

Example 5. **Union of open and closed halfspaces**

Define the feasible set

$$M = \{x \in \mathbb{R}^2 \mid -y \leq 0,\ \forall y \in Y(x)\},$$

where

$$Y(x) = \{y \in \mathbb{R} \mid x_1 + 1 - y \leq 0,\ y - x_2 + 1 \leq 0\}.$$

Then, we obtain

$$Y(x) = \begin{cases} \emptyset & \text{if } x_1 + 1 > x_2 - 1, \\ x_1 + 1 \leq y \leq x_2 - 1 & \text{if } x_1 + 1 \leq x_2 - 1 \end{cases}$$

and

$$M = \{x \in \mathbb{R}^2 \mid x_1 + 2 > x_2\} \cup \{x \in \mathbb{R}^2 \mid x_1 \geq -1,\ x_1 + 2 \leq x_2\}.$$

Hence, M is the union of an open and a closed halfspace (cf. Fig. 22.4).

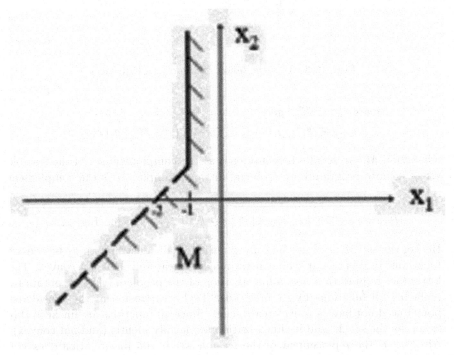

Fig. 22.4. M is the union of an open and a closed halfspace

Problems and difference to the finite case: According to this example, a feasible set of a generalized semi-infinite programming problem is *not closed*

in general. As already mentioned in Example 4, the feasible set M_{FP} of a finite programming problem is always the intersection of finitely many closed sets, i.e. M_{FP} is always a *closed* set. Several solution algorithms calculate a sequence of iteration points tending to a limit point which may be a solution point and, perhaps, a boundary point of the feasible set. In order to ensure the feasibility of the limit point, the closedness of the feasible set is assumed. However, if the feasible set is not closed (as it may happen in the semi-infinite case), then the limit point is, possibly, not feasible and, therefore, not a solution point. In that case the algorithm would calculate a sequence of feasible points which tends to a non-feasible limit point.

Example 6. **Local non-closedness**

As illustrated in Example 5, the feasible set of a semi-infinite programming problem may be the union of an open and a closed halfspace. In the current example the feasible set M is neither open nor closed as well, but M is not the union of halfspaces. Let

$$M = \{x \in \mathbb{R}^2 \mid x_2 - y \leq 0, \ \forall y \in Y(x)\}$$

and

$$Y(x) = \{y \in \mathbb{R} \mid y - 1 \leq 0, \ x_2 - x_1^4 - y \leq 0\}$$

which implies that

$$Y(x) = \begin{cases} \emptyset & \text{if } x_2 - x_1^4 > 1, \\ \{1\} & \text{if } x_2 - x_1^4 = 1, \\ [x_2 - x_1^4, 1] & \text{if } x_2 - x_1^4 < 1. \end{cases}$$

Then, M can be described as the union of an open and a closed set (cf. Fig. 22.5):

$$M = \{x \in \mathbb{R}^2 \mid x_2 - x_1^4 > 1\} \cup \{x \in \mathbb{R}^2 \mid x_1 = 0, \ x_2 \leq 1\}.$$

Problems and difference to the finite case: As in Example 5.

22.3 Applications

This section presents a selection of recently published applications from economics, finance and engineering which can be modelled and solved by using the calculus of semi-infinite programming. As already mentiond in the introductory section there exist several recent monographs and state-of-the-art volumes [8, 9, 22, 24] describing a multitude of possible applications of semi-infinite programming to the modelling and solution of real-life problems. Furthermore, we refer to the bibliography in Section 22.5 with a list of special references on applications of semi-infinite programming. In the current section we consider the following applications:

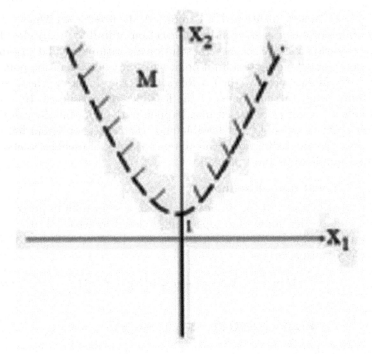

Fig. 22.5. M is neither open nor closed

Example 7: Application to Optimal Control Problems

Example 8: Stackelberg games and semi-infinite programming

Example 9: A portfolio problem and robust optimization

Example 10: A technical trading system for future contracts

Example 7. **Application to Optimal Control Problems (cf. [27])**

Many procedures in engineering and economics are dynamical (time-dependent) processes which can be modelled as so-called *optimal control problems* that represent a special class of mathematical programming problems. The dynamical part is often described by a system of ordinary or partial differential equations and the corresponding programming problem is defined in infinite-dimensional function spaces. Therefore, in order to solve such a problem numerically, one discretizes the original infinite-dimensional problem and obtains a *semi-infinite programming problem*. In the following we present the basic mathematical idea; for further details we refer to [27]. An optimal control problem consists of an objective integral function $F(u)$ as well as a dynamical system with a control function $u \in L$ (L is an appropriate function space) and can be described as follows:

(OCP) : $\underset{u \in L}{\text{minimize}} \; F(u)$

subject to $\begin{cases} a(t) \le u(t) \le b(t), \; \forall t \in [0, T] \\ c(x, t) \le S(u)(x, t) \le d(x, t), \; \forall t \in [0, T], \; \forall x \in \Omega. \end{cases}$

For each control $u \in L$ there exists a uniquely defined output-function

$$S : u \in L \mapsto S(u)$$

where

$$S(u) : (x, t) \in \Omega \subset I\!\!R^n \times [0, T] \mapsto S(u)(x, t).$$

Furthermore, there are constraints for $u(t)$ and $S(u)(x, t)$, e.g. in a simple case there could be box constraints of the type

$$a(t) \le u(t) \le b(t),$$
$$c(x, t) \le S(u)(x, t) \le d(x, t),$$

where all appearing functions are taken from the corresponding function spaces. Since u belongs to the infinite-dimensional function space L one has to introduce an appropriate discretization in order to compute the optimal control numerically. By defining a finite-dimensional subspace of piecewise constant controls

$$u_N(\alpha)(t) = \sum_{i=1}^{N} \alpha_i u_i(t), \quad \alpha \in I\!\!R^N,$$

the optimal control problem (OCP) is transformed into a *semi-infinite programming problem* depending on a finite dimensional vector $\alpha \in I\!\!R^N$:

$\underset{\alpha \in I\!\!R^N}{\text{minimize}} \; F_N(\alpha) = F(u_N(\alpha))$

subject to $\begin{cases} a(t) \le u_N(\alpha)(t) \le b(t), \; \forall t \in [0, T] \\ c(x, t) \le S(u_N(\alpha))(x, t) \le d(x, t), \; \forall t \in [0, T], \; \forall x \in \Omega. \end{cases}$

In [27] three applications of optimal control problems and their corresponding discretizations are discussed:

- Robot trajectory planning; where a robot arm follows a given path subject to certain conditions (see also [11]).
- Sterilization of food; where the heating process is optimized ensuring that the bacteria in food are killed and the damage of the nutrients (e.g. loss of vitamines) is minimal (see also [29]).
- Flutter control, where the flutter of aircraft wings have to be avoided in a certain range of velocities (see also [7]).

Example 8. **Stackelberg games and semi-infinite programming (cf. [30])**

Stackelberg games may appear in different contexts from economics; they can be seen as a decision process between an upper level (a leader) and finitely many parallel lower levels (several followers). In the upper level problem the leader chooses his strategy by optimizing a function f that depends on a parameter vector x; where in the lower level problem each follower i solves independently a parameter dependent problem $L^i(x)$, $i = 1, \ldots, l$ (each parameter value x represents a strategy of the leader). In particular, the decision of the upper level takes into account the parameter-dependent (i.e. x-dependent) solutions of the followers. Therefore, a Stackelberg game can be described as a bi-level programming problem consisting of the following two steps:

Lower level: Solve the parameter-dependent programming problems $L^i(x)$, $i = 1, \ldots, l$ of the followers.

Upper level: Solve the programming problem of the leader; his solution strategy will influence the follower's optimal strategies.

The solution of a Stackelberg game can be applied to an oligopoly where one enterprise is assumed to be dominant. There exists an exhaustive literature on bi- and multi-level programming problems as well as on bi-level programming and Stackelberg games (see e.g. [28] and the bibliography review [36]).

In order to see the relation between semi-infinite programming and Stackelberg games we consider an extended semi-infinite programming problem:

$$\text{(EGSIP)}: \qquad \text{minimize } f(x) \text{ subject to } x \in M$$

where

$$M = \{x \in \mathbb{R}^n \mid g_1(x, y) \le 0, \ldots, g_l(x, y) \le 0, \forall y \in Y(x)\}.$$

Instead of only one inequality constraint $g(x, y) \le 0$ we allow now finitely many constraints of this type. It is well-known that the optimal solution of a programming problem depends on the so-called active constraints, i.e. those which are fulfilled as equalities (cf. Section 22.2, Example 3). We define for a feasible point $\bar{x} \in M$ and for each $i \in \{1, \ldots, l\}$ the corresponding *active constraint set*:

$$Y_i(\bar{x}) = \{y \in Y(\bar{x}) \mid g_i(\bar{x}, y) = 0\}.$$

Obviously, for $\bar{x} \in M$ and $i \in \{1, \ldots, l\}$ each active constraint $\bar{y} \in Y_i(\bar{x})$ is a global solution of the corresponding programming problem

$$L^i(\bar{x}) \quad \text{maximize } g_i(\bar{x}, y) \quad \text{subject to } y \in Y(\bar{x}).$$

Hence, in order to check optimality conditions of (EGSIP) at a point $\bar{x}$ one has to solve *firstly* the k problems $L^i(\bar{x})$, $i = 1, \ldots, l$, which correspond to the parameter-dependent lower level problems of a Stackelberg game. Then, *secondly*, the solution of (EGSIP) corresponds to the upper level problem of a Stackelberg game.

Example 9. **A portfolio problem and robust optimization (cf. [2] and [30] for numerical treatment)**

Consider a portfolio consisting of n different parts j, $j = 1, \ldots, n$; e.g. stocks from n different enterprises. Let $y_j \geq 0$ be the investor's return that is obtained for each 100 \$-investment in part j, $j = 1, \ldots, n$ at the end of a given time-period. Then, the goal is to maximize the investor's return at the end of this time-period, e.g. to determine the amounts $x_j \geq 0$ to be invested in part j, $j = 1, \ldots, n$:

$$\text{maximize} \quad \sum_{j=1}^{n} x_j y_j \quad \text{subject to} \quad x \in M_1$$

where

$$M_1 = \left\{ x \in \mathbb{R}^n \ \Big| \ \sum_{j=1}^{n} x_j = 1, \ x_j \geq 0, \ j = 1, \ldots, n \right\}.$$

If the values of the investor's return parameters $y = (y_1, \ldots, y_n)$ are explicitly known, then, obviously, the optimal solution is to invest the whole amount in that part j^* which has the maximal value

$$y_{j^*} = \max\{y_j, \ j = 1, \ldots, n\}.$$

However, in general, the vector $y \in \mathbb{R}^n$ will not be known a priori and one has to assume that y varies in a set Y. Then, the corresponding programming problem is of *semi-infinite type*:

$$\text{maximize} \quad z \quad \text{subject to} \quad (x, z) \in M_2$$

where

$$M_2 = \left\{ (x, z) \in \mathbb{R}^n \times \mathbb{R} \ \left| \ \begin{array}{l} z - \sum_{j=1}^{n} x_j y_j \leq 0, \ \forall y \in Y \\[2mm] \sum_{j=1}^{n} x_j = 1, \ x_j \geq 0, \ j = 1, \ldots, n \end{array} \right. \right\}.$$

The latter problem can also be considered as a *robust optimization problem*. If parameter values (here: y_j, $j = 1, \ldots, n$) are not fixed or not known (uncertain), then a 'worst case' description may be that the corresponding constraint has to be fulfilled for *all* possible parameter values y from a corresponding set Y; that implies, obviously, a semi-infinite formulation:

$$g(x, y) \leq 0, \quad \forall y \in Y.$$

Robust optimization problems are extensively studied in [2], where, in particular, a close relationship to so-called *semi-definite programming problems* (cf. e.g. [6, 34]) is investigated.

Let us return to our portfolio problem. In [2], the set Y has an ellipsoidal form

$$Y = \left\{ y \in \mathbb{R}^n \;\middle|\; \sum_{j=1}^{n} \frac{(y_j - \bar{y}_j)^2}{\sigma_j^2} \leq \vartheta^2 \right\}$$

with so-called *nominal values* $\bar{y}_j$ and where σ_j, $j = 1, \ldots, n$ is a scaling parameter and ϑ measures the risk aversion of the decision maker. It is shown in [2] that for

$$\left. \begin{array}{l} \bar{y}_j = 1.15 + \frac{0.05j}{n} \\ \sigma_j = \frac{0.05}{3n} \sqrt{2n(n+1)j} \end{array} \right\} \quad j = 1, \ldots, n$$

$$\vartheta = 1.5$$

one obtains an optimal investor's return by investing the same amount in each part, i.e. $x_j = \frac{1}{n}$, $j = 1, \ldots, n$.

If, additionally, the risk aversion depends on x ($\vartheta = \vartheta(x)$), e.g. the risk aversion may increase if the values of x are different from $\frac{1}{n}$), then we obtain a *generalized semi-infinite programming problem*:

$$\text{maximize } z \text{ subject to } (x, z) \in M_3$$

where

$$M_3 = \left\{ (x, z) \in \mathbb{R}^n \times \mathbb{R} \;\middle|\; \begin{array}{l} z - \sum_{j=1}^{n} x_j y_j \leq 0, \; \forall y \in Y(x) \\[2mm] \sum_{j=1}^{n} x_j = 1, \; x_j \geq 0, \; j = 1, \ldots, n \end{array} \right\}$$

and

$$Y(x) = \left\{ y \in \mathbb{R}^n \;\middle|\; \sum_{j=1}^{n} \frac{(y_j - \bar{y}_j)^2}{\sigma_j^2} \leq (\vartheta(x))^2 \right\}.$$

Example 10. **A technical trading system for future contracts (cf. [33]; continuation of Example 1 in Section 22.1)**

The paper [33] deals with the numerical treatment of an asset price model using a deterministic (i.e. non-stochastic) approach. According to the hypothesis

– that the current price of the asset reflects completely the past and
– that the market responds immediately to relevant new information,

the modelling of an asset price means modelling of relevant new information. This could be done by using a class of linear dynamical systems (cf. Example 7) under uncertainty; here, the investor's return is associated with the change of the asset price divided by the asset price. A standard way is to decompose this return into two parts: the predictable (deterministic) return and the nonanticipated (stochastic) return. The latter part can be modelled by an adequate class of input perturbations that is characterized by a parameter vector p whose components are restricted to corresponding lower and upper bounds. Each parameter vector value $\bar{p}$ defines a trajectory of the asset price. For the choice of the class of input perturbations one assumes that the information to be observed is discrete, e.g. by observing the price of the asset at finitely many timepoints in a pre-determined time interval. Stochastic models use distributions for describing the uncertainty; however, many of them assume mathematical properties which, in general, can hardly be verified under real conditions. The use of distributions can be avoided by solving a corresponding minimax-programming problem which calculates an optimal value p^* for the parameter vector. This minimax problem is a *semi-infinite programming problem* (in [33] two different models are discussed, one of them is related to the classical one of Vasicek [35]). The so-obtained deterministic model results a unique trajectory of the asset price and allows the prediction of future prices for the next time interval. As already mentiond in Section 22.1, Example 1, the authors apply their model to the DAX future and obtain a trading system that gives selling and buying signals for the next day based on the final courses of the DAX of the current day. It is shown that the system works successfully; it is highly active and values are changing quickly according to the conditions of the actual stock market. We also refer to the related paper [18].

22.4 Conclusions and Open Questions

This paper illustrates a possible bridge between a modern field of continuous nonlinear programming, the so-called (generalized) semi-infinite programming, and its applications to the modelling and solution of real-life problems from economics, finance and engineering. As we have seen, there already exists a multitude of applications and it is our hope that there will be more mathematicians, economists and engineers in the future entering this exciting field linking mathematical programming methods and instruments from finance and cconomics. Several ideas of this paper merit further investigations; the outlined results suggest the following conclusions and open questions.

1. The topological structure of the feasible set M of a semi-infinite programming problem may be characterized by the following specific properties:

- Local disjunctive structure: M can locally be described as the union of finitely many halfspaces (Example 4).
- M need not be a closed set; in particular, M can locally be described as the union of finitely many closed and finitely many open subsets (Examples 5 and 6).
- In general, the convexity of the constraints which are describing M does *not* imply the convexity of the set M (Example 4).
- The local linearization of M does not provide a corresponding convex linearized set (Example 4).

2. These specific geometric features imply that known solution methods have a qualitatively different computational performance; hence, successful solution algorithms for semi-infinite programming problems can only be obtained after a deep understanding of the topology of this class of problems. We refer to [23] for an exhaustive survey on solution methods for semi-infinite programming problems. However, as we have seen in Example 8, there exists a close relation between semi-infinite programming and bi- or multi-level programming (or, more general, parametric programming). Then, the question arises whether one can apply solution methods from bi-level programming which exploit the specific bi-level structure to the solution of (SIP) and (GSIP) (cf. [30]). Is it possible to adjust these known methods to the geometry of the current class of problems? Furthermore, which relations exist to corresponding methods from parametric programming, e.g. to path-following methods?

3. There are still existing many open questions with respect to the local and global structure of the class of (generalized) semi-infinite programming problems. These questions are dealing with

- generic conditions; which of the phenomena discussed above appear generically; which of the non-generic ones can be excluded by assuming additional conditions?
- Constraint qualifications and
- first and second order optimality conditions.
- Under which conditions can the original problem be transformed locally in a finite programming problem in order to apply solution methods from finite programming (as it can be done e.g. when applying the so-called reduction approach, cf. [12]).

4. Applications: the modelling of a particular real-life problem from economics, finance or engineering may lead to a particularly structured (SIP) or (GSIP); e.g. to a linear problem, a convex problem, or a problem where some of these specific topological features considered do not appear. How can these particular structures be exploited for the design of solution methods?

5. What are the relations between the geometric phenomena, the consequences for the use of solution methods and, on the other hand, the application context from economics, finance and engineering? Are there, on the practitioner's side, existing interpretations of the features discussed?

22.5 Bibliography

Finally, in this section we present a list of several articles which are dealing with possible applications of semi-infinite programming to real-life problems. Although this list cannot be complete, it is our hope that it illustrates several interesting bridges between the theory of semi-infinite programming and its applications to the modelling and solution of problems in economics, finance and engineering. Those articles which are marked by an "*" are also included in the list of references at the end of this paper.

Applications to economics:

Charnes A, Clower RW, Kortanek KO (1967) Effective control through coherent decentralization with preemptive goals. Econometrica 35:294–320

Davis OA, Kortanek KO (1971) Centralization and decentralization: the political economy of public school systems. American Economic Review 61:456–462

Game and duality theory:

Charnes A, Eisner MJ, Kortanek KO (1970) On weakly balanced games and duality theory. Cahiers du Centre d'Etude de Recherche Operationnell 12:7–21

Semi-infinite linear production situations:

Fragnelli V, Patrone F, Sideri E, Tijs S (1999) Balanced games arising from infinite linear models. Math. Methods of Oper. Research 50:385–397

Timmer J, Tijs S, Llorca N (2000) Games arising from infinite production situations. International Game Theory Review 2:97–106

Timmer J, Tijs, S, Llorca N, Sánchez-Soriano J (2001) The Owen set and the core of semi-infinite production situations. In: Goberna MA, Lopez MA (eds) Semi-infinite Programming – Recent Advances. Kluwer, Boston, 365-386

Semi-infinite programming as a tool for pricing stock options:

Smith JE (1995) Generalized Chebyshev inequalities: theory and applications in decision analysis. Operations Research 43:807–825

Applications to geometric programming (Cobb-Douglas production functions):

Rajgopal J, Bricker DL (1990) Posynomial geometric programming as a special case of semi-infinite linear programming. J. Optim. Theory Appl. 66:455–475

Applications to portfolio problems:

*Ben-Tal A, Nemirovski A (1999) Robust solutions of uncertain linear programs. Operations Research Letters 25:1–13

*Stein O (to appear) Bi-level Strategies in Semi-infinite Programming. Kluwer, Boston

*Tichatschke R, Kaplan A, Voetmann T, Böhm M (2002) Numerical treatment of an asset price model with non-stochastic uncertainty. TOP 10, 1:1–30

Applications of mathematical programming to Stackelberg games:

*Stein O (to appear) Bi-level Strategies in Semi-infinite Programming. Kluwer, Boston

Shimizu K, Aiyoshi E (1981) A new computational method for Stackelberg and min-max problems by use of a penalty method. IEEE Transactions on Automatic Control AC-26:460–466

*Shimizu K, Ishizuka Y, Bard JF (1997) Nondifferentiable and Two-Level Mathematical Programming. Kluwer, Boston

*Vicente LN, Calamai PH (1994) Bilevel and multilevel programming: a bibliography review. J. of Global Optimization 5:291–306

Applications to control problems:

*Sachs EW (1998) Semi-infinite programming in control. In: Reemtsen R, Rückmann J-J (eds) Semi-infinite Programming. Kluwer, Boston, 389-411

References

1. Ben-Israel A, Charnes A, Kortanek KO (1971) Asymptotic duality in semi-infinite programming and the convex core topology. Rendiconti Di Matematica 4:751–761
2. Ben-Tal A, Nemirovski, A (1999) Robust solutions of uncertain linear programs. Operations Research Letters 25:1–13
3. Bonnans JF, Cominetti R, Shapiro A (1999) Second order optimality conditions based on parabolic second order tangent sets. SIAM J. Optimization 9:466–492
4. Bonnans JF, Shapiro A (1998) Optimization problems with perturbations, a guided tour. SIAM Review 40,2:228–264
5. Bonnans JF, Shapiro A (2000) Perturbation Analysis of Optimization Problems. Springer, New York
6. de Klerk E (2002) Aspects of Semidefinite Programming. Interior Point Algorithms and Selected Applications. Kluwer, Boston
7. Fahl M, Sachs EW (1997) Modern optimization methods for structural design under flutter constraints. Technical report, University of Trier, Germany
8. Goberna MA, López MA (1998) Linear Semi-infinite Optimization. Wiley, Chichester
9. Goberna MA, López MA (eds) (2001) Semi-infinite Programming – Recent Advances. Kluwer, Boston
10. Graettinger TJ, Krogh BH (1988) The acceleration radius: a global performance measure for robotic manipulators. IEEE J. of Robotics and Automation 4:60–69
11. Haaren-Retagne E (1992) Semi-Infinite Programming Algorithms for Robot Trajectory Planning. PhD Thesis, University of Trier, Germany
12. Hettich R, Kortanek KO (1993) Semi-infinite programming: theory, methods and applications. SIAM Review 35,3:380–429
13. Hettich R, Still G (1991) Semi-infinite programming models in robotics. In: Guddat J et al. (eds) Parametric Optimization and Related Topics II, Math. Res. 62, Akademie-Verlag, Berlin, 112–118

14. Hettich R, Still G (1995) Second order optimality conditions for generalized semi-infinite programming problems. Optimization 34:195–211

15. Jongen HTh, Rückmann J-J, Stein O (1998) Generalized semi-infinite optimization: a first order optimality condition and examples. Math. Programming 83:145–158

16. Kaplan A, Tichatschke R (1997) On a class of terminal variational problems. In: Guddat J, Jongen HTh, Nožička F, Still G, Twilt F (eds) Parametric Optimization and Related Topics IV, Peter Lang Verlag, Frankfurt a.M., 185–199

17. Klatte D (1994) Stable local minimizers in semi-infinite optimization: regularity and second-order conditions. J. Comp. Appl. Math. 56:137–157

18. Kortanek KO, Medvedev G (1999) Models for estimating the structure of interest rates from observations of yield curves. In: Avellaneda M (ed) Quantitative Analysis in Financial Markets, World Sci., 53–120

19. Kortanek KO, Moulin P (1998) Semi-infinite programming in orthogonal wavelet filter design. In [24], 323–360

20. Krabs W (1987) On time-minimal heating or cooling of a ball. In: International Series of Numerical Mathematics 81, Birkhäuser, Basel, 121–131

21. Polak E (1987) On the mathematical foundations of nondifferentiable optimization in engineering design. SIAM Review 29:21–89

22. Polak E (1997) Optimization. Algorithms and Consistent Approximations. Springer, New York

23. Reemtsen R, Görner S (1998) Numerical methods for semi-infinite programming: a survey. In [24], 195–275

24. Reemtsen R, Rückmann J-J (eds) (1998): Semi-Infinite Programming. Kluwer, Boston

25. Rückmann J-J, Shapiro A (1999) First-order optimality conditions in generalized semi-infinite programming. J. Optim. Theory Appl. 101:677–691

26. Rückmann J-J, Shapiro A (2001) Second-order optimality conditions in generalized semi-infinite programming. Set-Valued Analysis 9,1–2:169–186

27. Sachs EW (1998) Semi-infinite programming in control. In [24], 389–411

28. Shimizu K, Ishizuka Y, Bard JF (1997) Nondifferentiable and Two-Level Mathematical Programming. Kluwer, Boston

29. Silva CLM, Oliveira FAR, Hendrickx M (1993) Modelling optimum processing conditions for the sterilization of prepackaged foods. Food Control 4,2:67–78

30. Stein O (to appear) Bi-level Strategies in Semi-infinite Programming. Kluwer, Boston

31. Stein O, Still G (2000) On optimality conditions for generalized semi-infinite programming problems. J. Optim. Theory Appl. 104:443–458

32. Stein O, Still G (submitted) Solving semi-infinite optimization problems with interior point techniques

33. Tichatschke R, Kaplan A, Voetmann T, Böhm M (2002) Numerical treatment of an asset price model with non-stochastic uncertainty. TOP 10,1:1–30

34. Vandenberghe L, Boyd S (1998) Connections between semi-infinite and semi-definite programming. In [24], 277–294

35. Vasicek O (1977) An equilibrium characterization of the term structure. J. of Financial Economics 5:177–188

36. Vicente LN, Calamai PH (1994) Bilevel and multilevel programming: a bibliography review. J. of Global Optimization 5:291–306

37. Weber G-W (1999) Generalized Semi-infinite Optimization and Related Topics. Habilitation thesis, Darmstadt University of Technology

Index